EXERCISE WORKBOOK for Beginning AutoCAD® 2004

by

Cheryl R. Shrock

Chairperson

Drafting Technology

Orange Coast College, Costa Mesa, Ca.

INDUSTRIAL PRESS INC.

New York

Limits of Liability and disclaimer of Warranty
The author and publisher make no warranty of any kind, expressed or implied, with regard to the documentation contained in this book.

First Industrial Press edition published July, 2004
ISBN 0-8311-3198-5

Exercise Workbooks written by Cheryl R. Shrock:

Advanced AutoCAD **2000**	ISBN 0-8311-3193-4
Beginning AutoCAD **2000, 2000i & LT**	ISBN 0-8311-3194-2
Advanced AutoCAD **2000, 2000i & LT**	ISBN 0-8311-3195-0
Beginning AutoCAD **2002**	ISBN 0-8311-3196-9
Advanced AutoCAD **2002**	ISBN 0-8311-3197-7
Beginning AutoCAD **2004**	ISBN 0-8311-3198-5
Advanced AutoCAD **2004**	ISBN 0-8311-3199-3

New Exercise Workbooks Coming Fall 2004

Beginning AutoCAD **2005**	ISBN 0-8311-3200-0
Advanced AutoCAD **2005**	ISBN 0-8311-3201-9

For information about these workbooks,
visit www.industrialpress.com

For information about Cheryl Shrock's online courses,
visit www.shrockpublishing.com

Table of Contents

Introduction

Lesson 1

Lesson 2

Lesson 27

Lesson 28

Lesson 29

Lesson 30

APPENDIX

Index

INTRODUCTION

About this workbook
Exercise Workbook for Beginning AutoCAD® 2004 is designed for classroom instruction or self-study. There are 30 lessons. Each lesson starts with step by step instructions followed by exercises designed for practicing the commands you learned within that lesson.

You may find the order of instruction in this workbook somewhat different from most textbooks. The approach I take is to familiarize you with the drawing commands first. After you are comfortable with the drawing commands, you will be taught to create your own setup drawings. This method is accomplished by supplying you with drawings "Workbook Helper" and "9A Helper". These drawings are preset and ready for you to open and use. For the first 8 lessons you should not worry about settings, **you just draw.**

I realize that not everyone will agree with this approach and if this is the case, you may want to change the order in which the lessons are learned. I have had success with this method because my students feel less intimidated and more confident. This feeling of confidence increases student retention. Learning should be fun not a headache.

The exercises in the workbook, that include printing, are designed for a Hewlett Packard 4MV printer capable of printing a 17 X 11 drawing. These exercises can be amended to match your printer or plotter specifications. To configure your printer, refer to Appendix A, "Add a Printer / Plotter. But it is important to note that you can configure a printer / plotter even though your computer is not attached to it. I advise you to configure the HP4MV, to complete the lessons within the workbook, even though you will not use this printer for actual printing.

How to get the drawings listed above?
Please download the 2 files mentioned above from our website,
www.shrockpublishing.com

About the Author
Cheryl R. Shrock is a Professor and Chairperson of Computer Aided Design at Orange Coast College in Costa Mesa, California. She is also an Autodesk® registered author / publisher. Cheryl began teaching CAD in 1990. Previous to teaching, she owned and operated a commercial product and machine design business where designs were created and documented using CAD. This workbook is a combination of her teaching skills and her industry experience.

"Sharing my industry and CAD knowledge has been the most rewarding experience of my career. Students come to learn CAD in order to find employment or to upgrade their skills. Seeing them actually achieve their goals, and knowing I helped, is a real pleasure. If you read the lessons and do the exercises, I promise, you will not fail."

Cheryl R. Shrock

Configuring your system

AutoCAD ® 2004 allows you to customize it's configuration. While you are using this workbook, it is necessary for you to make some simple changes, to your configuration, so our configurations are the same. This will ensure that the commands and exercises work as expected. The following instructions will walk you through those changes.

A. First start AutoCAD® 2004.
   1. Click "Start" button in the lower left corner of the screen.
   2. Choose "Programs / Autodesk / AutoCAD 2004 or LT / AutoCAD 2004 or LT
   3. You should see a blank screen. (If the "Create a new drawing" dialog box appears, select "Cancel" and continue.

B. At the bottom of the screen there is a white rectangular area called the "*Command Line*". Type: ***Options*** then press the **<enter>** key. (not case sensitive)

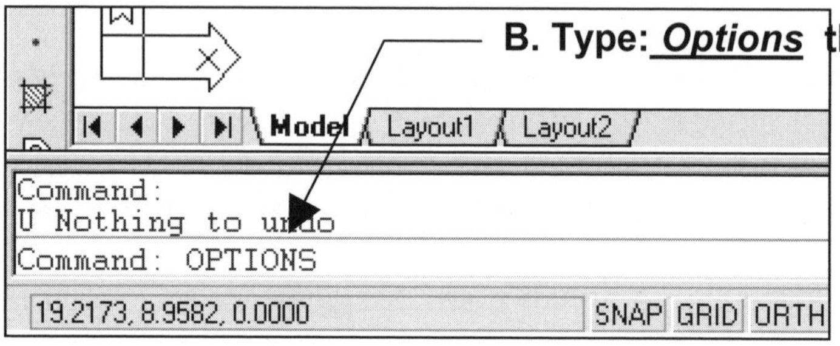

B. Type: *Options* then press <enter>

Model / Layout1 / Layout2

Command:
U Nothing to undo
Command: OPTIONS

19.2173, 8.9582, 0.0000 SNAP GRID ORTH

C. Select the ***Display*** tab and change the settings on your screen to match the dialog box below. Pay special attention to the settings with an ellipse around it.

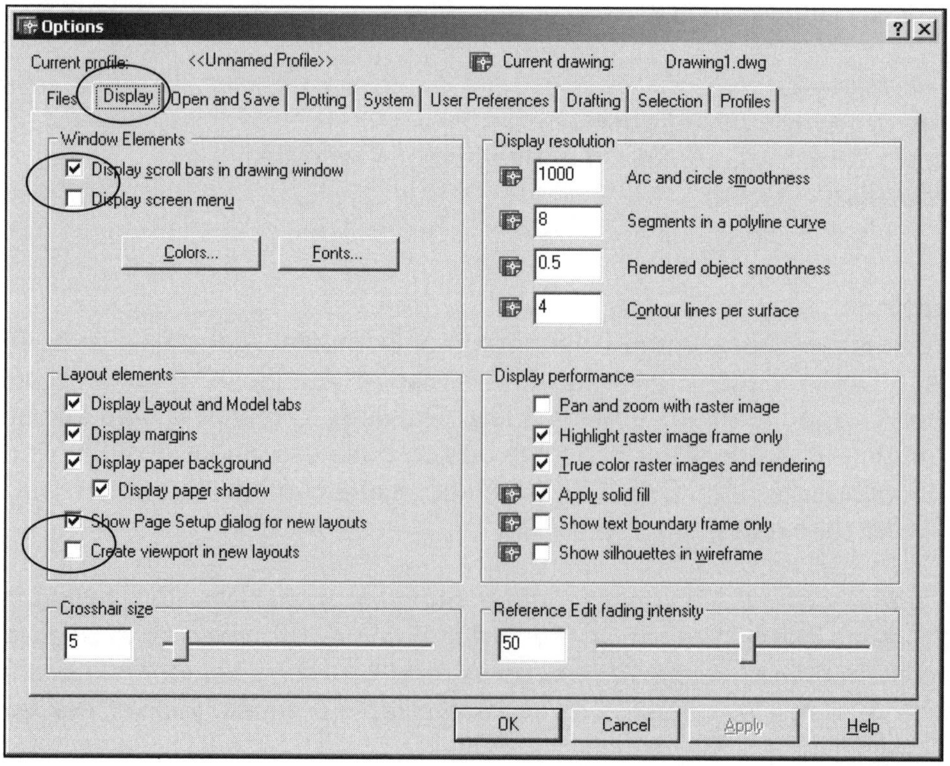

Options

Current profile: <<Unnamed Profile>> Current drawing: Drawing1.dwg

Files | Display | Open and Save | Plotting | System | User Preferences | Drafting | Selection | Profiles

Window Elements
- [x] Display scroll bars in drawing window
- [ ] Display screen menu

Colors... Fonts...

Display resolution
- 1000 Arc and circle smoothness
- 8 Segments in a polyline curve
- 0.5 Rendered object smoothness
- 4 Contour lines per surface

Layout elements
- [x] Display Layout and Model tabs
- [x] Display margins
- [x] Display paper background
  - [x] Display paper shadow
- [x] Show Page Setup dialog for new layouts
- [ ] Create viewport in new layouts

Display performance
- [ ] Pan and zoom with raster image
- [x] Highlight raster image frame only
- [x] True color raster images and rendering
- [x] Apply solid fill
- [ ] Show text boundary frame only
- [ ] Show silhouettes in wireframe

Crosshair size: 5

Reference Edit fading intensity: 50

OK Cancel Apply Help

E. Select the ***Open and Save*** tab and change the settings on your screen to match the dialog box below.

Options

Current profile: <<Unnamed Profile>> Current drawing: Drawing2.dwg

Files | Display | Open and Save | Plotting | System | User Preferences | Drafting | Selection | Profiles

File Save
Save as:
AutoCAD 2004 Drawing (*.dwg)
☑ Save a thumbnail preview image
50 Incremental save percentage

File Safety Precautions
☑ Automatic save
10 Minutes between saves
☑ Create backup copy with each save
☐ Full-time CRC validation
☐ Maintain a log file
ac$ File extension for temporary files
Security Options...
☐ Display digital signature information

File Open
9 Number of recently-used files to list
☑ Display full path in title

External References (Xrefs)
Demand load Xrefs:
Enabled with copy
☑ Retain changes to Xref layers
☑ Allow other users to Refedit current drawing

ObjectARX Applications
Demand load ObjectARX apps:
Object detect and command invoke
Proxy images for custom objects:
Show proxy graphics
☑ Show Proxy Information dialog box

OK | Cancel | Apply | Help

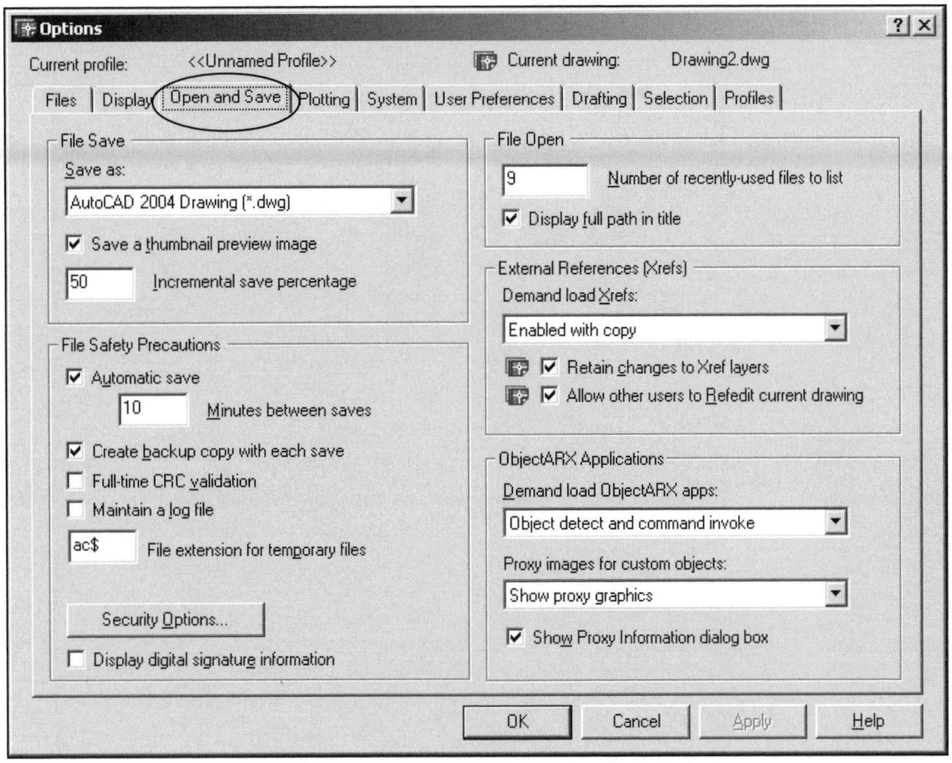

F. Select the ***Plotting*** tab and change the settings on your screen to match the dialog box below.

IMPORTANT: Add this printer.
See Appendix A for instructions
(Don't worry, it is not difficult)

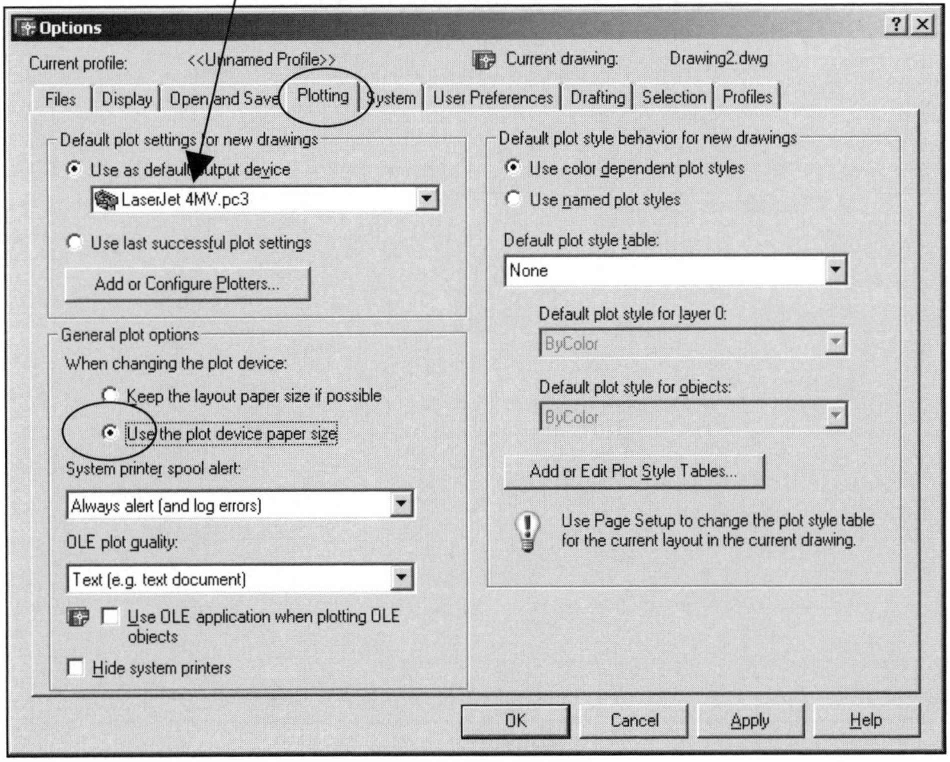

Options

Current profile: <<Unnamed Profile>> Current drawing: Drawing2.dwg

Files | Display | Open and Save | Plotting | System | User Preferences | Drafting | Selection | Profiles

Default plot settings for new drawings
◉ Use as default output device
LaserJet 4MV.pc3
○ Use last successful plot settings
Add or Configure Plotters...

General plot options
When changing the plot device:
○ Keep the layout paper size if possible
◉ Use the plot device paper size
System printer spool alert:
Always alert (and log errors)
OLE plot quality:
Text (e.g. text document)
☐ Use OLE application when plotting OLE objects
☐ Hide system printers

Default plot style behavior for new drawings
◉ Use color dependent plot styles
○ Use named plot styles
Default plot style table:
None
Default plot style for layer 0:
ByColor
Default plot style for objects:
ByColor
Add or Edit Plot Style Tables...
Use Page Setup to change the plot style table for the current layout in the current drawing.

OK | Cancel | Apply | Help

G. Select the ***System*** tab and change the settings on your screen to match the dialog box below.

Options

Current profile: <<Unnamed Profile>>   Current drawing: Drawing1.dwg

Files | Display | Open and Save | Plotting | System | User Preferences | Drafting | Selection | Profiles

Current 3D Graphics Display
GSHEIDI10   Properties...

Current Pointing Device
Current System Pointing Device
Accept input from:
Digitizer only
Digitizer and mouse

Layout Regen Options
Regen when switching layouts
Cache model tab and last layout
Cache model tab and all layouts

dbConnect Options
Store Links index in drawing file
Open tables in read-only mode

General Options
Single-drawing compatibility mode
Display OLE properties dialog
Show all warning messages
Beep on error in user input
Load acad.lsp with every drawing
Allow long symbol names
Startup: Show Startup dialog box

Live Enabler Options
Check Web for Live Enablers
5 Maximum number of unsuccessful checks

OK | Cancel | Apply | Help

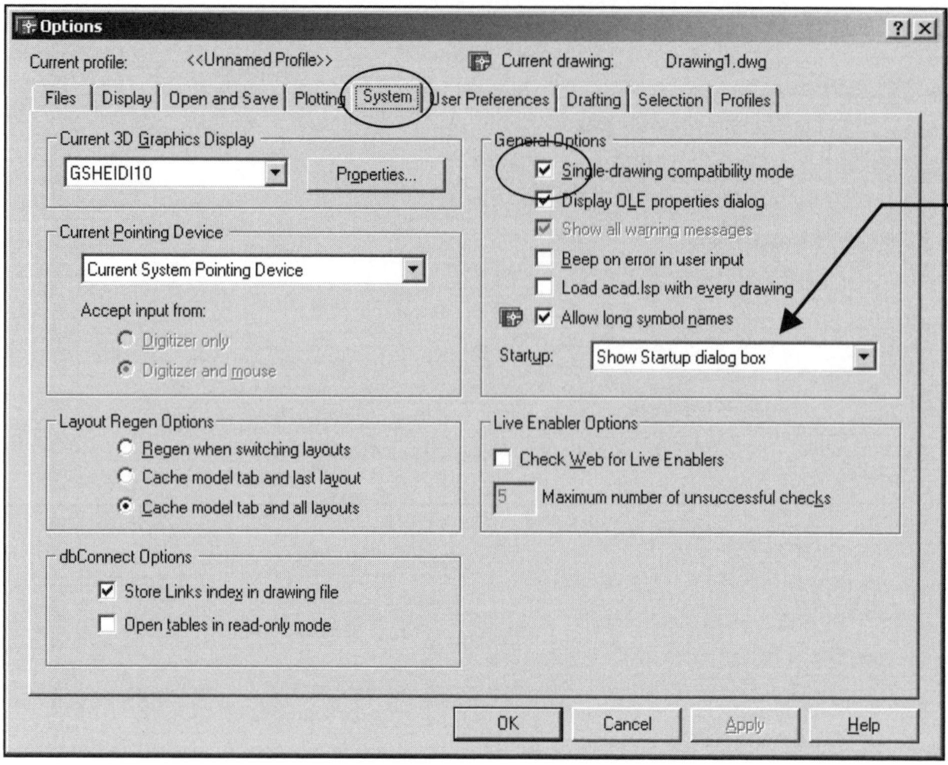

Select "Show Startup dialog box"

H. Select the ***User Preferences*** tab and change the settings on your screen to match the dialog box below.

Options

Current profile: <<Unnamed Profile>>   Current drawing: Drawing1.dwg

Files | Display | Open and Save | Plotting | System | User Preferences | Drafting | Selection | Profiles

Windows Standard Behavior
Windows standard accelerator keys
Shortcut menus in drawing area
Right-click Customization...

Drag-and-drop scale
Default settings when units are set to unitless:
Source content units:
Inches
Target drawing units:
Inches

Hyperlink
Display hyperlink cursor and shortcut menu
Display hyperlink tooltip

Priority for Coordinate Data Entry
Running object snap
Keyboard entry
Keyboard entry except scripts

Object Sorting Methods
Object selection
Object snap
Regens
Plotting

Associative Dimensioning
Make new dimensions associative

Hidden Line Settings...   Lineweight Settings...

OK | Cancel | Apply | Help

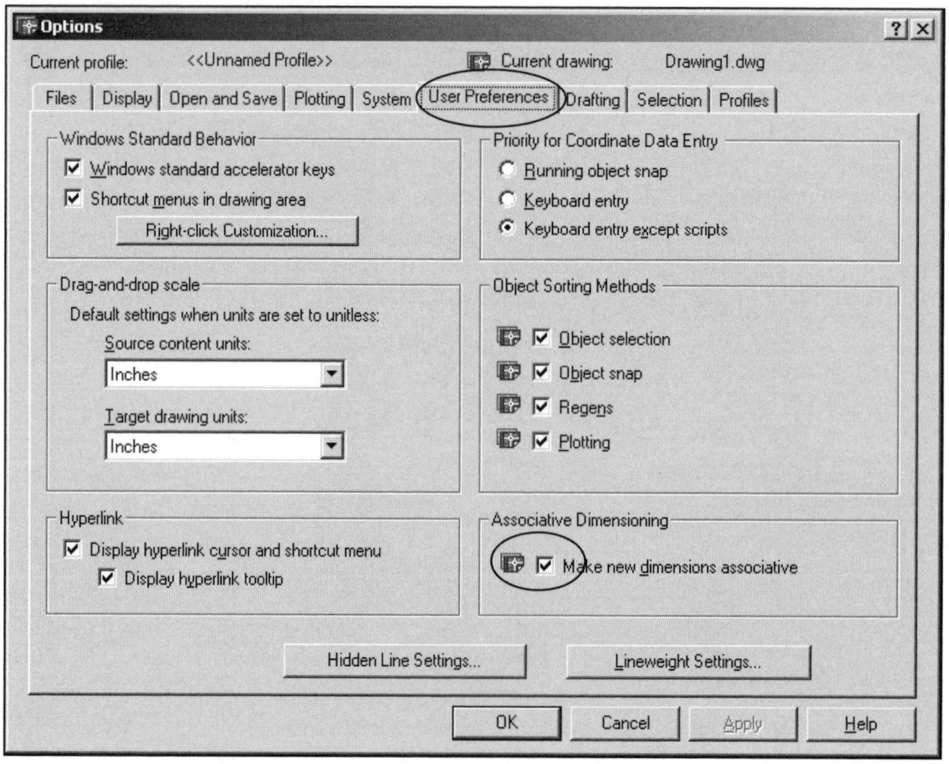

I. Select the ***Right-click Customization..*** box and change the settings on your screen to match the dialog box below.

Select "Right-click customization" button

Options
Current profile: CHERYL
Current drawing: Drawing1.dwg
Files | Display | Open and Save | Plotting | System | User Preferences | Drafting | Selection | Profiles
Windows Standard Behavior
Windows standard accelerator keys
Shortcut menus in drawing area
Right-click Customization...
Priority for Coordinate Data Entry
Running object snap
Keyboard entry
Keyboard entry except scripts

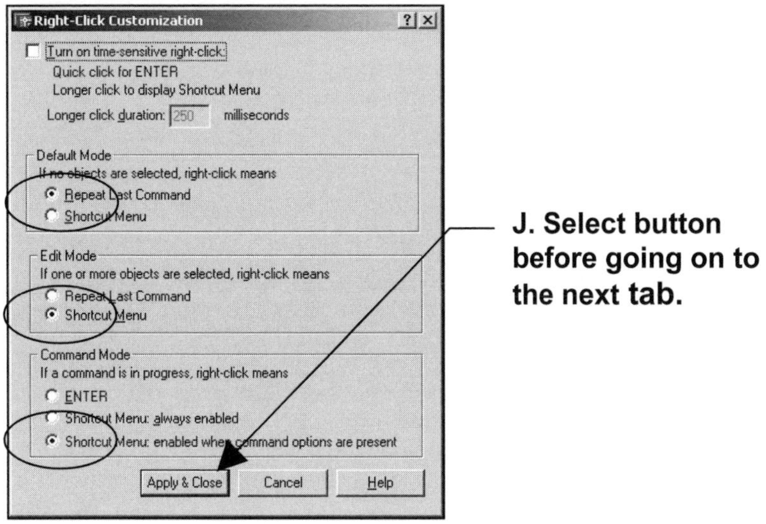

Right-Click Customization
Turn on time-sensitive right-click:
Quick click for ENTER
Longer click to display Shortcut Menu
Longer click duration: 250 milliseconds
Default Mode
If no objects are selected, right-click means
Repeat Last Command
Shortcut Menu
Edit Mode
If one or more objects are selected, right-click means
Repeat Last Command
Shortcut Menu
Command Mode
If a command is in progress, right-click means
ENTER
Shortcut Menu: always enabled
Shortcut Menu: enabled when command options are present
Apply & Close | Cancel | Help

J. Select button before going on to the next tab.

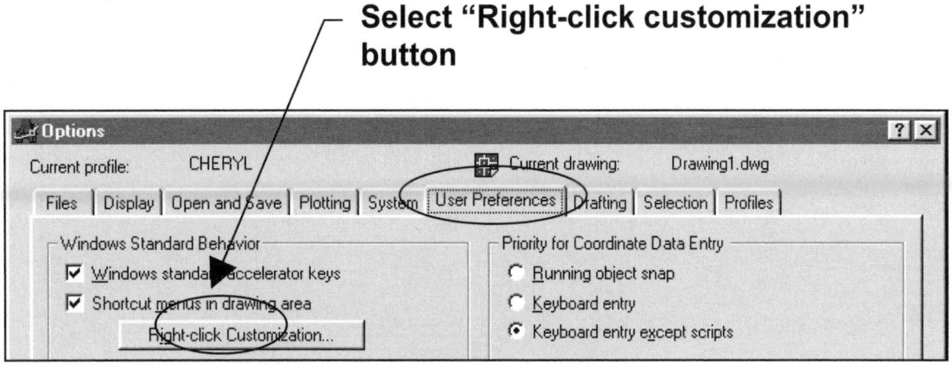

J. Select the ***Apply & Close*** button, shown above, before going on to the next tab.

K. Select the ***Drafting*** tab and change the settings on your screen to match the dialog box below.

Options
Current profile: <<Unnamed Profile>>
Current drawing: Drawing1.dwg
Files | Display | Open and Save | Plotting | System | User Preferences | Drafting | Selection | Profiles
AutoSnap Settings
Marker
Magnet
Display AutoSnap tooltip
Display AutoSnap aperture box
AutoSnap marker color:
Red
AutoTrack Settings
Display polar tracking vector
Display full-screen tracking vector
Display AutoTrack tooltip
Alignment Point Acquisition
Automatic
Shift to acquire
AutoSnap Marker Size
Aperture Size

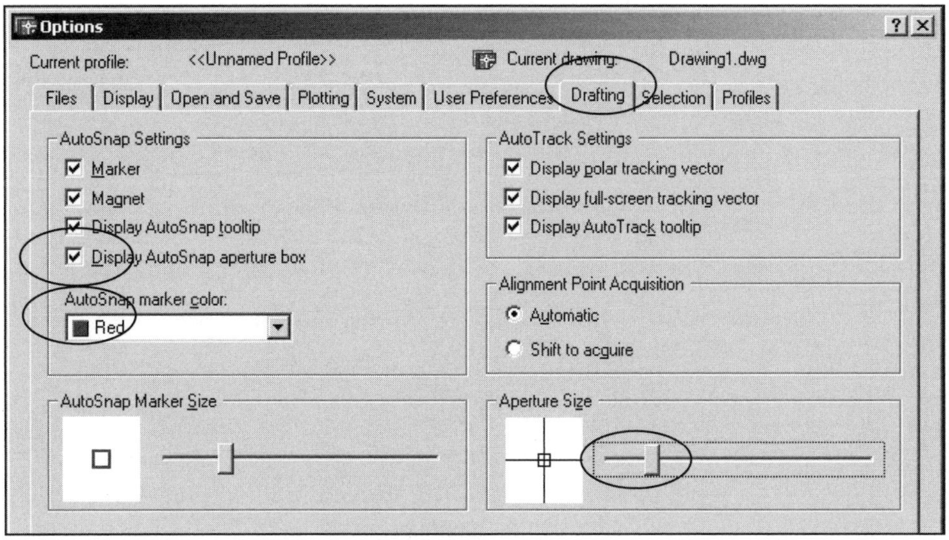

L. Select the ***Selection*** tab and change the settings on your screen to match the dialog box below.

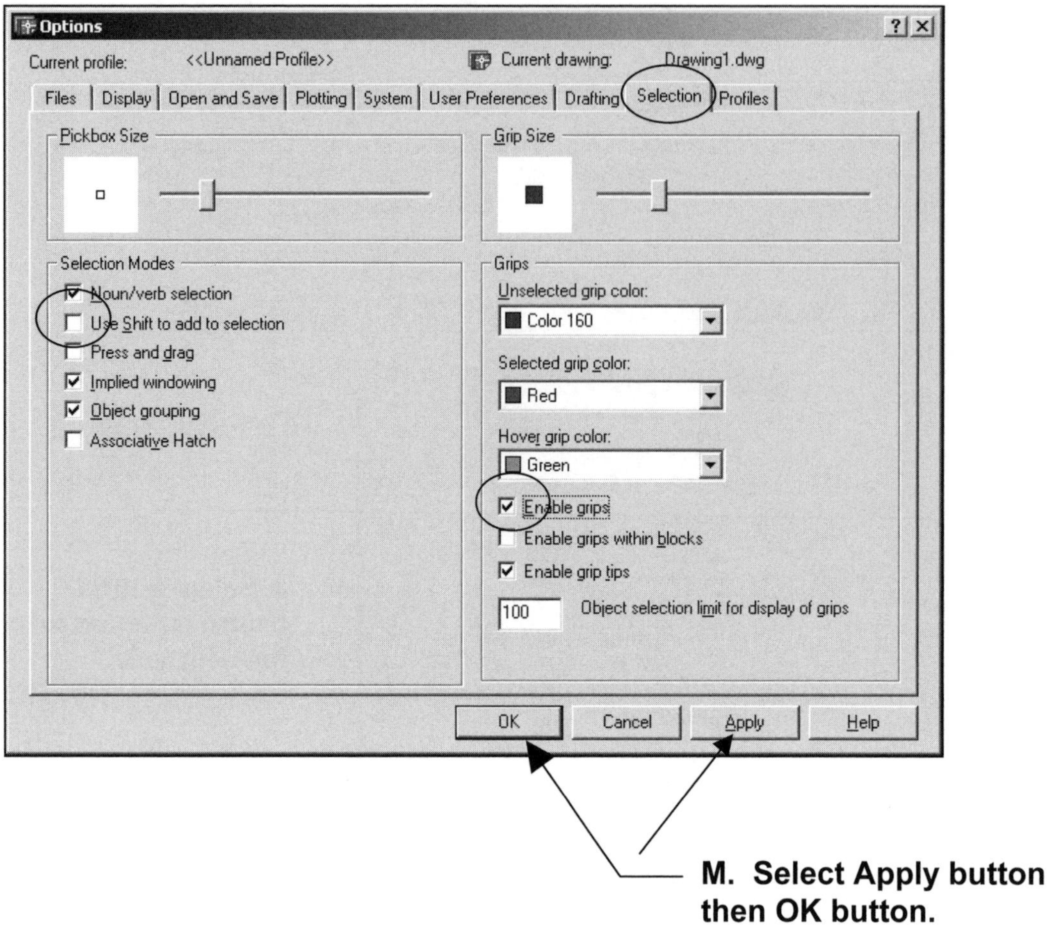

M. Select Apply button then OK button.

M. Select the ***Apply*** button then the ***OK*** button.

Customizing your Wheel Mouse

A Wheel mouse has two or more buttons and a small wheel between the two top side buttons. The default functions for the two top buttons are as follows:
Left Hand button is for **input**
Right Hand button is for **Enter** or the **shortcut menu**.

You will learn more about this later. But for now follow the instructions below.

Using a Wheel Mouse with AutoCAD®

To get the most out of your Wheel Mouse set the **MBUTTONPAN** setting to **"0"** as follows:

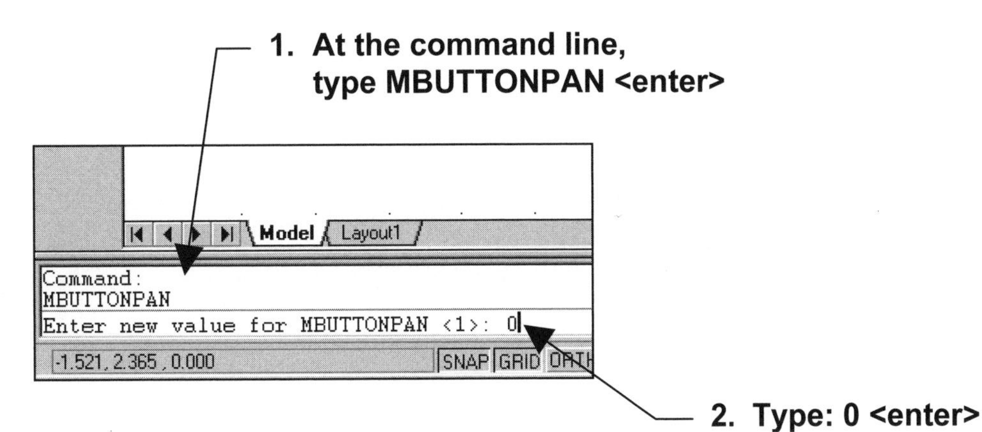

After you understand the function of the "Mbuttonpan" variable, you can decide whether you prefer the setting "0" or "1" as described below.

MBUTTONPAN setting 0:

ZOOM Rotate the wheel forward to zoom in
Rotate the wheel backward to zoom out

OBJECT SNAP Object Snap menu will appear when you press the wheel

MBUTTONPAN setting 1:

ZOOM Rotate the wheel forward to zoom in
Rotate the wheel backward to zoom out

ZOOM EXTENTS Double click the wheel

PAN Press the wheel and drag

NOTES:

LEARNING OBJECTIVES

After completing this lesson, you will be able to:

1. Understand basic computer terms.
2. Understand the term "CAD".
3. Determine what computer to purchase.
4. Know the system requirements for AutoCAD.
5. Start AutoCAD four ways.

LESSON 1

Part 1. UNDERSTANDING COMPUTERS

A BRIEF HISTORY OF COMPUTERS AND SOFTWARE.

The first computers were developed in the 1950s, shortly after the transistor was invented. In the mid 1960s General Motors, Boeing and IBM began developing CAD programs, but the development was slowed by the high cost of computer hardware and programming.

In 1971, Ted Hoff developed the first microprocessor. All circuitry of the central processing unit (CPU) was now on one chip. This started the era of the personal computer (PC). In the 1980s, additional improvements to the microprocessor changed the mainframe computers to powerful desktop models.

Of course, computer software was advancing along with the computer hardware. CAD started as a simple drafting tool and has now evolved into a powerful design tool. CAD has progressed from two-dimensional (2-D) to three-dimensional (3-D), to surface modeling and to solid modeling with animation. Each generation becoming more powerful and more user friendly.

HARDWARE

Microprocessor

The complex procedure that transforms raw input data into useful information for output is called processing. The **processor** is the "brain" of the computer. The processor interprets and carries out instructions. In Personal Computers (PCs) the processor is a single chip plugged into a circuit board. This chip is called a **microprocessor**.

Central Processing Unit (CPU)

The CPU is the term used for the computer's processor. The CPU contains the intelligence of the machine. It is where the calculations and decisions are made.

Memory (RAM)

Your CPU needs memory to hold pieces of information while it works. While this information remains in memory, the CPU can access it directly. This memory is called **random access memory (RAM)**. RAM holds information only while the power is on. When you turn off or reset the computer, the information disappears.
The more RAM a computer has, the quicker it works and the more it can do.
The most common unit of measurement for computer memory is the **byte**. A **byte** can be described as the amount of memory it takes to store a single character. A **kilobyte (KB)** equals 1,024 bytes. A **Megabyte (MB)** equals 1,024 kilobytes, or 1,048,576 bytes. So a computer with 64 MB of memory actually has (64 X 1,048,576) 67,108,864 bytes. This is equal to approximately 1024 pages of information.

Input / Output devices

Input devices accept data and instructions from the user. The most common input devices are the keyboard, mouse and scanner. Output devices return processed data back to the user. The most common output devices are the monitor, printer and speakers.

Storage

The purpose of storage is to hold data that the computer isn't using. When you need to work with a set of data, the computer retrieves the data from storage and puts it into memory. When it no longer needs the data, it puts it back into storage. There are 2 advantages to storage. First, there is more room in storage and second, storage retains its contents when the computer is turned off. Storage devices include: Hard disks, floppy disks, zip disks, CDR/W, etc.

SOFTWARE

Operating Systems

When you turn on the computer, it goes through several steps to prepare itself for use.

The first step is a self-test. This involves:

a. Identifying the devices attached to it. (Such as the monitor, mouse and printer)
b. Counts the amount of memory available.
c. Checks to see if the memory is functioning properly.

The second step is searching for a special program called the Operating System. When the computer finds the operating system, it loads it into memory. (remember RAM) The operating system enables the computer to:

a. communicate with you.
b. use devices such as the disk drives, keyboard and monitor.

The operating system is now ready to accept commands from you. The operating system continues to run until the computer is turned off. Examples of operating systems are: Windows 98, Windows NT, ME, 2000, XP, OS/2, Unix and more.

Note: 1. Apple / Macintosh computers have their own operating system.
2. AutoCAD 2004 will not work with Windows 98 or Apple / Macintosh.

Application software

The operating system is basically for the computer. The Application Software is for the user. Application Software is designed to do a specific task.
There are basically four major categories:
Business, Utility, Personal, and Entertainment.

Business application software would be desktop publishing, spreadsheet programs, database software and graphics. *AutoCAD is a "graphics" business application software.*

Utility application software helps you maintain your computer. You would use a utility program to recover an accidentally deleted file, improve the efficiency of your computer and help you move, copy or delete files. *Norton Utilities is an example of a "utility application" software.*

Personal application software is basically what it sounds like. This software is designed for your personal needs, such as: balancing your checkbook, making an address book, creating a calendar and many more tasks.

Entertainment application software are video games, puzzles, flight simulators and even educational programs.

COMPUTER SIZES AND CAPABILITIES

Computers are divided into five catagories:
Supercomputer, Mainframe, Minicomputer, Workstation and Personal Computer.

Supercomputers are the most powerful. These computers process huge amounts of information very quickly. For example, scientists build models of complex processes and simulate the processes on a supercomputer.

Mainframe computers are the largest. These computers are designed to handle tremendous amounts of input, output and storage. For example, the government uses mainframe computers to handle the records for Social Security.

Minicomputers are smaller than mainframe computers but bigger than personal computers. They do not handle as much as the mainframe computers but they are less expensive. A company that needs the features of a mainframe but can't afford such a large computer, may choose a minicomputers.

Workstations resemble a personal computer but are much more powerful. Their internal construction is different than a PC. Workstations use a different CPU design called reduced instruction set computing (RISC), which makes the instructions process faster. Workstations using the UNIX operating system are generally used by scientists and engineers.

Personal computers (PC), originally named microcomputers, are small computers that usually reside on a desktop. This category would include Laptops.

What is a Clone?

In 1981, IBM called its first microcomputer the IBM PC. Many companies copied this design and they functioned just like the original. These copies were called "clones" or "compatibles". The term PC is now used to describe this family of computers.
Note: the Apple/Macintosh computer is neither an IBM or a compatible. It should not be called a PC.

Part 2. What is CAD?

Computer Aided Design (CAD) is simply, **design** and **drafting** with the aid of a computer. Design is creating a real product from an idea. Drafting is the production of the drawings that are used to document a design. CAD can be used to create 2D or 3D computer models. A CAD drawing is a file that consists of numeric data, in binary form that will be saved on to a disk.

Why should you use CAD?
Traditional drafting is repetitious and can be inaccurate. It may be faster to create a simple "rough" sketch by hand but larger more complex drawings with repetitive operations, are drawn more efficiently using CAD.

Why use AutoCAD?
AutoCAD is a computer aided design software developed by Autodesk Inc. AutoCAD was first introduced in 1982. By 2000, it is estimated that there were over 4 million AutoCAD users worldwide.

What this means to you is that many employers are in need of AutoCAD operators. In addition, learning AutoCAD will give you the basics for learning other CAD packages because many commands, terms and concepts are used universally.

Part 3. Buying your first computer

Buying your first computer is not an easy task. Here are a few tips.

1. Make a list of tasks for which you will use your computer?
    a. Select the software for those tasks.
    b. Select the computer that will run that software.

2. Talk to other computer owners.
    a. Listen to their good and bad experiences.

3. Educate yourself.
    a. Go to your local library and spend an evening reading through the computer magazines. Most are written with the novice in mind.

4. Decide how much you can afford.
    a. Remember, you can always upgrade or add components later.

5. Find the right company to buy from. A great deal can turn into a bad investment if you can't get help when you need it.
    a. Ask about their customer service and support.
    b. How long is the warranty?

Part 4. AutoCAD 2004 system requirements

Operating system:

Windows NT 4.0, ME, 2000, XP (Windows 95 and 98 can't be used)

RAM and Hard Disk Space

128 MB of RAM minimum, 256 MB recommended
130MB of hard disk space
64MB of swap space
100MB free disk space in system folder

Hardware (required)

Pentium III or later with 800 MHz processor or better
Mouse or other pointing device
1024 x 768 video graphics display with true color
CD-ROM drive for initial installation only
Disk drive (3-1/2 floppy, Zip or CDW) for saving files.

Hardware (optional)

Printer or Plotter
Serial or Parallel port (for peripheral devices)
Sound card with speakers

Miscellaneous (required)

Microsoft Internet Explorer 6.0

STARTING AutoCAD

To Start AutoCAD, select the **START** button / **PROGRAMS** / **Autodesk / AutoCAD 2004 or LT** / **AutoCAD 2004 or LT**.

If one of the dialog boxes, shown below, does not appear automatically, refer to page Intro-4 item G to change your setting.

I prefer these dialog boxes for students new to AutoCAD. But after you become an "expert" you may disable this option.

Notice the four buttons located in the upper left corner of this dialog box. Each button provides a different way to start a drawing. A brief description of each is listed below.

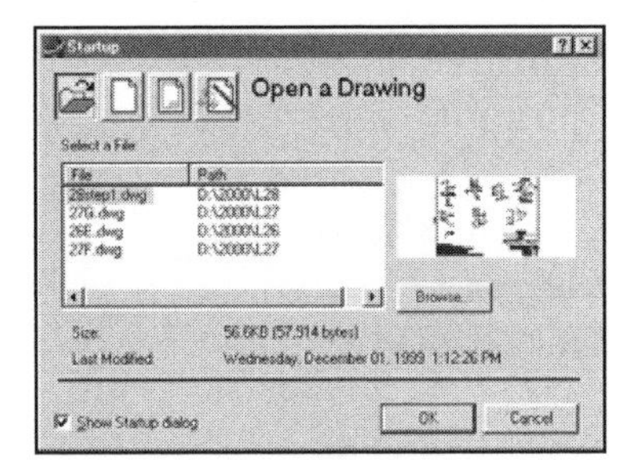

Open a Drawing
allows you to select a drawing from a list of the most recently opened drawings or select the "Browse" button to search for more drawing files. After you select the file desired, select the OK button. The file selected will appear on your screen. (This option is only active when you first enter AutoCAD. Normally you will use **File / Open** refer to page 2-14)

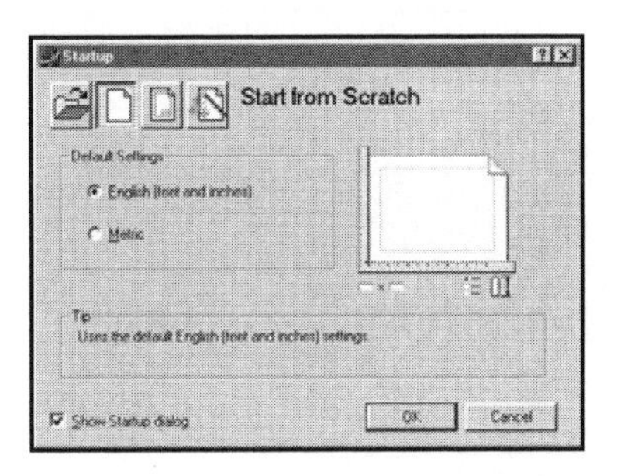

Start from Scratch
allows you to begin a new drawing from scratch. Starting from scratch means, all settings are preset by AutoCAD. You must select which measurement system on which to base your new drawing, English or Metric.

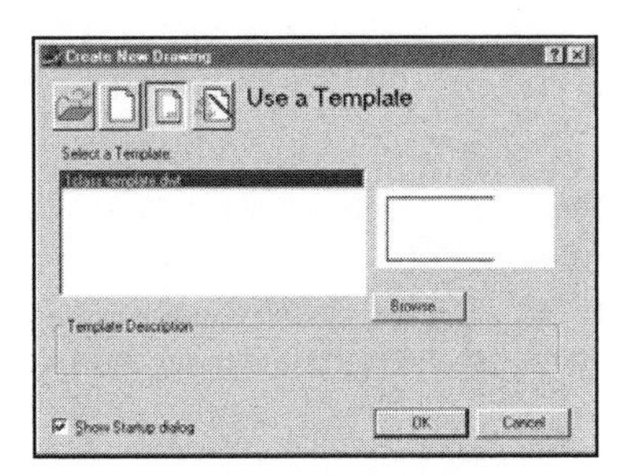

Use a Template
Allows you to choose a previously created template. You can choose one of the templates supplied with AutoCAD or create your own.
Note:
We will be creating a Template in Lesson 2.

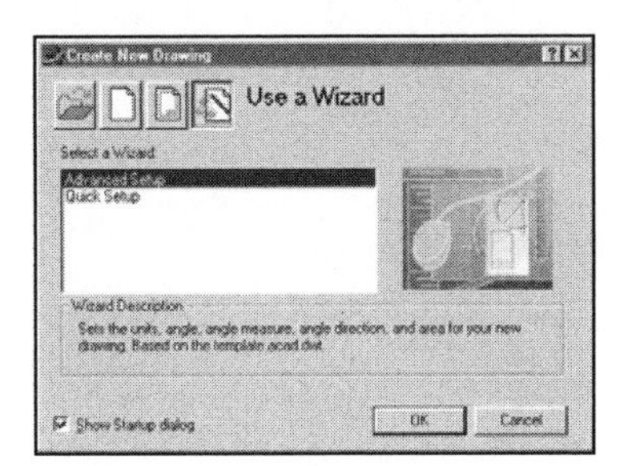

Use a Wizard
Allows you to start a new drawing using either the "Quick" or "Advanced" setup wizard. The wizard sets the units, angle, angle measurement, angle direction and area for your new drawing. (You will learn all of these settings in Lesson 9)

Using AutoCAD's HELP system

This workbook was created to make your AutoCAD learning experience fun and uncomplicated. To accomplish this I do not include every little detail about each command. I teach you the "meat and potatoes" of AutoCAD. If you would like to learn more about a command or system variable try AutoCAD's HELP system.

HOW TO OPEN THE HELP SYSTEM

Method 1.
1. Start a command.
2. Press the F1 key.

Method 2.
1. Select HELP menu at the top of the screen.
2. Select HELP from the drop down menu.

Method 3.
1. Press F1
2. Click on the blinking "? AutoCAD 2004 Help" button on the task bar.

USING THE HELP SYSTEM

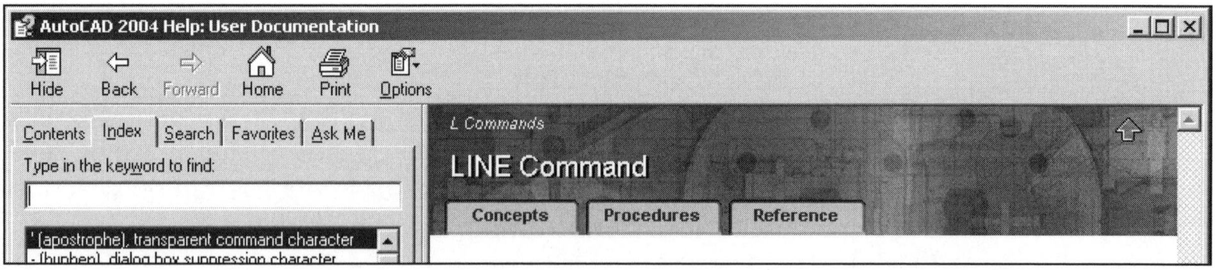

The Contents tab Organizes by topic like a table of contents in a book.

The Index tab Alphabetical listing of topics. Type the first few letters of the word. As you type, the list jumps to the closest match.

The Search tab Find keywords. Type a word in the text box.

The Favorites tab You may collect your favorite help topics here. First, display the topic. Then click the Favorites tab and click Add. To display a help topic from the Favorites list, double click it.

The Ask Me tab Ask a question and hopefully get an answer. Type a question or phrase and press <enter>

Concepts The overall description.

Procedures How to do it.

Reference Related commands.

Active Assistance

The Active Assistance window provides a brief explanation about the AutoCAD command you are using as you work.

You may "close" or "open" Active Assistance.
Temporarily close: select the Close button [X].

Turn Off: See "On Demand" below.

Open it: double click on the Active Assistance **icon** in the lower right corner of the task bar or select "Active Assistant" from the HELP menu.

You may select "when" and "how" the Active Assistant displays.
Displaying the Active Assistance is displayed is a personal preference. When you are first learning AutoCAD you may want it to display often. After you are more experience, you may want it displayed less frequently. The settings below allow you to select your preference.

Place the cursor on the Active Assistance, press the right mouse button and then select "Settings".

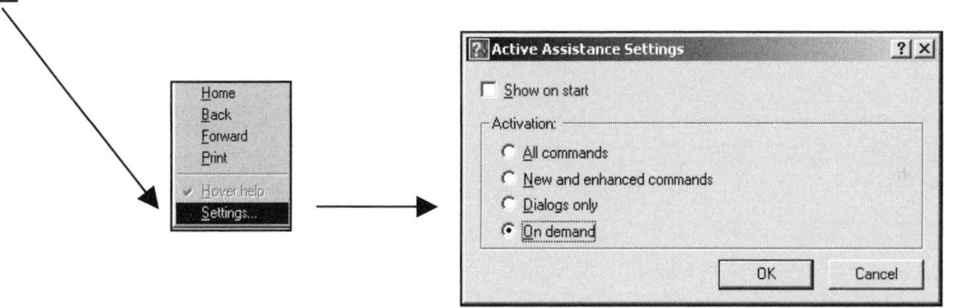

Show on start: Starts Active Assistance when AutoCAD starts. If this is off, you must open Active Assistance when you need it. See "Open it" above.

All Commands: Automatically opens the Active Assistance window when any command is activated

New and enhanced commands: Automatically opens the Active Assistance window when any new or enhanced commands are activated.

Dialogs only: Automatically opens the Active Assistance window when a dialog box is displayed. You cannot close the Active Assistance window while a dialog box is displayed.

On Demand: Turns Active Assistance OFF. You must open the Active Assistance when you need it. See "Open it" above.

NOTES:

LEARNING OBJECTIVES

After completing this lesson, you will be able to:

1. Create a template.
2. Getting familiar with the AutoCAD Window.
3. Understand the use of the function keys.
4. Select commands using the Pull-down Menu Bar, Toolbars or typing at the Command Line.
5. Recognize a dialog box.
6. Open, Close and Move a toolbar.
7. Draw, Erase and Select Lines.
8. Clear the screen.
9. Save a drawing.
10. Open an existing drawing.
11. Exit AutoCAD.

LESSON 2

CREATE A TEMPLATE

The first item on the learning agenda is how to create a template file from the **"Workbook Helper.dwg".** Go to the website **www.shrockpublishing.com** and download the files for workbooks 2004 and save them to a disk.

Now we will create a template. This will be a very easy task.

1. Start AutoCAD:
   Start button / Programs / Autodesk / AutoCAD 2004 or LT / AutoCAD 2004 or LT

   Note: If a dialog box appears select the "Cancel" button.

2. Select **File / Open**

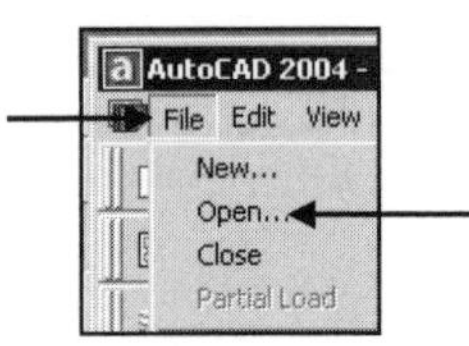

3. Select the **Directory** in which the downloaded files were saved. (Click on the ▾)

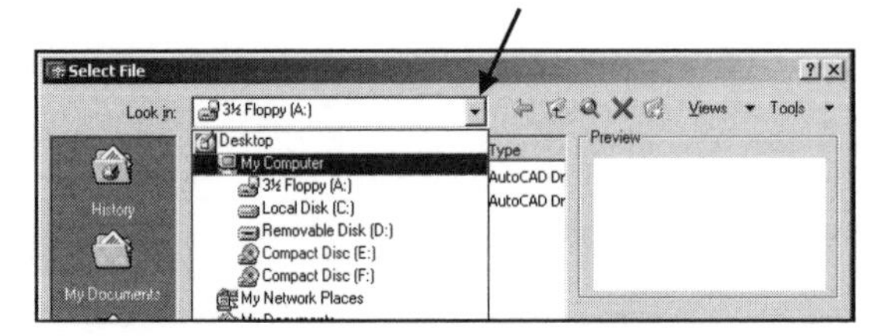

4. Select the file "**Workbook Helper.dwg**" and then "**Open**" button.

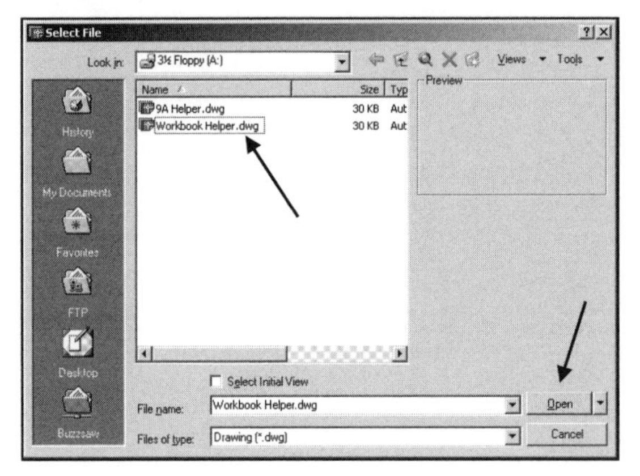

Notice the 3 letter extension for a "drawing" file is ".dwg"

5. Select **"File / Save As..."**

6. Select the "**Files of type**:" down arrow ▾ to display different saving formats. Select "**AutoCAD Drawing Template (*.dwt)**"

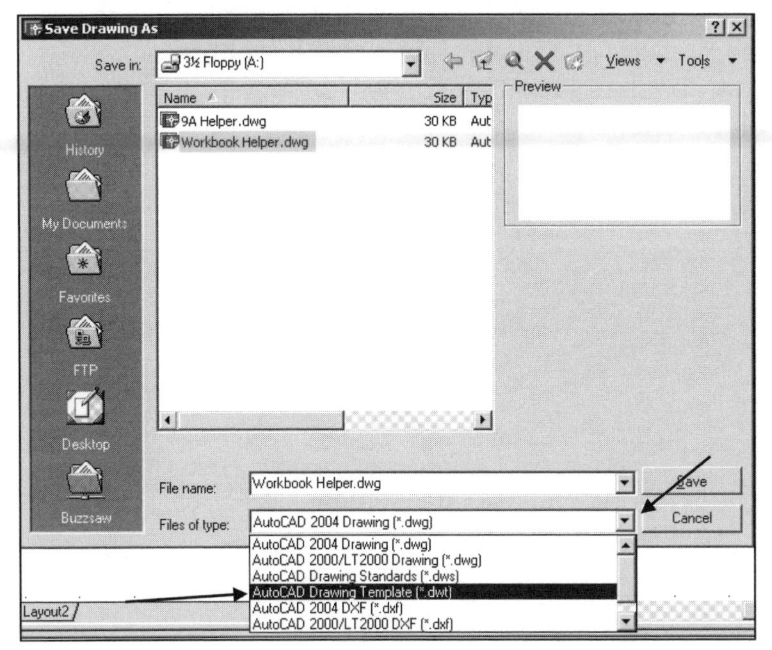

Notice the 3 letter extension for Template is ".dwt"

A list of all the AutoCAD templates will appear.

7. Type the new name "**1Workbook Helper**" in the "**File name**:" box and then select the "**Save**" button.

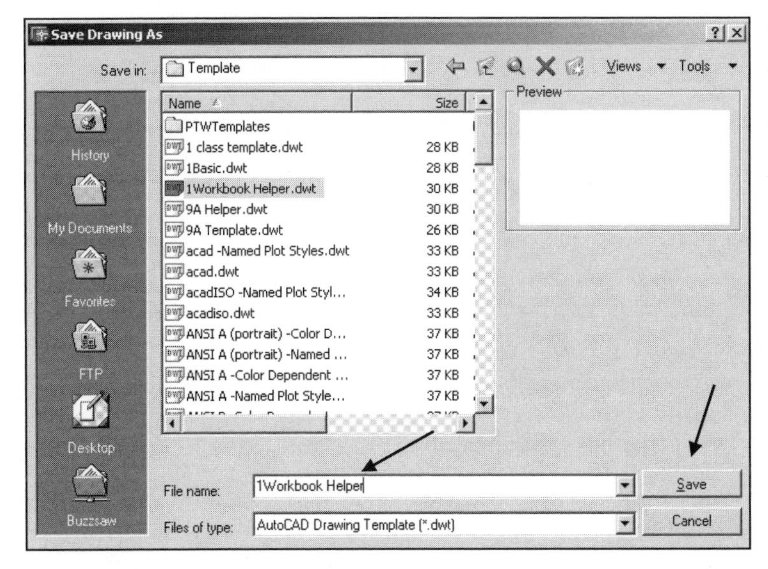

The "1" in the name places the file at the top of the list. AutoCAD lists numbers first and then alphabetical.

Notice it was not necessary to type the extension .dwt because "Files of type" was previously selected.

8. Type a description and the select the "OK" button.

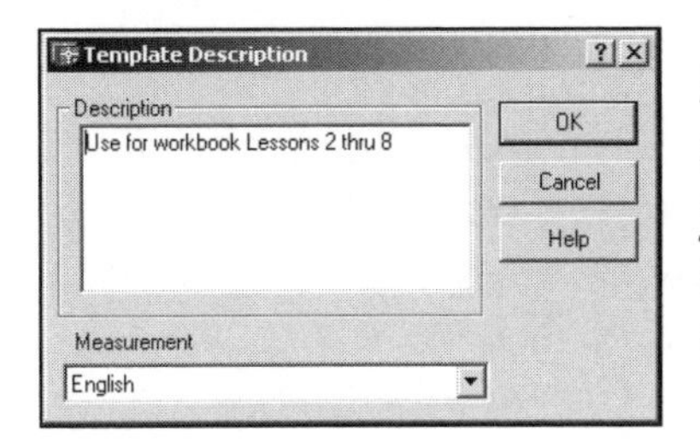

Now you have a template to use for lessons 2 through 8. At the beginning of each of the exercises you will be instructed to open this template. Using a template as a master setup drawing is good CAD management.

OPENING A TEMPLATE

The template that you created on the previous page will be used for lessons 2 through 8. It will appear as a blank screen but there are many variables that have been preset. This will allow you to start drawing immediately. You will learn how to set those variables before you complete this workbook but for now you will concentrate on learning the AutoCAD commands and, hopefully, have some fun.

Let's start by opening the "1Workbook Helper.dwt" template.

1. Select **FILE / NEW**.

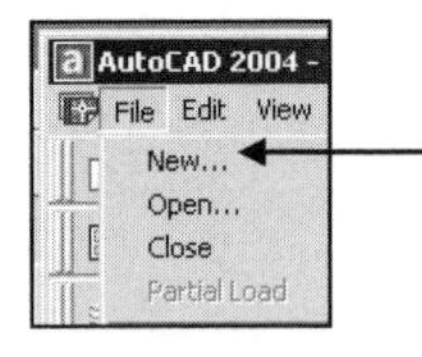

2. Select the **Use a Template** box. (third from the left)

3. Select **1workbook helper.dwt** from the list of templates.

   ***(NOTE:** If you do not have this template, refer to page 2-2)*

4. Select the **OK** button

***NOTE:** If you find that you have more than one drawing open it is important that you have configured your AutoCAD software to only allow one drawing open at one time. It will be less confusing for now. When you progress to the Advanced workbook, you will configure AutoCAD for multiple open drawings. But for now refer to Intro-4 for "Single drawing compatibility mode" setting under "General Options". Check this option box.*

GETTING FAMILIAR WITH THE AUTOCAD WINDOW

Before you can start drawing you need to get familiar with the AutoCAD window. In the following lessons, I will be referring to all of the areas described below. So it is important for you to understand each of them. But remember, this page will always be here for you to refer back to.

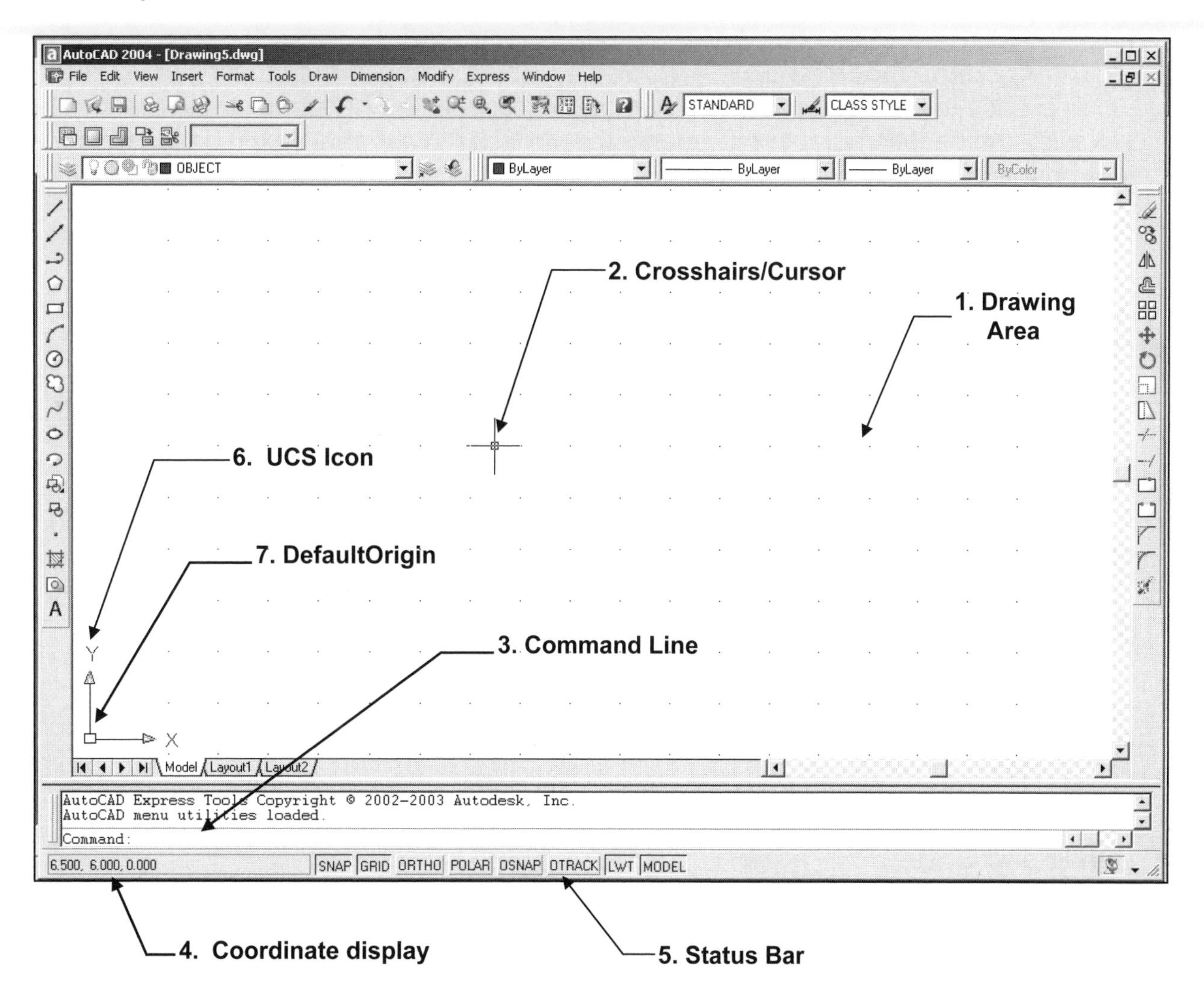

1. **DRAWING AREA**
   Location: The large area in the center of the screen.
   This is where you will draw. This area represents the piece of paper.
   The color of this area can be changed using Tools / Option / Display / Color.
   The default color for 2004 is white. The default color for 2004 LT is black.

2. **CROSSHAIRS / CURSOR**
   Location: Can be anywhere in the Drawing Area.
   The movement of the cursor is controlled by the movement of the pointing device such as a mouse. You will use the cursor to locate points, make selections and draw objects. The size can be changed using Tools / Options / Display / Crosshair Size.

3. **COMMAND LINE**
Location: The three lines at the bottom of the screen.
This is where you enter commands and Autocad will prompt you to input information.

4. **COORDINATE DISPLAY** (F6)
Location: Lower left corner

In the **Absolute mode (coords = 1)**: displays the location of the crosshairs / cursor in reference to the Origin. The first number represents the horizontal movement (Xaxis), the second number represents the vertical movement (Yaxis) and the third number is the Zaxis which is used for 3D.

In the **Relative Polar mode (coords = 2)**: displays the distance and angle of the cursor from the last point entered. (Distance<Angle)

5. **STATUS BAR**
Location: Below the Command Line.
Displays your current settings. These settings can be turned on and off by clicking on the word (Snap, Grid, Ortho, etc.) or by pressing the function keys, F1, F2, etc. See button descriptions below.

[SNAP] (F9)
Increment Snap controls the movement of the cursor. If it is off the cursor will move smoothly. If it is ON, the cursor will jump in an incremental movement.
The increment spacing can be changed, at any time using **Tools / Drafting Settings / Snap and Grid**. The default spacing is .250.

[GRID] (F7)
The grid (dots) is merely a visual "drawing aid". The default spacing is 1 unit.
You may change the grid spacing at any time using: **Tools / Drafting Settings / Snap and Grid.**

[ORTHO] (F8)
When Ortho is ON, cursor movement is restricted to horizontal or vertical. When Ortho is OFF, the cursor moves freely.

[POLAR] (F10)
POLAR TRACKING creates "Alignment Paths" at specified angles.
(More detailed information on page 11-3)

OSNAP (F3)
RUNNING OBJECT SNAP (More detailed information on page 4-3 & 4-4)
Specific Object Snaps can be set to stay active until you turn them off.

OTRACK (F11)
OBJECT SNAP TRACKING
Creates "Alignment Paths" at precise positions using object snap locations.

LWT
LINEWEIGHT. Displays the width assigned to each object. (More information on page 9-7)

MODEL
Switches your drawing between paperspace and modelspace.
(More information in Lesson 26)

6. **UCS ICON (User Coordinate System)**
Locaton: Lower left corner of the screen. The UCS icon indicates the location of the Origin. The UCS icon appearance can be changed using: **View / Display / Icon / Properties**.

7. **ORIGIN**
The location where the X, Y and Z axes intersect. 0,0,0
(More information in Lesson 9)

FUNCTION KEYS

F1	Help	Explanations of Commands
F2	Flipscreen	Toggles from Text Screen to Graphics Screen.
F3	Osnap	Toggles Osnap On and Off.
F4	Tablet	Toggles the Tablet On and Off.
F5	Isoplane	Changes the Isoplane from Top to Left to Right.
F6	Coordinate Display	Changes the display from Absolute / Off / Relative Polar.
F7	Grid	Toggles the Grid On or Off
F8	Ortho	Toggles Ortho On or Off
F9	Snap	Toggles Increment Snap on or off
F10	Polar	Toggles Polar Tracking On or Off.
F11	Otrack	Toggles Object Snap Tracking On and Off.

SPECIAL KEY FUNCTIONS

Escape Key Cancels the current command, menu or Dialogue Box.

Enter Key Ends a command; or will repeat the previous command if the command line is blank.

Space Bar Same as the Enter Key except when entering text.

PULL-DOWN "MENU BAR"

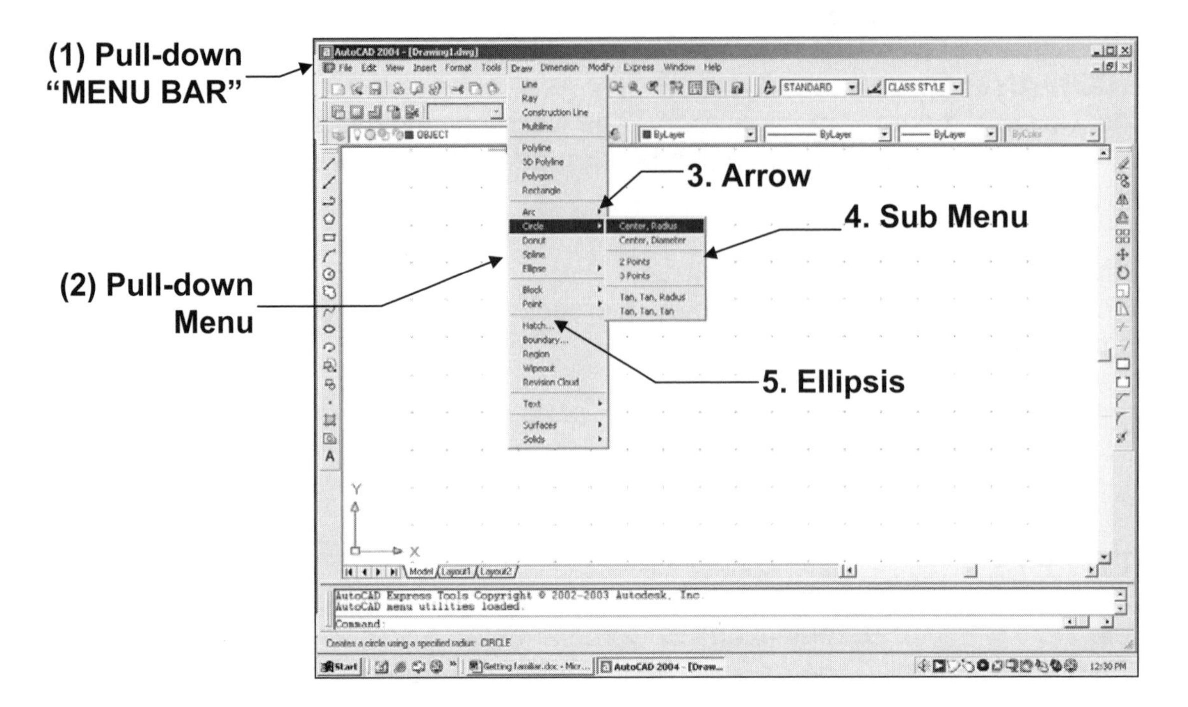

(1) The pull-down "MENU BAR" is located at the top of the screen.

AutoCAD 2004 - [Drawing1.dwg]

File Edit View Insert Format Tools Draw Dimension Modify Express Window Help

By selecting any of the words in the MENU BAR, a **(2) Pull-down menu** appears.
If you select a word from the pull-down menu that has an **(3) Arrow ➤** a
(4) Sub Menu will appear. (Example : Draw / Circle)
If you select a word with **(5) Ellipsis ...** a dialog box will appear.
(Example: Draw / Hatch…)

DIALOG BOX

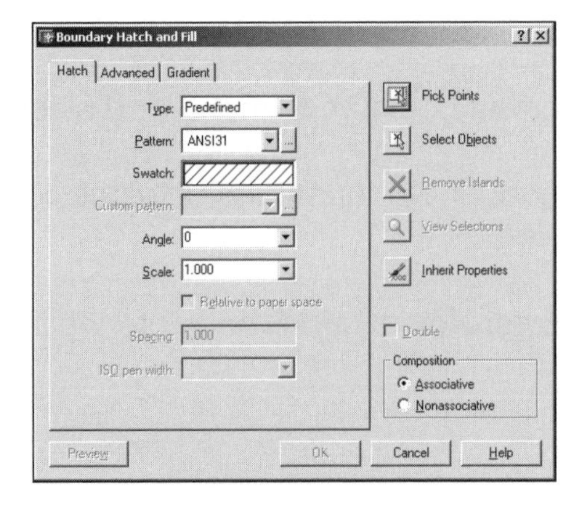

Many commands have **multiple options** and require you to make selections. These commands will display a dialog box. Dialog boxes, such as the ***Hatch*** dialog box shown above, make selecting and setting options easy.

TOOLBARS

AutoCAD provides several toolbars to access frequently used commands.

The **(1)Standard, (2)Object Properties, (3) Draw, and (4) Modify** toolbars are displayed by default.
Toolbars contain ***icon buttons (5).*** These icon buttons can be selected to Draw or Edit objects and manage files.

If you place the pointer on any icon and wait a second, a **tool tip (6)** will appear and a **help message (7)** will appear at the bottom of the screen.

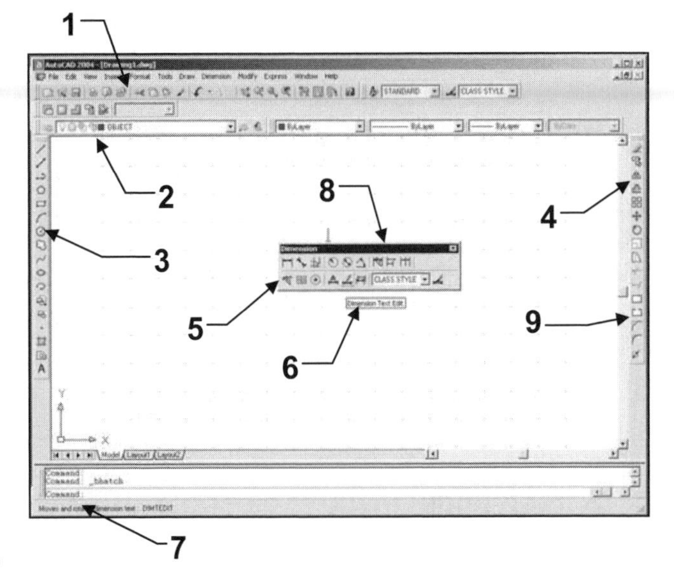

Toolbars can be ***"floated"*** or ***"docked".***

Floating toolbars (8) move freely in the drawing area and can be resized.
To move; place the pointer on the toolbar title then hold the left mouse button down, drag to the new location and release the mouse button.
To resize; place the pointer on the right or bottom edge of the toolbar. When the pointer changes to a double ended arrow, hold the left mouse button down and drag. When desired size is achieved, release the mouse button.

Docked toolbars (9) locked into place along the top, bottom or sides of the AutoCad Window.
To dock; place the pointer on the toolbar title, hold the left mouse button down and drag to the top, bottom, or either side of the AutoCAD window. When the outline of the toolbar appears, release the mouse button.

OPEN OR CLOSE TOOLBARS

Many other toolbars are available by selecting **View / Toolbars** from the Pull-down menu. Select the "Toolbars" tab. A list of available toolbars will appear.
(A check mark indicates the toolbars that are **"open"**.)

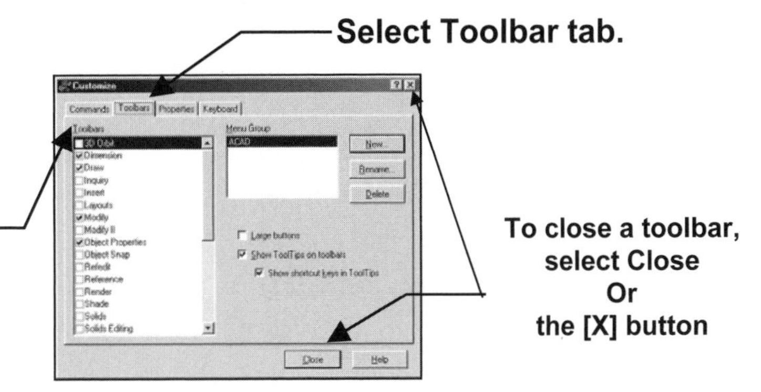

METHODS OF ENTERING COMMANDS

AutoCAD has 5 different methods of entering commands. All 5 methods will accomplish the same end result. AutoCAD allows you to use the method you prefer. The following are descriptions of all 5 methods and an example of how each one would be used to start a command such as the Line command.

1. **Pull down Menu** (page 2-8) **(Select Draw / Line)**
   The Pull down menus are activated by: **a**. moving the cursor to the Menu Bar **b**. Highlighting a Menu header **c**. Pressing the left mouse button.

2. **Tool Bars** (page 2-9) **(Select the Line icon from the Draw toolbar)**
   Move the cursor to an icon on a toolbar and press the left mouse button.

3. **Keyboard (Type L and <enter>)**
   Type the command on the command line.

4. **Screen (side) menu (Select the Draw 1 menu and then the Line command)**
   a. Move the cursor to the Screen Menu, **b**. highlight a Menu option **c**. Press the left mouse button. *Note: AutoCAD does not normally display this menu. If you want to activate this menu select Tools / Options / Display. In the "Window Elements" area, select the "Display Screen Menu" box. When activated, the Screen menu is located on the right side of the screen.*

5. **Tablet Menu (Move pointer to Line command box and press)**
   The Tablet Menu can only be used with a Digitizer. Move the mouse on the tablet overlay to the desired box and press the mouse button.

What is a SHORTCUT menu?
In addition to the methods listed above, AutoCAD has shortcut menus. Shortcut Menus give you quick access to command options. Shortcut Menus are only available when brackets [ ] enclose the options, on the command line. (Example below)
To activate a Shortcut Menu, press the right mouse button.

Example:

Select Draw / Circle / Center, Radius
_circle Specify center point for circle or **[3P/2P/Ttr (tan tan radius)]:**

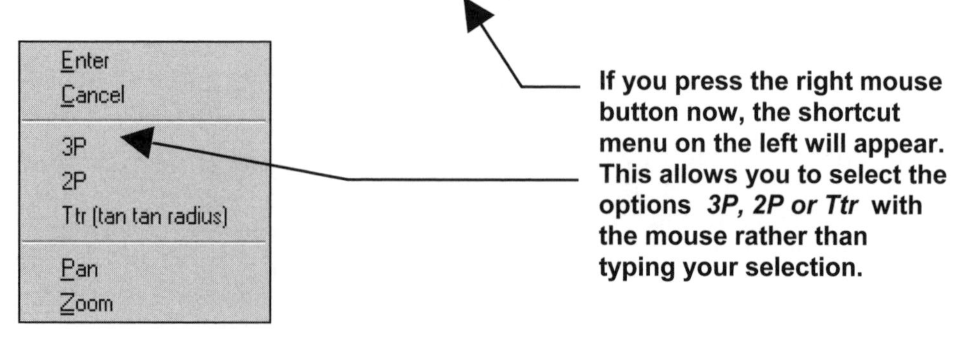

Selecting the **Shortcut menus** is the AutoCAD's **"Heads Up"** drawing method within AutoCAD. Using this method should improve your efficiency and productivity.

DRAWING LINES

A ***LINE*** *can be one segment or a series of connected segments. Each segment is a individual object.*

One segment **Series of connected segments**

Start the Line command by using one of the following methods:

Type = L <enter>
PULLDOWN MENU = DRAW / LINE
TOOLBAR = DRAW

Lines are drawn by specifying the locations for the endpoints.
Move the cursor to the location of the **"first"** endpoint then press the left mouse button.
Move the cursor again to the **"next"** endpoint and press the left mouse button.
Continue locating **"next"** endpoints until you want to stop.

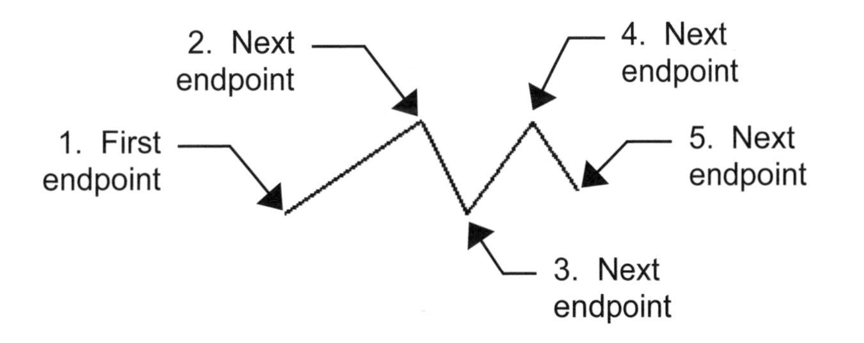

There are 3 ways to **Stop** drawing a line: 1. Press <enter> key. 2. Press <Space Bar> or 3. Press the right mouse button then select enter from the short cut menu.

To draw perfectly **Horizontal** or **Vertical** lines select the **ORTHO** mode by clicking on the **ORTHO button** on the **Status Bar** or pressing **F8.**

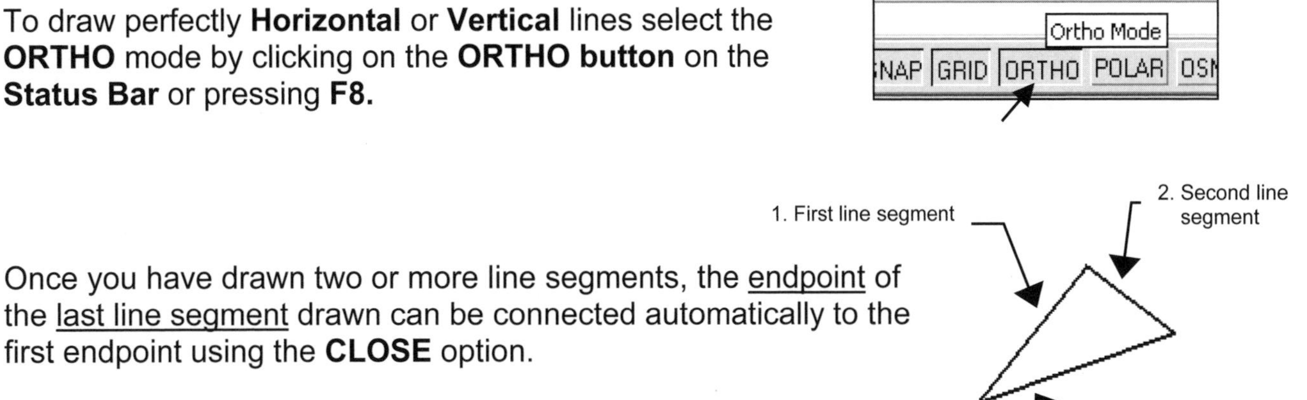

Once you have drawn two or more line segments, the endpoint of the last line segment drawn can be connected automatically to the first endpoint using the **CLOSE** option.

To use this option, draw two or more line segments, then type **C** <enter>.

ERASE

There are 3 methods to erase (or delete) objects from the drawing. Again, AutoCAD has multiple methods. You decide which one you prefer to use. They all work equally well.

METHOD 1.
Select the Erase command and then the objects.

1. Start the command by using one of the following:

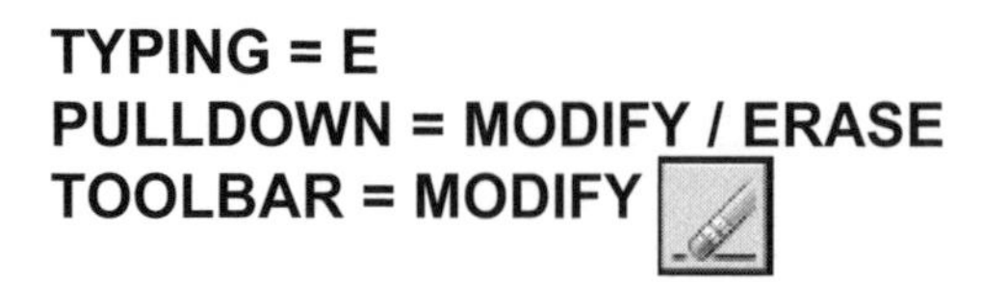

TYPING = E
PULLDOWN = MODIFY / ERASE
TOOLBAR = MODIFY

2. Select objects: ***pick one or more objects***
   Select objects: ***press <enter> and the objects will disappear***

METHOD 2.
Select the Objects and then the Erase command from the shortcut menu.

1. Select the object(s) to be erased
2. Press the right mouse button.
3. Select **"Erase"** from the short-cut menu.

METHOD 3.
Select the Objects and then the Delete key

1. Select the object(s) to be erased
2. Press the Delete key.

NOTE: Very important
If you want the objects to return, press **U <enter>** or **Ctrl + Z** or the **Undo arrow icon.**

This will **"Undo"** the effects of the last command.

METHODS OF SELECTING OBJECTS

Most AutoCAD commands prompt you to "select objects". This means, select the objects that you want the command effect.

There are 2 methods. **Method 1. Pick**, is very easy and should be used if you have only 1 or 2 objects to select. **Method 2. Window**, is a little more difficult but once mastered it is extremely helpful and time saving. Practice the examples shown below.

Method 1. PICK : When the command line prompt reads, "Select Objects", place the cursor (pick box) on top of the object and click the left mouse button. The selected object will change in appearance. This appearance change is called "dithered". This gives you a visual notice of which objects have been selected.

Method 2. WINDOW: Crossing and Window

Crossing:
Place your cursor in the area up and to the right of the objects that you wish to select (P1) and press the left mouse button. Then move the cursor down and to the left of the objects (P2) and press the left mouse button again. **Only** objects that this window **crosses or completely encloses** will be selected.

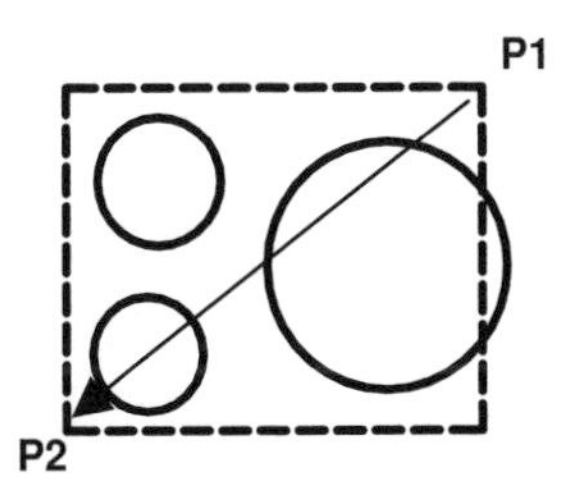

In the example on the right, all 3 circles have been selected. (The 2 small circles are ***completely enclosed*** and
the large circle is ***crossed*** by the window.)
Note: Crossing windows are identified by a ***dashed*** line appearance.

Window:
Place your cursor in the area up and to the left of the objects that you wish to select (P1) and press the left mouse button. Then move the cursor down and to the right of the objects (P2) and press the left mouse button. **Only** objects that this window **completely encloses** will be selected.

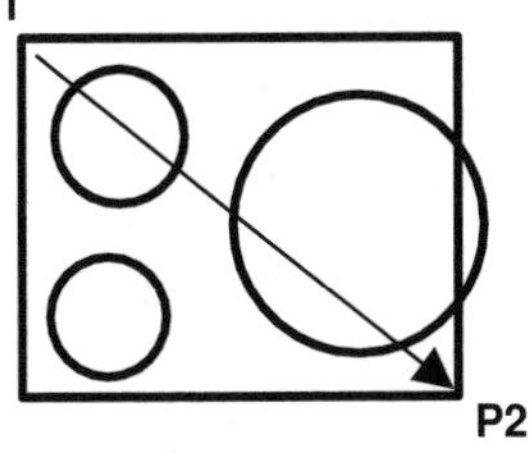

In the example on the right, only 2 circles have been selected. (The large circle is ***not*** completely enclosed.) This Window is identified by a ***continuous (solid)*** line appearance.

Note: if these windows do not show up on your screen, it means that your "implied windowing" is turned off. Select Tools / Options / Selection tab. In the section "Selection Modes", on the left, place a check mark in the "implied windowing" box. Now select "Apply" and "OK" at the bottom of the dialog box.

STARTING A NEW DRAWING (or clearing the screen)

1. Start the command using one of the following methods:

TYPING: **NEW or CTRL + N**
PULLDOWN: **FILE / NEW**
TOOLBAR: **STANDARD**

The Dialog box shown below should appear.

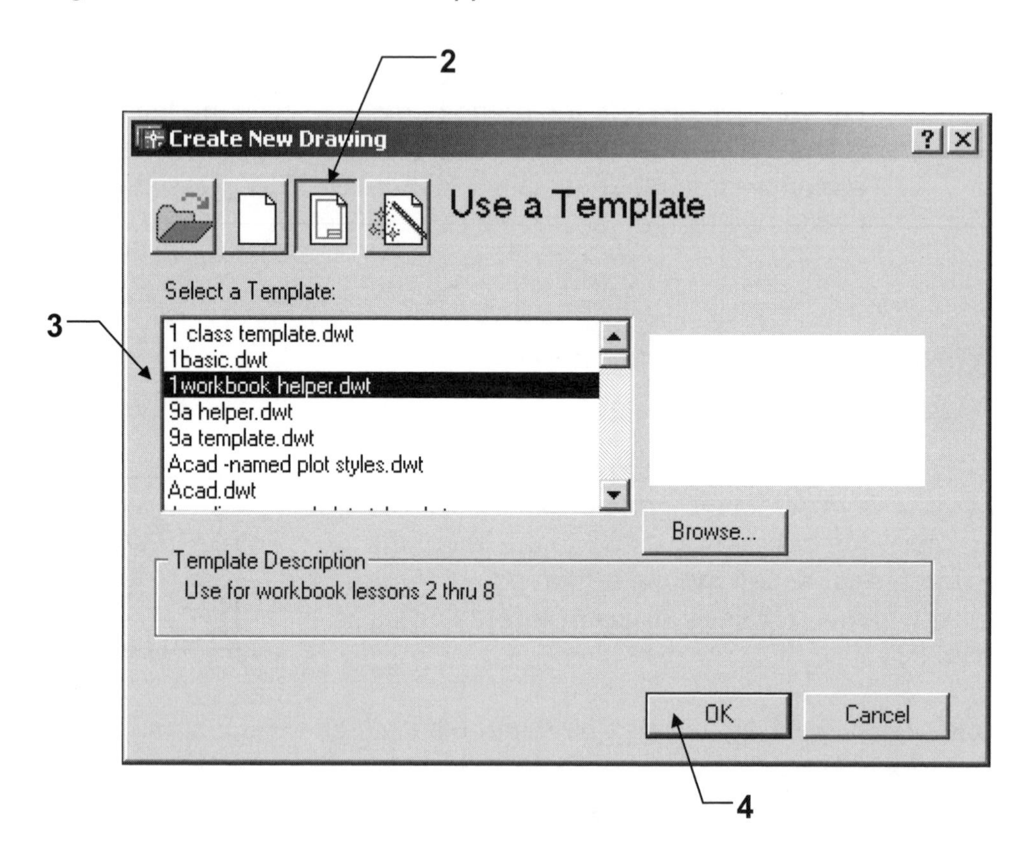

2. Select the **Use a Template** box. (third from the left)
3. Select **1workbook helper.dwt** from the list of templates.
(NOTE: If you do not have this template, refer to page 2-2)
4. Select the **OK** button (bottom right).

NOTE: It is important that you have configured your AutoCAD software to only allow one drawing open. It will be less confusing for now. When you progress to the Advanced workbook, you will configure AutoCAD for multiple open drawings. Refer to Intro-4 for “Single drawing compatibility mode” setting.

SAVING A DRAWING

After you have completed a drawing, it is very important to save it. Learning how to save a drawing correctly is more important than making the drawing. If you can't save correctly, you will lose the drawing and hours of work.

There are 2 commands for saving a drawing: ***Save*** *and* ***Save As****. I prefer to use* ***Save As****. The* ***Save As*** *command always pauses to allow you to choose where you want to store the file and what name to assign to the file. This may seem like a small thing, but it has saved me many times from saving a drawing on top of another drawing by mistake. The* ***Save*** *command will automatically save the file either back to where you retrieved it or where you last saved a previous drawing. Neither may be the correct destination. So play it safe, use* ***Save As*** *for now.*

1. Start the command by using one of the following methods:

TYPING:	**SAVEAS**
PULLDOWN:	**FILE / SAVEAS**
TOOLBAR:	**No icon for Save As, only for Save (Don't use it)**

This Dialog box should appear:

Descriptions on page 2-13

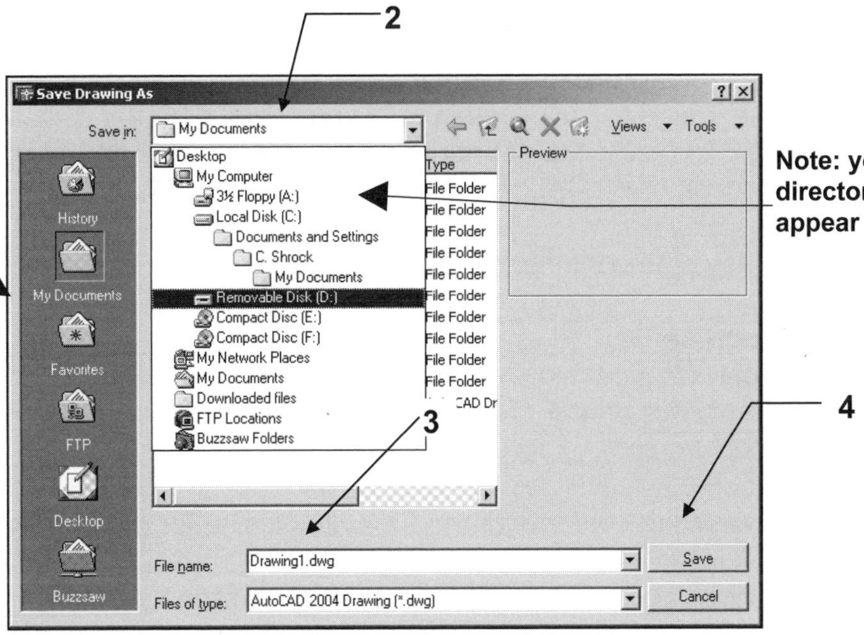

Note: your directories may appear different

2. Select the appropriate drive and directory from the **"SAVE IN"** box. (This is where your drawing will be saved)
3. Type the new drawing file name in the **"FILE NAME"** box.
4. Select the **"SAVE"** box.

BACK UP FILES

When you save a drawing file, Autocad creates a file with a **.dwg** extension. For example, if you save a drawing as **House**, Autocad saves it as **House.dwg**. The next time you save that same drawing, Autocad replaces the old with the new but renames the old version to **House.bak**. The old version is now a back up file.
(Only 1 backup file is stored)

How to open a back up file.
You can not open a **.bak** file. It must first be renamed to a **.dwg** file extension.

How to view the list of back up files.
Type "*.bak" in the "file name" box and <enter>. A list of the backup (.bak) files, within the chosen directory, will appear.

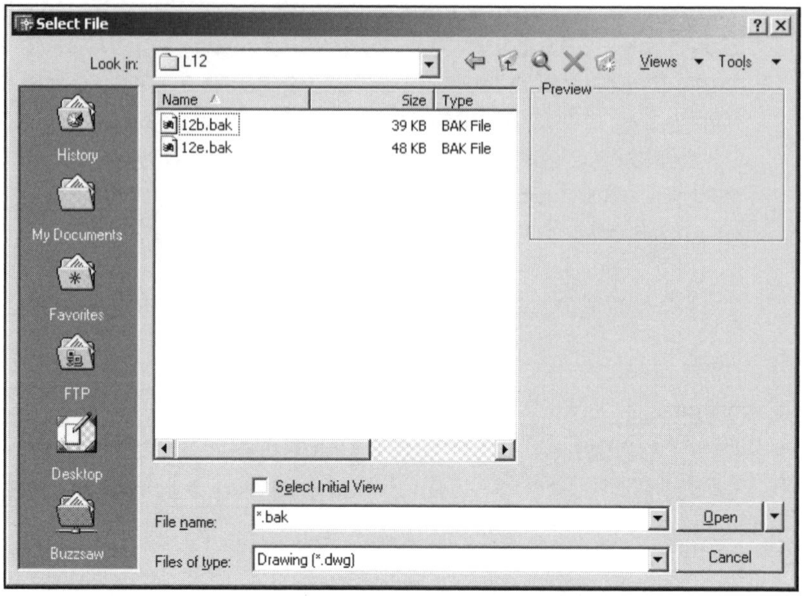

<u>The following is information only. We will not be using these in this workbook.</u>

History: Displays shortcuts to the files most recently accessed from the dialog box. Note: The shortcuts remain until you remove them. This list can get very long. To delete the shortcuts go to: Windows / Application Data / Autodesk / AutoCAD / Recent / Save Drawing As.

Desktop: Displays the contents of your desktop.

Personal / My Documents:
Displays the contents of the Personal or My Documents folder for the current user profile. The name of this location depends on your operating system version.

Favorites: Displays the contents of the Favorites folder for the current user profile.

Buzzsaw: Provides access to projects hosted by Buzzsaw.com—a business-to-business marketplace for the building design and construction industry.

RedSpark: Provides access to projects hosted by RedSpark—a business-to-business marketplace for the manufacturing industry.

FTP: Displays the FTP sites that are available for browsing in the standard file selection dialog box.

OPENING AN EXISTING DRAWING FILE

1. Start the command by using one of the following methods.

TYPING:	**OPEN or CTRL + O**
PULLDOWN:	**FILE / OPEN**
TOOLBAR:	**STANDARD**

The Dialog box shown below should appear.

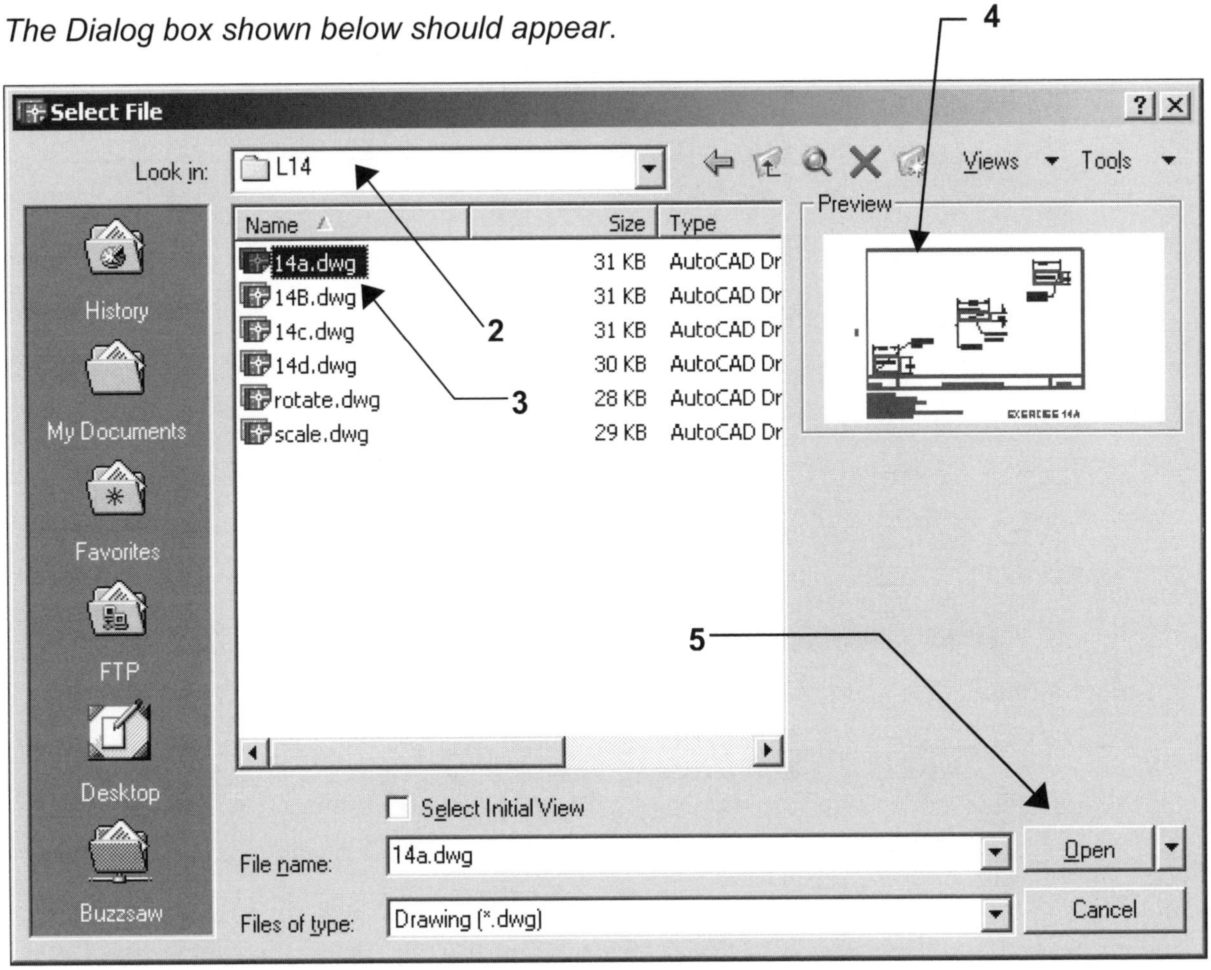

2. Select the **Drive and Directory** from the **"LOOK IN"** Box.
3. Select the drawing file from the list. (You may double click on the file name to automatically open the drawing)
4. Displays a "Thumbnail Preview Image" of the selected file.
5. Select the **OPEN** button.

NOTE: It is important that you have configured your AutoCAD software to only allow one drawing open. It will be less confusing for now. When you progress to the Advanced workbook, you will configure AutoCAD for multiple open drawings. Refer to Intro-4 for "Single drawing compatibility mode" setting.

EXITING AUTOCAD

1. Start the command by using one of the following methods:

TYPING: **EXIT OR QUIT**
PULLDOWN: **FILE / EXIT (Safest method)**
TOOLBAR: **NONE**

This Pull-down menu should appear when you select "File / Exit:

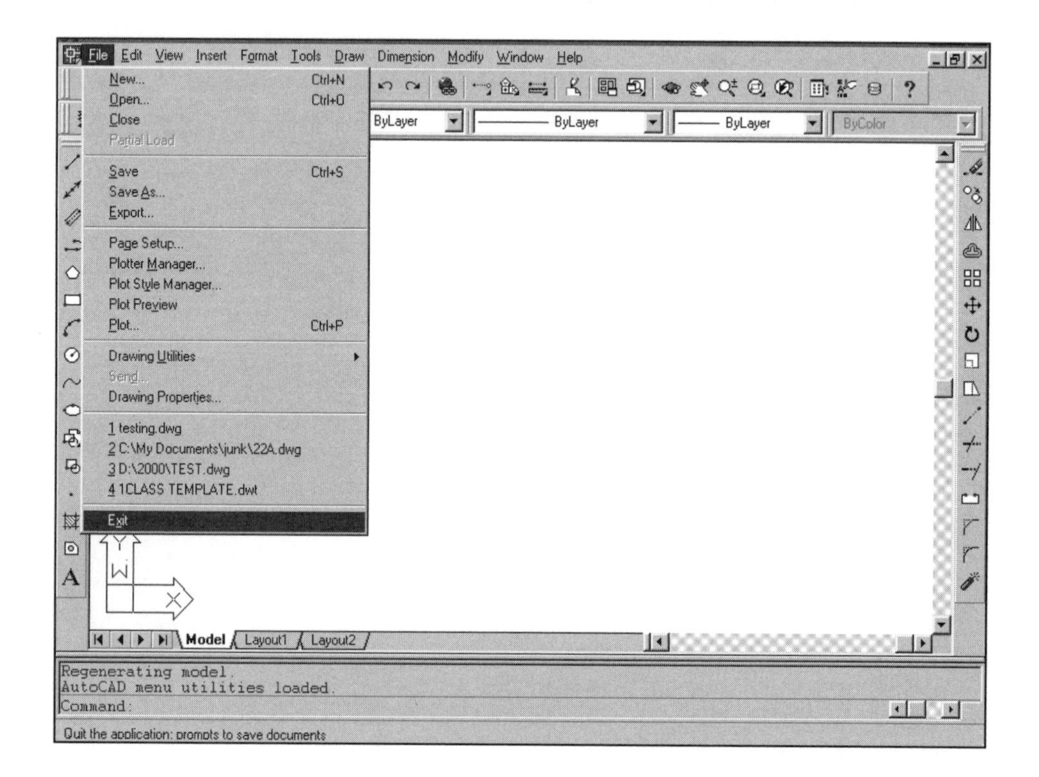

If any changes have been made to the drawing since the last save, the dialog box below will appear asking if you want to ***SAVE THE CHANGES***?

You will have to select ***YES, NO*** *or* ***CANCEL***.

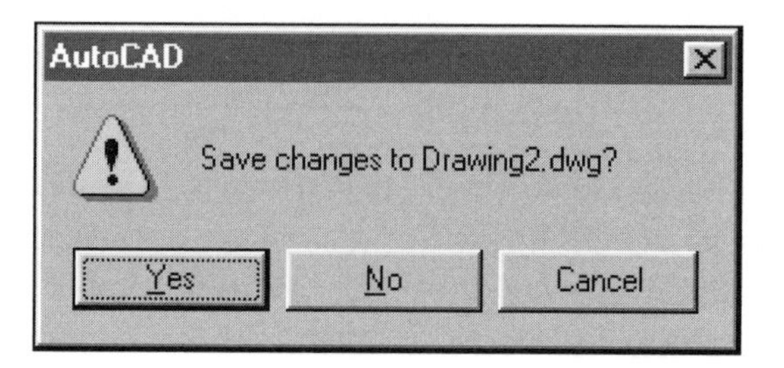

EXERCISE 2A

INSTRUCTIONS:

1. Start a **New** file (refer to 2-15) and select **1workbook helper.dwt**.
2. **Draw** the objects below using:
   LINE command
   Ortho (f8) **ON** for **Horizontal** and **Vertical** lines
   Ortho (f8) **OFF** for lines drawn on an **Angle.**
   Increment Snap (f9) **ON**
   Osnap (f3) **OFF**
3. **Save** this drawing using:
   File / Save as / **EX2A**

EXERCISE 2B

INSTRUCTIONS:

1. Using drawing **EX2A**, **ERASE** the missing lines.
2. **Save** this drawing using:
   File / Save as / **EX2B**

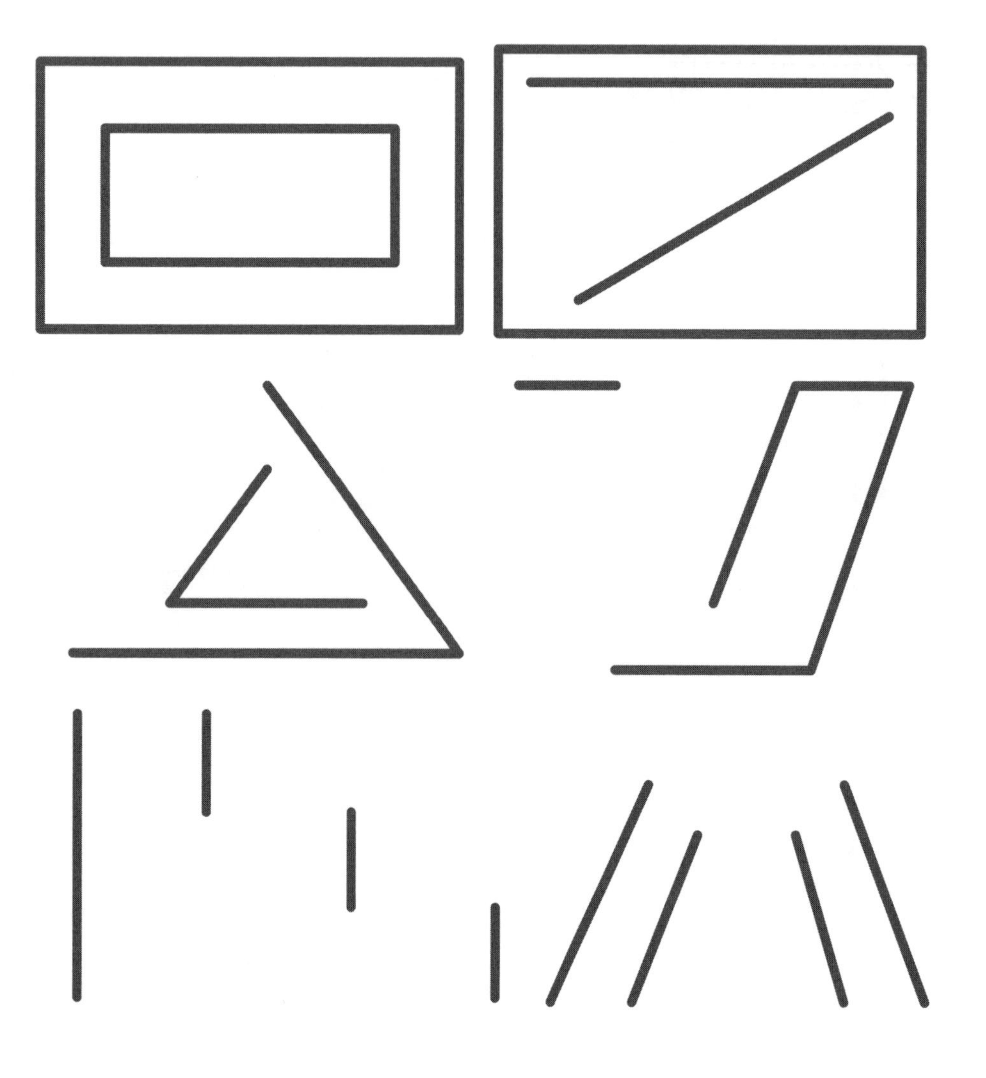

EXERCISE 2C

INSTRUCTIONS:

1. Start a **New** file and select **1workbook helper.dwt**.
2. **Draw** the objects below using:
   Draw / Line
   Ortho (f8) **ON** for **Horizontal** and **Vertical** lines
   Ortho (f8) **OFF** for lines drawn on an **Angle.**
   Increment Snap (f9) **ON**
   Osnap (f3) **OFF**
3. **Save** this drawing using:
   File / Save as / **EX2C**

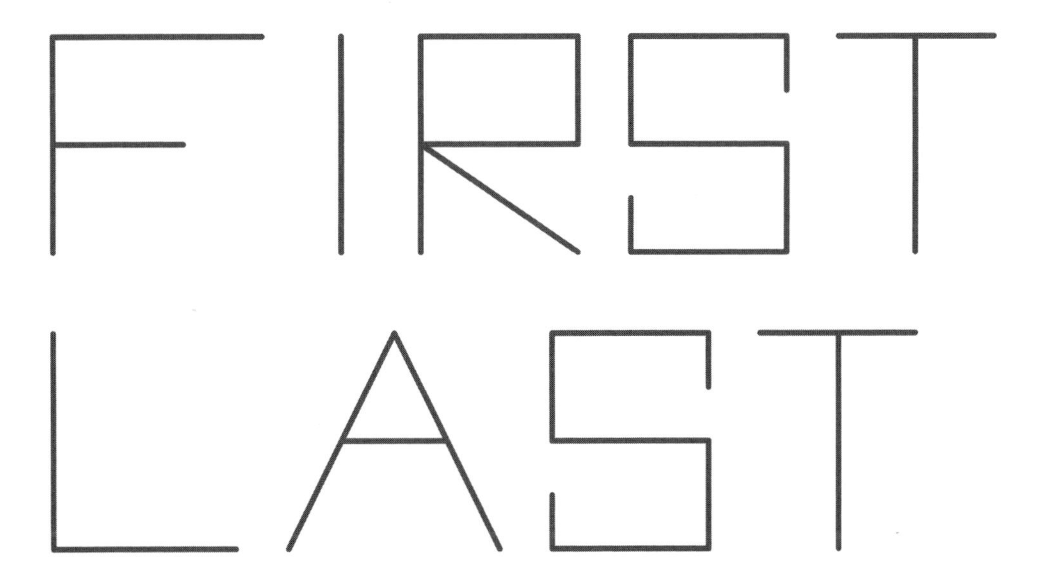

EXERCISE 2D

INSTRUCTIONS:

1. Start a **New** file and select **1workbook helper.dwt**.
2. **Draw** the objects below using:
   Draw / Line
   Ortho (f8) **ON** for **Horizontal** and **Vertical** lines
   Ortho (f8) **OFF** for lines drawn on an **Angle.**
   Increment Snap (f9) **ON**
   Osnap (f3) **OFF**
3. **Save** this drawing using:
   File / Save as / **EX2D**

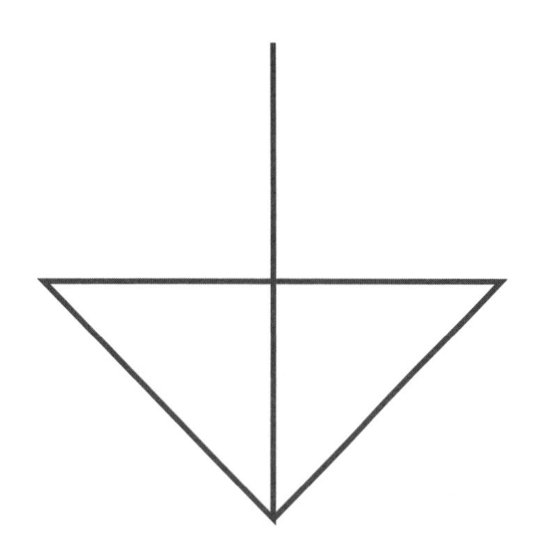
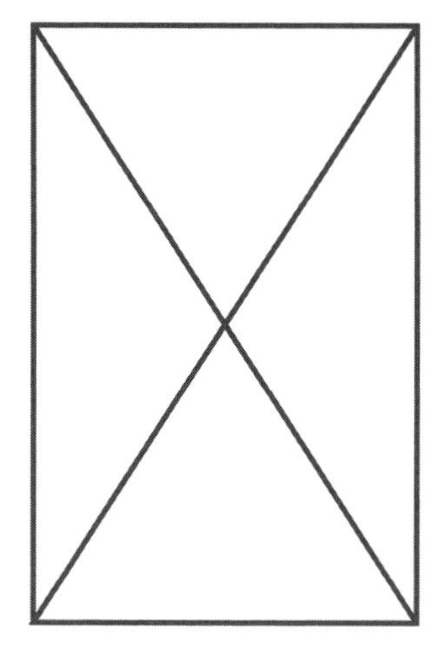

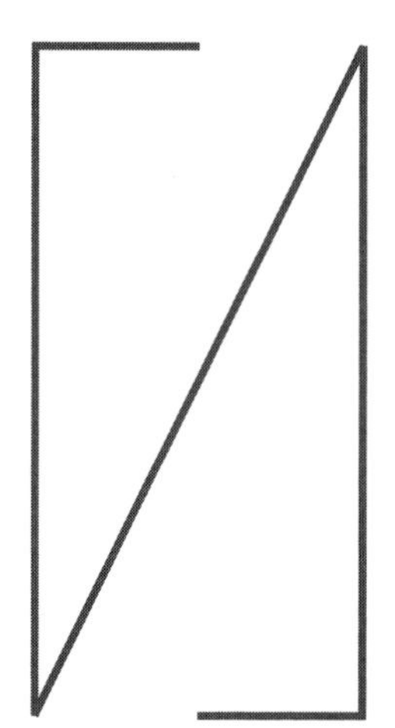

LEARNING OBJECTIVES

After completing this lesson, you will be able to:

1. Create a Circle using 6 different methods.
2. Create a Rectangle with width, chamfers or fillets.
3. Set Grids and Increment Snap using the Drafting Settings option.
4. Change current Layers.

LESSON 3

CIRCLE

*There are 6 options to create a circle. The default option is "**Center, radius**".*
(Probably because that is the most common method of creating a circle.)
*We will try the **"Center, radius"** option first.*

1. Start the **Circle** command by using one of the following:
   TYPING = C
   PULLDOWN = DRAW / CIRCLE
   TOOLBAR = DRAW

2. The following will appear on the command line:

Command: _circle Specify center point for circle or [3P/2P/Ttr (tan tan radius)]:

3. Locate the center point, for the circle, by moving the cursor to the desired location in the drawing area and press the left mouse button.

4. Now move the cursor away from the center point and you should see a circle forming.

5. When it is approximately the size desired, press the left mouse button, or if you want the exact size, type the radius, then press <enter>.

Note: To use one of the other methods described below, first select the Circle command, then press the right mouse button. A "short cut" menu will appear. Select the method desired by placing the cursor on the option and pressing the left mouse button. Or you can type 3P or 2P or T, then press <enter> (The short cut menu is simple and more efficient)

Center, Radius: (Default option)
1. Specify the center (P1) location
2. Specify the Radius (P2)

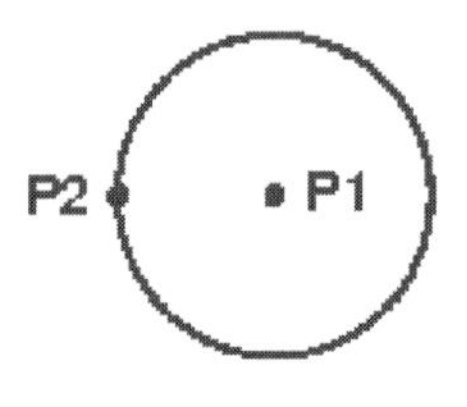

Center, Diameter:
1. Specify the center (P1) location
2. Select the Diameter option using the shortcut menu or type "D" <enter>
3. Specify the Diameter (P2).

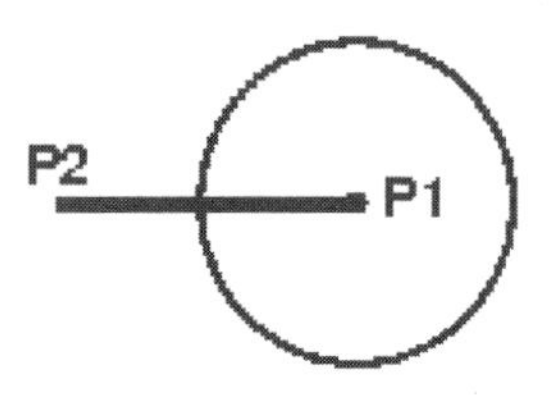

2 Points:

1. Select the 2 point option using the short cut menu or type 2P <enter>
2. Specify the 2 points (P1 and P2) that will determine the Diameter .

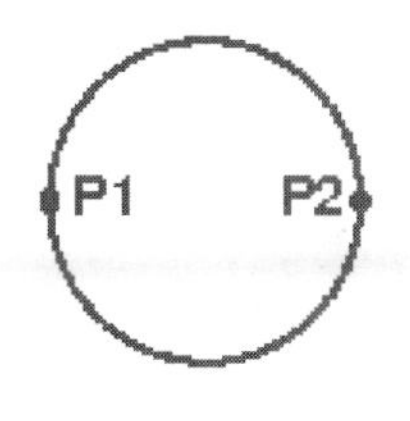

3 Points:

1. Select the 3 Point option using the short cut menu or type 3P <enter>
2. Specify the 3 points (P1, P2 and P3) on the circumference.
   The Circle will pass through all three points.

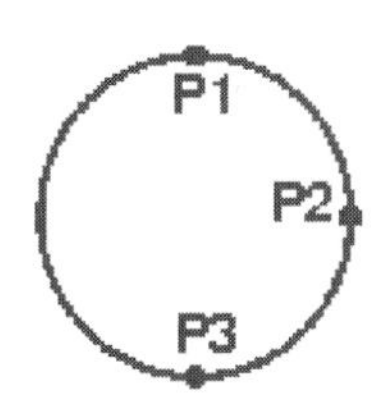

Tangent, Tangent, Radius:

1. Select the Tangent, Tangent, Radius option using the short cut menu or type T <enter>
2. Select two objects (P1 and P2) for the Circle to be tangent to by placing the cursor on the object and pressing the left mouse button
3. Specify the radius.

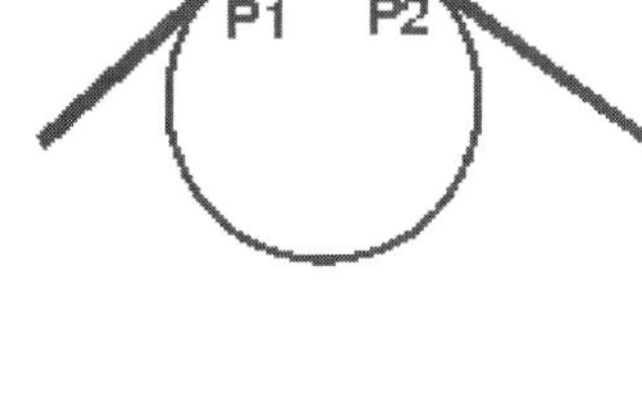

Tangent, Tangent, Tangent:

1. Select the Tangent, Tangent, Tangent option using the pull down menu. This option is not available in the short cut menu. (I don't know why, so don't ask)
2. Specify three objects (P1, P2 and P3) for the Circle to be tangent to by placing the cursor on the object and pressing the left mouse button. (The diameter will be calculated by the computer.)

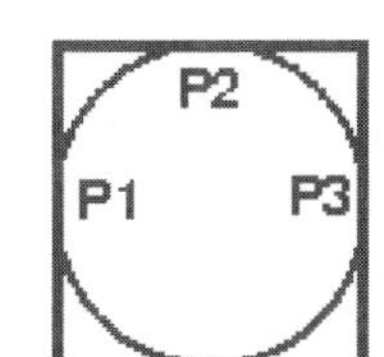

RECTANGLE

To create a rectangle you must specify two diagonal corners.
The rectangle can be any size and the sides are always drawn horizontal and vertical.
A Rectangle is one object, not four separate lines.

1. Start the **RECTANGLE** command by using one of the following:

    TYPING = REC
    PULLDOWN = DRAW / RECTANGLE
    TOOLBAR = DRAW

2. The following will appear on the command line:

Command: _rectang
Specify first corner point or [Chamfer/Elevation/Fillet/Thickness/Width]:

3. Specify the location of the first corner by moving the cursor to a location (P1) and then press the left mouse button.

    The following will appear on the command line:

    Specify other corner point or [Dimension]:

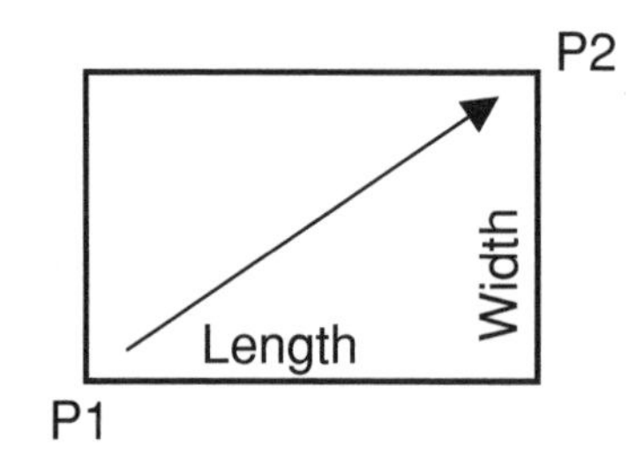

4. Specify the location of the **diagonal** corner (P2) by moving the cursor diagonally from the first corner (P1) and pressing the left mouse button.

    --OR—

    Type **D** <enter> (or use the short cut menu)
    Specify length for rectangle <0.000>: ***Type length <enter>***
    Specify width for rectangle <0.000>: ***Type width <enter>***
    Specify other corner point or [Dimension]: ***move the cursor up, down, right or left to specify where you want the second corner relative to the first corner.***

OPTIONS:

To select one of the following options, use the short cut menu or type the uppercase letter, such as "C" for Chamfer.

CHAMFER Automatically draws all 4 corners with chamfers. (All the same size)
FILLET Automatically draws all 4 corners with fillets. (All the same size)
WIDTH Sets the width of the rectangle lines. (Note: Do not confuse this with the Length and Width. This makes the lines appear to have width.)
ELEVATION Used in 3D only.
THICKNESS Used in 3D only.

DRAFTING SETTINGS

The **DRAFTING SETTINGS** dialog box allows you to set the **INCREMENT SNAP** and **GRID SPACING**. You may change the Increment Snap and Grid Spacing at anytime while creating a drawing. The settings are only drawing aids to help you visualize the size of the drawing and control the movement of the cursor.

INCREMENT SNAP controls the movement of the cursor. If it is **OFF** the cursor will move smoothly. If it is **ON**, the cursor will jump in an *incremental* movement. This incremental movement is set by changing the **"Snap X and Y spacing"** .

GRID is the dot matrix in the drawing area. Grid dots will not print. It is only a visual aid. The Grid dot spacing is set by changing the **"Grid X and Y spacing".**

1. Select **DRAFTING SETTINGS** by using one of the following:

   TYPING = DS or DSETTINGS
   PULL-DOWN = TOOLS / DRAFTING SETTINGS
   TOOLBAR = NONE

2. The dialog box shown below will appear.

3. Select the **"Snap and Grid"** tab.

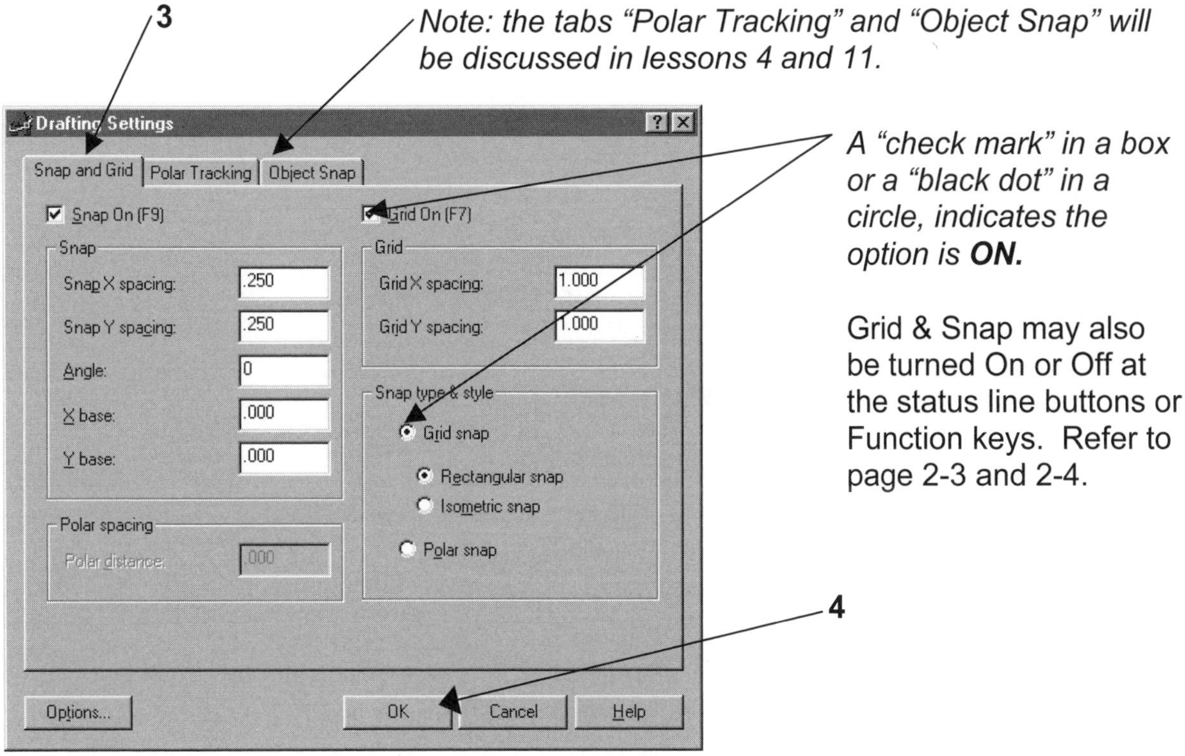

*A "check mark" in a box or a "black dot" in a circle, indicates the option is **ON.***

Grid & Snap may also be turned On or Off at the status line buttons or Function keys. Refer to page 2-3 and 2-4.

4. Make your changes and select the **OK** button to save them.
   If you select the **CANCEL** button, your changes will **not** be saved.

LAYERS

A **LAYER** is like a transparency. Have you ever used an overhead light projector? Remember those transparencies that are laid on top of the light projector? You could stack multiple sheets but the projected image would have the appearance of one document. Layers are basically the same. Multiple layers can be used but it is still only one drawing.

This example shows 3 layers.
One for annotations (text), one for dimensions and one for objects.

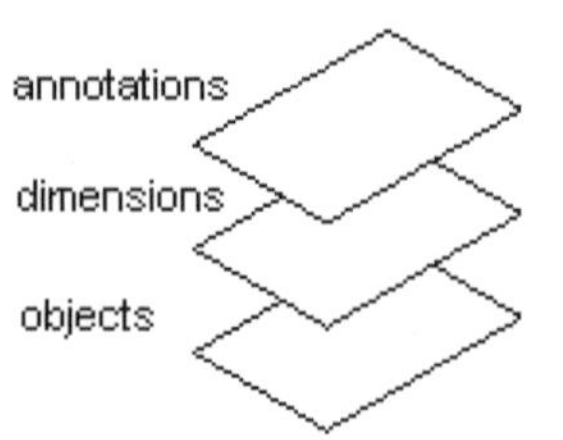

It is good "drawing management" to draw related objects on the same layer. For example, in an architectural drawing, you could have the walls of a floor plan on one layer and the Electrical and Plumbing on two other layers. These layers can then be Thawed (ON) or Frozen (OFF) independently. If a layer is Frozen, it is not visible. When you Thaw the layer it becomes visible again. This will allow you to view or make plots with specific layers visible or invisible.
(You will learn more about layers in lesson 26)

SELECTING A LAYER - Method 1. (Method 2 on next page)

1. Display the LAYER CONTROL DROP-DOWN LIST below by clicking on the ▼ down arrow.

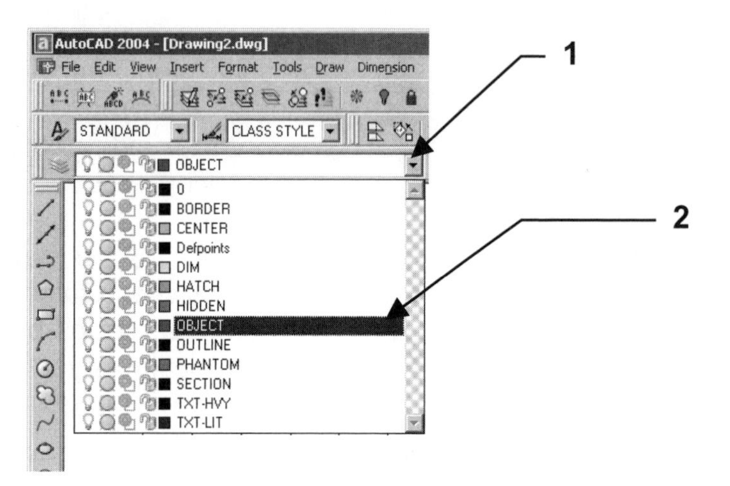

2. Click on the LAYER **NAME** you wish to select. The Layer selected will become the **CURRENT** layer and the drop-down list will disappear.

SELECTING A LAYER - Method 2.

1. Select the Layer command using one of the following:

   TYPE = LA
   PULLDOWN = FORMAT / LAYER
   TOOLBAR = OBJECT PROPERTIES 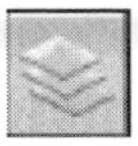

2. The dialogue box below will appear.

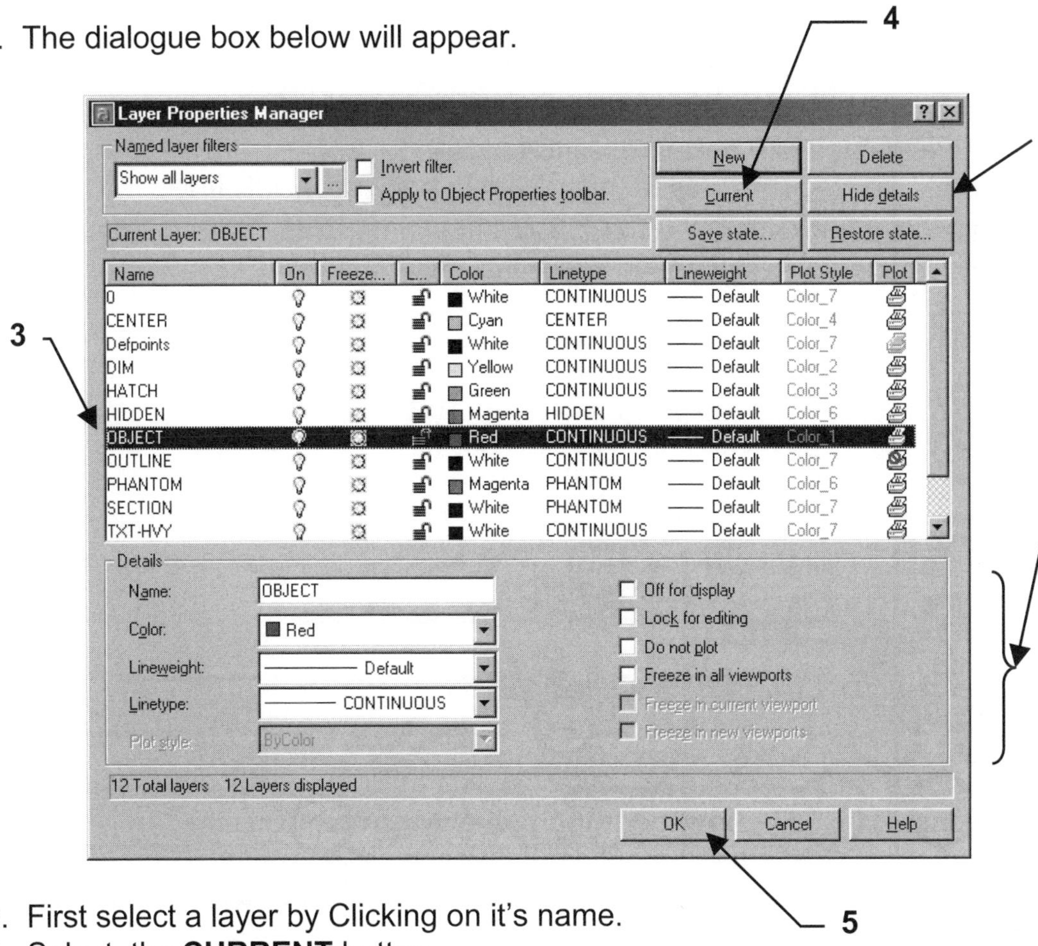

See "Note About Details" below.

3. First select a layer by Clicking on it's name.
4. Select the **CURRENT** button.
5. Then select the **OK** button.

The layer you have just selected is now the **CURRENT** layer. This means that the next object drawn will reside on this layer and will have the same properties as this layer. Notice that the layers shown above have specific colors, linetypes and lineweights. These are called Properties.

Note about "Details":
If your dialog box does not have the "**Details**" area at the bottom, select the "Show details" button. When you select the "Show details" button, it will change to a "Hide details" button, as shown above.

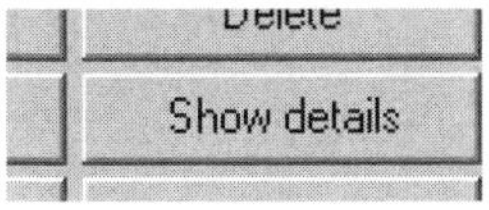

EXERCISE 3A

INSTRUCTIONS:

1. Start a **New** file and select **1workbook helper.dwt**
2. **Draw** the **LINES** below using:
   Draw / Line
   Ortho (f8) **ON** (to help you draw horizontal lines)
   Increment Snap (f9) **ON**
3. **Change** to the appropriate **layer** before drawing each line.
4. **Save** this drawing using:
   File / Save as / **EX3A**

Layer HIDDEN

Layer OBJECT

Layer PHANTOM

Layer SECTION

Layer TXT-HVY

Layer TXT-LIT

Layer DIM

Layer CENTER

Layer HATCH

EXERCISE 3B

INSTRUCTIONS:

1. Start a **New** file and select **1workbook helper.dwt**
2. Change the **GRID SPACING** to .40 and **SNAP** to .20 using: **TOOLS / DRAFTING SETTINGS**
3. Draw the objects below, use the **layers** indicated.
4. **Save** this drawing using:
   File / Save as / **EX3B**

LAYER = OBJECT

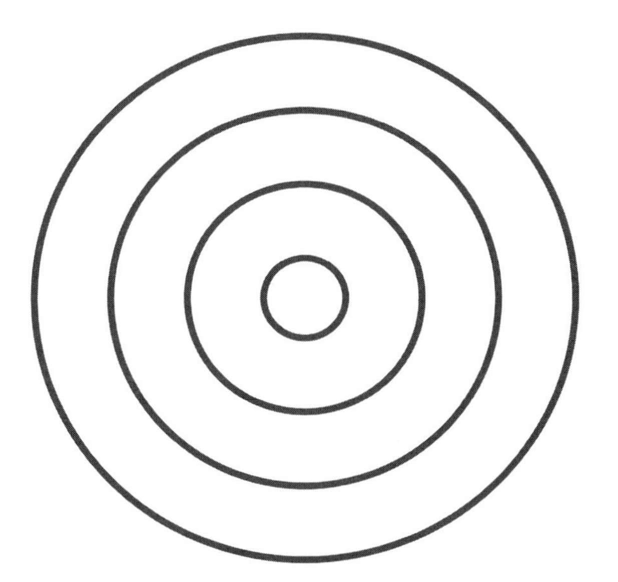

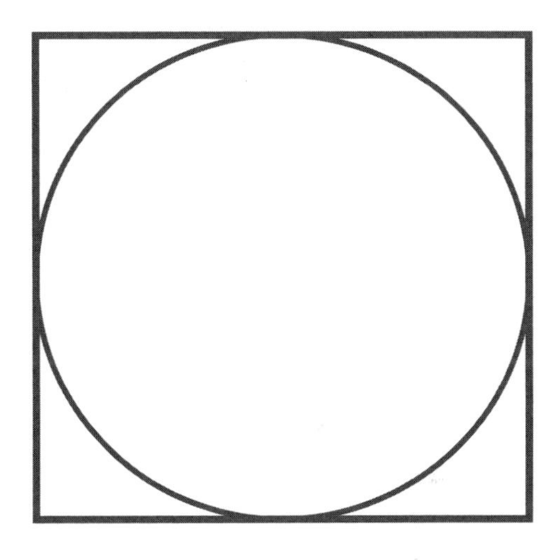

LAYER = HIDDEN

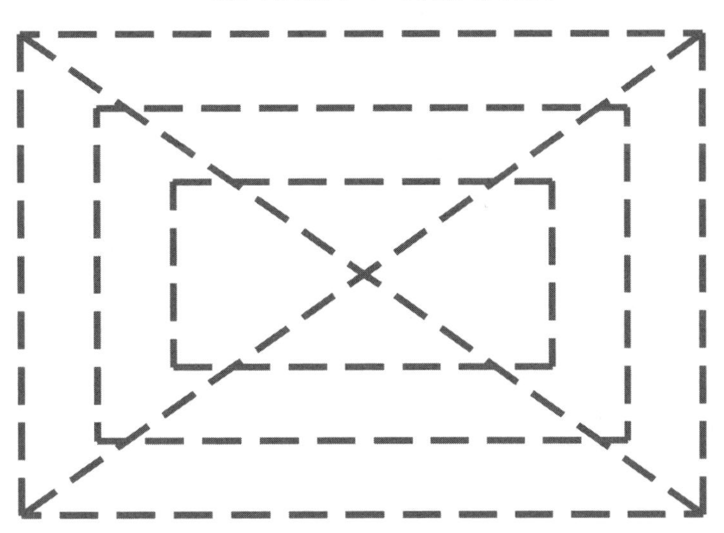

EXERCISE 3C

INSTRUCTIONS:

1. Start a **New** file and select **1workbook helper.dwt**.
2. Draw the **RECTANGLES** below using the options:
   CHAMFER, FILLET and **WIDTH**
3. **Save** this drawing as: **EX3C**

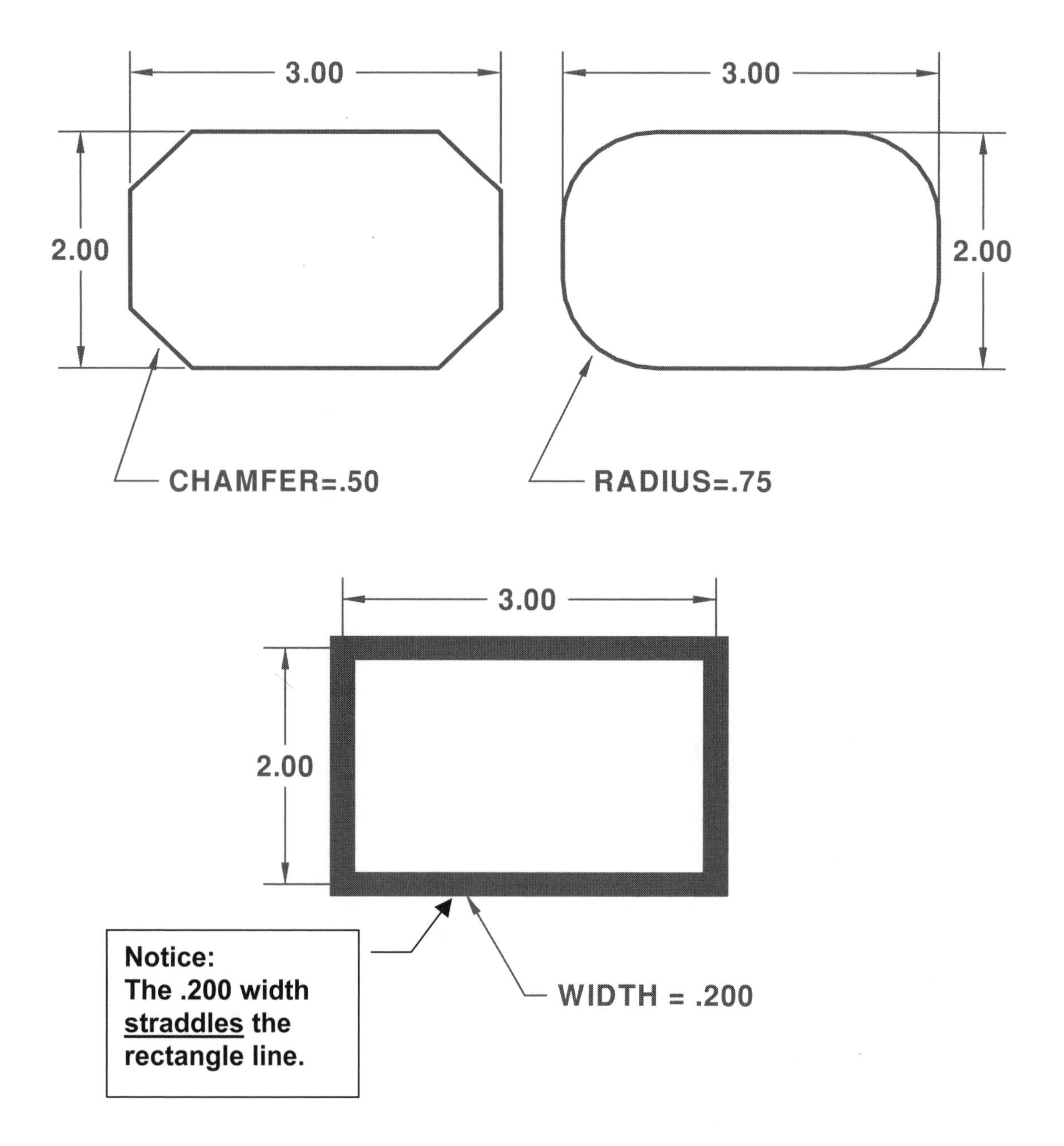

EXERCISE 3D

INSTRUCTIONS:

1. Start a **New** file and select **1workbook helper.dwt.**
2. Draw the house below using at least <u>4 different layers</u>.
3. You can change the **GRID** and **INCREMENT SNAP** settings to whatever you like.
4. You decide when to turn Ortho and Snap On or Off.

   Have some fun with this one!

5. Save this drawing as: **EX3D**

NOTES:

LEARNING OBJECTIVES

After completing this lesson, you will be able to:

1. Understand the function of Object Snap.
2. Use 7 Object Snap modes.
3. Operate the Running Snap function.
4. Toggle the Running Snap function On and Off.
5. Use the Zoom options to view the drawing.
6. Understand the basic concept of Setting up your drawing.
7. Change the drawing paper size.
8. Select the Units of Measurement to draw with.

LESSON 4

OBJECT SNAP

In Lesson 2 you learned about Increment Snap. Increment Snap enables the cursor to move in an incremental movement. So you could say your cursor is "snapping to increments" preset by you.

Now you will learn about Object Snap. If Increment Snap, snaps to increments, what do you think Object Snap, snaps too? That's right, object snap enables you to snap to "objects", in very specific and accurate locations. For example: the endpoint of a line or the center of a circle.

Selecting an Object Snap option using the Toolbar: select View / Toolbars.
Select the toolbar tab and then the Object Snap box. (Close the Customize dialog box)

Selecting an Object Snap option using a Popup Menu: *(I prefer this method)*
Method 1: Press the wheel and the Object Snap menu will appear.
*(**Note**: The command **"Mbuttonpan"** must be set to **0.** Refer to Intro-7)*
Method 2: While holding down the shift key, press the right mouse button and the Object Snap menu will appear.

OBJECT SNAP OPTIONS: *(Note: Refer to Lesson 5 for more Object Snap selections.)*

ENDpoint	Snaps to the closest endpoint of a Line, Arc or polygon segment. Place the cursor on the object close to the end.
MIDpoint	Snaps to the middle of a Line, Arc or Polygon segment. Place the cursor anywhere on the object.
INTersection	Snaps to the intersections of any two objects. Place the Pick box directly on top of the intersection or select one object and then the other and Autocad will locate the intersection.
CENter	Snaps to the center of an Arc, Circle or Donut. Place the cursor on the object, or the approximate center location.
QUAdrant	Snaps to a 12:00, 3:00, 6:00 or 9:00 o'clock location on a circle. Place the cursor on the circle near the desired quadrant location.
PERpendicular	Snaps to a point perpendicular to the object selected. Place the cursor anywhere on the object.
TANgent	Calculates the tangent point of an Arc or Circle. Place the cursor on the object as near as possible to the expected tangent point. (Note: nothing happens until you select the next point.)

How to use OBJECT SNAP

The following is an example of attaching a line segment to previously drawn vertical lines. The new line will start from the upper endpoint (P1) to the midpoint (P2) to the lower endpoint (P3).

1. Select the Line command
2. Draw two vertical lines as shown below.
3. Select the Line command again.
4. Select the “Endpoint” object snap option using one of the methods listed on the previous page.
5. Place the cursor close to the upper endpoint of the left hand line (P1). *(Notice that a square appears at the end of the line, an “endpoint” tool tip should appear and the cursor snaps to the endpoint like a magnet. This is what “object snap” is all about. You are snapping the cursor to a previously drawn object.)*
6. Press the left mouse button to attach the new line to the endpoint of the previously drawn line. (Do not end the Line command yet)
7. Now select the “Midpoint” object snap option.
8. Move the cursor to approximately the middle of the right hand vertical line (P2). A triangle and a “midpoint” tool tip appear and the cursor should snap to the middle of the line like a magnet.
9. Press the left mouse button to attach the new line to the midpoint of the previously drawn line. (Do not end the Line command yet)
10. Select the “endpoint” object snap option.
11. Move the cursor close to the lower endpoint of the left hand vertical line (P3).
12. Press the left mouse button to attach the new line to the endpoint of the previously drawn line.
13. Disconnect by pressing <enter>

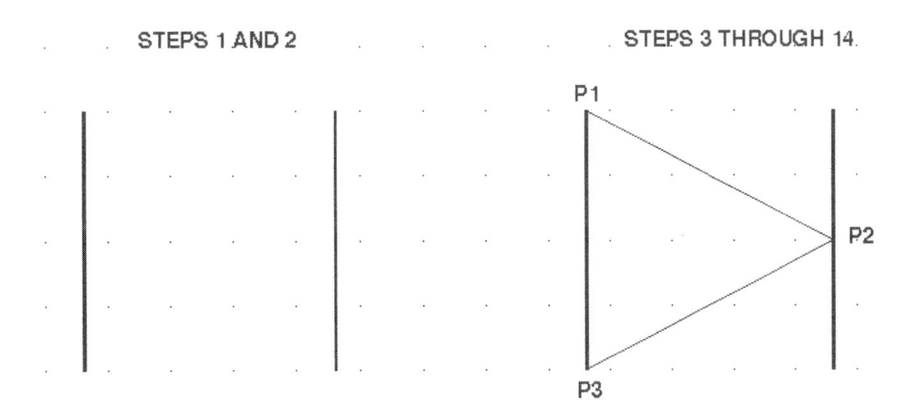

RUNNING OBJECT SNAP

RUNNING OBJECT SNAP is a method of **presetting** the **object snap options** so specific options, such as center, endpoint or midpoint, stay **active** until you **de-activate** them. When Running Object Snap is active, markers are displayed automatically as you move the cursor near the object and the cursor is drawn, to the object snap location, like a **magnet**.

For example, if you need to snap to the endpoint of 10 lines, you could preset the running object snap **endpoint** option. Then when you place the cursor near any one of the lines, a marker will appear at the endpoint and the cursor will automatically snap to the endpoint of the line. You then can move on to the next and the next and the next. Thus eliminating the necessity of invoking the object snap menu for each endpoint.

Running Object Snap can be toggled **ON** or **OFF** using the **F3** key or **clicking** on the **OSNAP** button on the status bar.

<u>Setting Running Object Snap</u>

1. Select the **Running Object Snap** option using one of the following:

    TYPE = OSNAP
    PULL DOWN = TOOLS / DRAFTING SETTINGS
    Right Click on the OSNAP tile, on the Status Bar, and select SETTINGS.

    (The dialog box below will appear.)

2. Select the **OBJECT SNAP** tab.

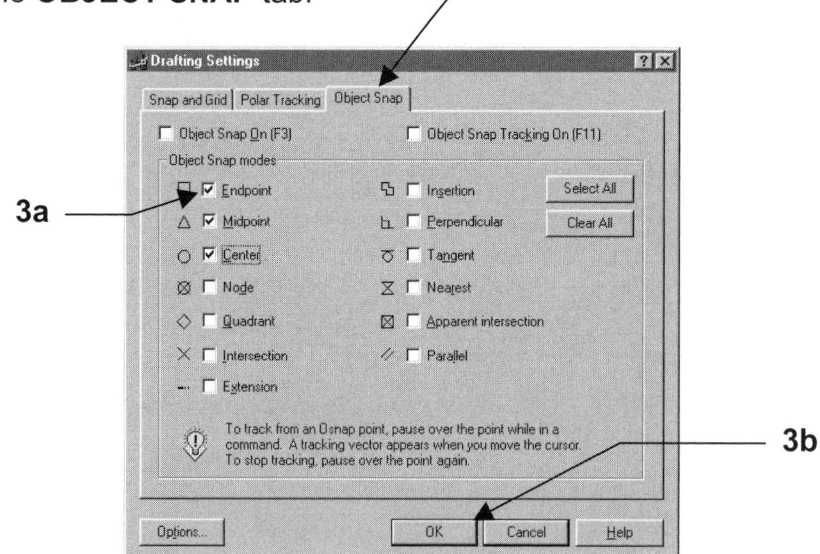

3. Select the Object Snap desired (3a) and then OK. (3b)

Note: Do not preset more than 3 object snaps, you will lose control of the cursor.

DRAWING SET UP

When drawing with a Computer, you must "set up your drawing area" just as you would on your drawing board if you were drawing with pencil and paper. You must decide what size your paper will be, what type of scale you will use (feet and inches or decimals, etc) and how precise you need to be. In CAD these decisions are called "Setting the **Drawing Limits, Units** and **Precision".**

DRAWING LIMITS

Consider the drawing limits as the size of the paper you will be drawing on. You will first be asked to define where the lower left corner should be placed, then the upper right corner, similar to drawing a Rectangle. An 11 x 17 piece of paper would have a **lower left corner** of 0,0 and an **upper right corner** of 17, 11. *(17 is the horizontal measurement or X-axis and 11 is the vertical measurement or Y-axis).*

HOW TO SET THE DRAWING LIMITS

1. Select the **DRAWING LIMITS** command using one of the following:

   TYPE = LIMITS
   PULLDOWN = FORMAT / DRAWING LIMITS
   TOOLBARS = NONE

2. The following will appear on the command line:

   Command: '_limits
   Reset Model space limits:
   Specify lower left corner or [ON/OFF] <0.000,0.000>:

3. Type the X,Y coordinates **0, 0** for the lower left corner location of your piece of paper then press <enter>.

4. The command line will now read:

   Specify upper right corner <12.000,9.000>:

5. Type the X,Y coordinates **17, 11** for the upper right corner of your piece of paper then press <enter>.

6. **This next step is very important:** Select **VIEW / ZOOM / ALL** to make the screen display the new drawing limits.

A personal note: *I love drawing limits. I can remember when I used to draw with pencil and paper and I would find that I had under estimated the size of the paper needed. With drawing limits you merely type a new size and magically the paper gets bigger or smaller. So now you can look forward to beautiful drawings with unlimited space. We will learn how to print any size drawing later in this workbook. Be patient for now.*

continued next page....

DRAWING SET UP continued

UNITS AND PRECISION

You now need to select what unit of measurement you want to work with.
Such as: Decimal (0.000) or Architectural (0'-0").
Next you should select how precise you want the measurements. This means, do you want the measurement rounded off to a 3 place decimal or the nearest 1/8".

HOW TO SET THE UNITS AND PRECISION.

1. Select the **UNITS** command using one of the following:

TYPE = UNITS
PULLDOWN = FORMAT / UNITS
TOOLBAR = NONE

(the dialogue box below will appear)

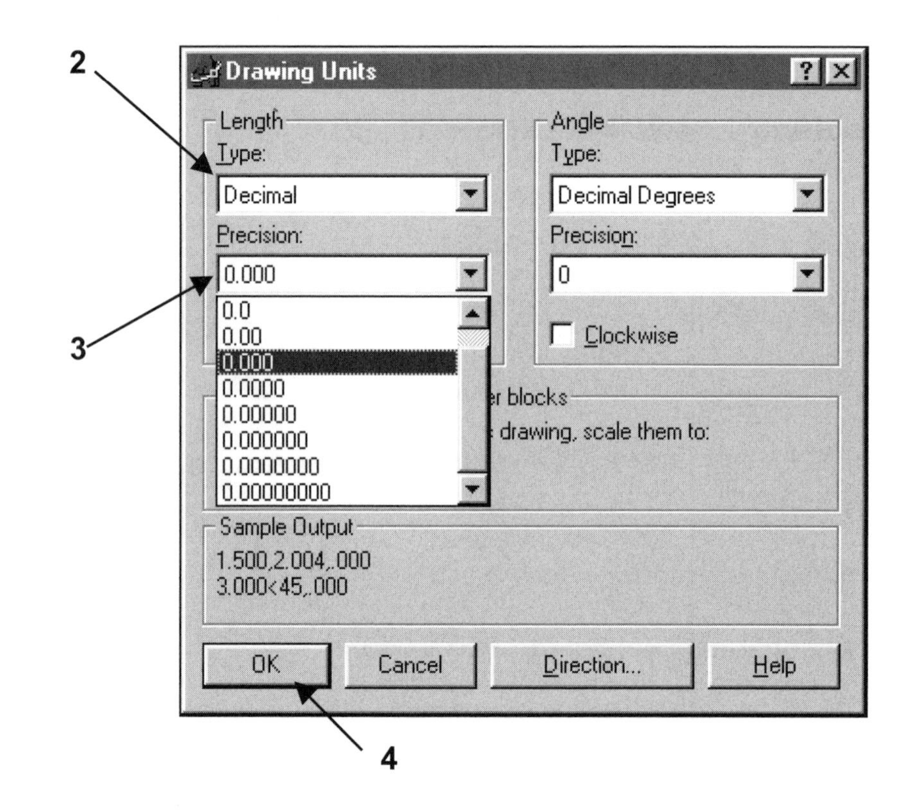

2. Select the appropriate **TYPE** such as: decimals or architectural.

3. Select the appropriate **PRECISION** associated with the "type".

4. Select the **OK** button to save your selections.

Easy, yes?

ZOOM

The **ZOOM** command is used to move closer or farther away to an object.

The following is an example of **Zoom / Window** to zoom in closer to an object.

1. Select the Zoom command by using one of the following:

   TYPING = Z
   PULLDOWN = VIEW / ZOOM
   TOOLBAR = STANDARD

2. Select the **"Window"** option and draw a window around the area you wish to magnify by moving the cursor to the lower left area of the object(s) and left click. Then move the cursor diagonally to form a square shape around the objects, and left click again. ***(Do not hold the left mouse button down while moving the cursor, just click it at the first and diagonal corners of the square shape)***

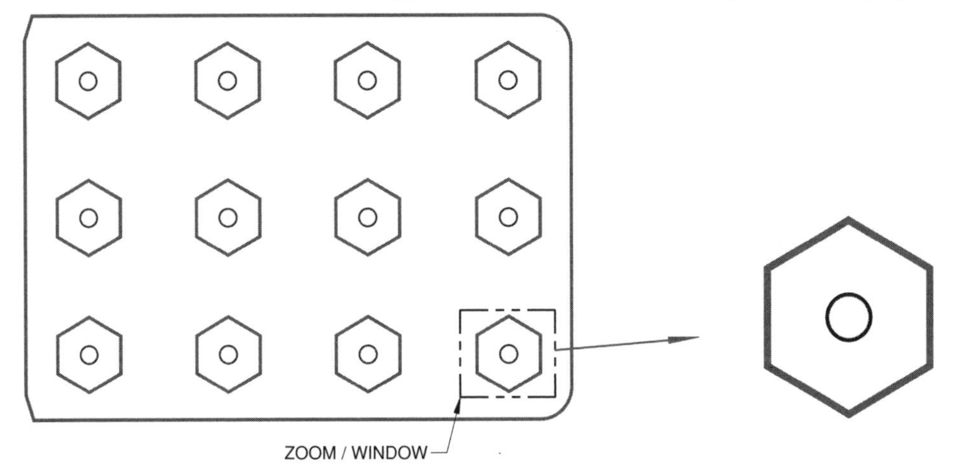

Additional Zoom options described below. (Try them)

1. **WINDOW** = zoom in on an area by specifying a window (rectangle) around the area.

2. **All** = Changes the screen to the size of the drawing limits. If you have objects outside of the drawing limits, Zoom/All will display them too.

3. **PREVIOUS** = returns the screen to the previous display. (Limited to 10 previous displays)

4. **IN or OUT** = moves in 2X or out 2X

5. **REAL TIME** = Interactive Zoom. You can zoom in or out by moving the cursor vertically up or down while pressing the left mouse button. To stop, press the Esc key.

6. **EXTENTS** = Displays all objects in the drawing file, using the smallest window possible.

EXERCISE 4A

INSTRUCTIONS:

1. Start a **New** file and select **1workbook helper.dwt**.

2. Using **FORMAT / UNITS:**
   set the units to **FRACTIONS**
   set the precision to **1/2**"

3. Using **FORMAT / DRAWING LIMITS** set the drawing limits to:
   Lower left corner = **0,0**
   Upper right corner = **20, 15**

4. Use **VIEW / ZOOM / ALL** to make the screen adjust to the new limits

5. Turn **OFF** the **GRIDS** (F7) **SNAP** (F9) and **ORTHO** (F8)
   (Your screen should be blank and your crosshair should move freely)

6. Draw the objects below using:
   DRAW / CIRCLE (CENTER, RADIUS) and **LINE**
   OBJECT SNAP = CENTER and TANGENT

Very Important: Use the Tangent option at each end of the line. AutoCAD needs to be told that you want each end of the line to be tangent to a circle.

7. **Save** this drawing as **EX4A**

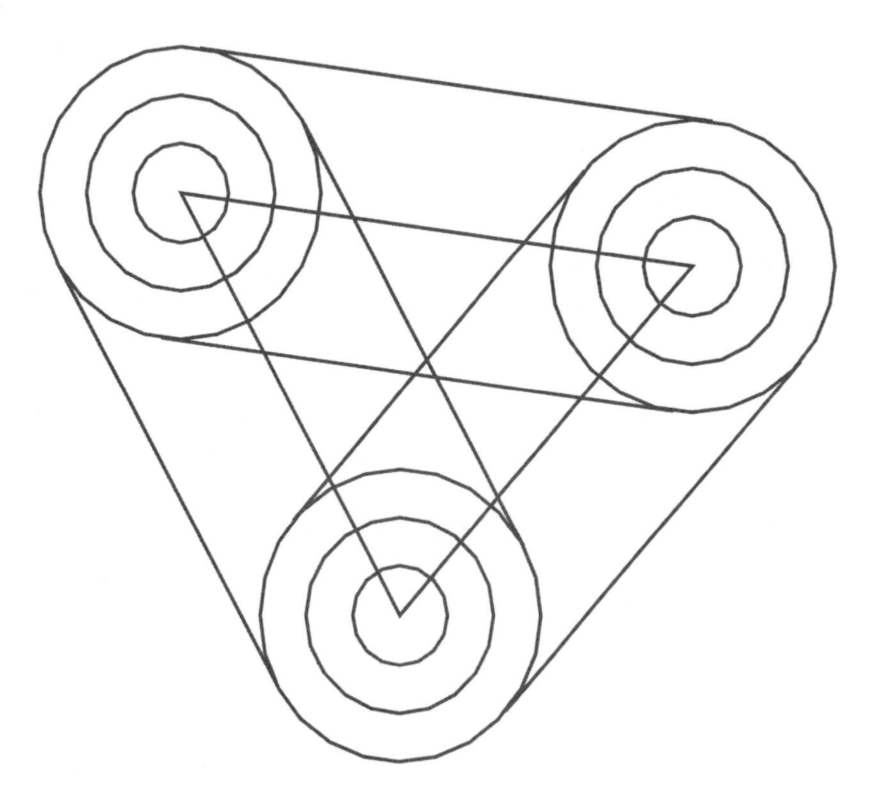

EXERCISE 4B

INSTRUCTIONS:

1. Start a **New** file and select **1workbook helper.dwt**.

2. Using **FORMAT / UNITS:**
   Set the units to **FRACTIONS**
   Set the precision to **1/4**"

3. Using **FORMAT / DRAWING LIMITS** set the drawing limits to:
   Lower left corner = **0,0**
   Upper right corner = **12, 9**

4. Use **VIEW / ZOOM / ALL** to make the screen adjust to the new limits.

5. Turn **OFF** the **GRIDS** (F7) **SNAP** (F9) and **ORTHO** (F8)
   (Your screen should be blank and your crosshair should move freely)

6. Draw the objects below using:
   DRAW / CIRCLE (CENTER, RADIUS) and **LINE**
   OBJECT SNAP = QUADRANT

7. **Save** this drawing as: **EX4B**

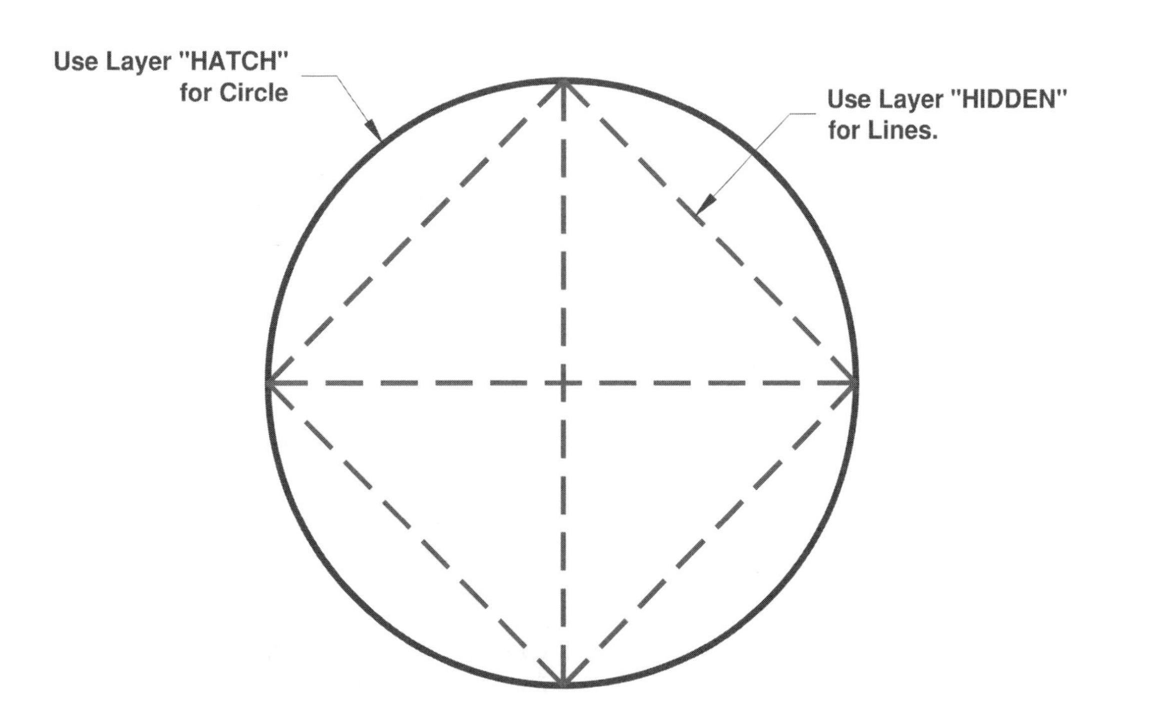

EXERCISE 4C

INSTRUCTIONS:

1. Start a **New** file and select **1workbook helper.dwt**.

2. Using **FORMAT / UNITS:**
   Set the units to **ARCHITECTURAL**
   Set the precision to **1/2**"

3. Using **FORMAT / DRAWING LIMITS** set the drawing limits to:
   Lower left corner = **0, 0**
   Upper right corner = **25, 20**

4. Use **VIEW / ZOOM / ALL** to make the screen adjust to the new limits.

5. Turn **OFF** the **GRIDS** (F7) **SNAP** (F9) and **ORTHO** (F8)
   (Your screen should be blank and your crosshair should move freely)

6. Draw the objects below using:
   DRAW / LINE
   OBJECT SNAP = PERPENDICULAR

7. **Save** this drawing as: **EX4C**

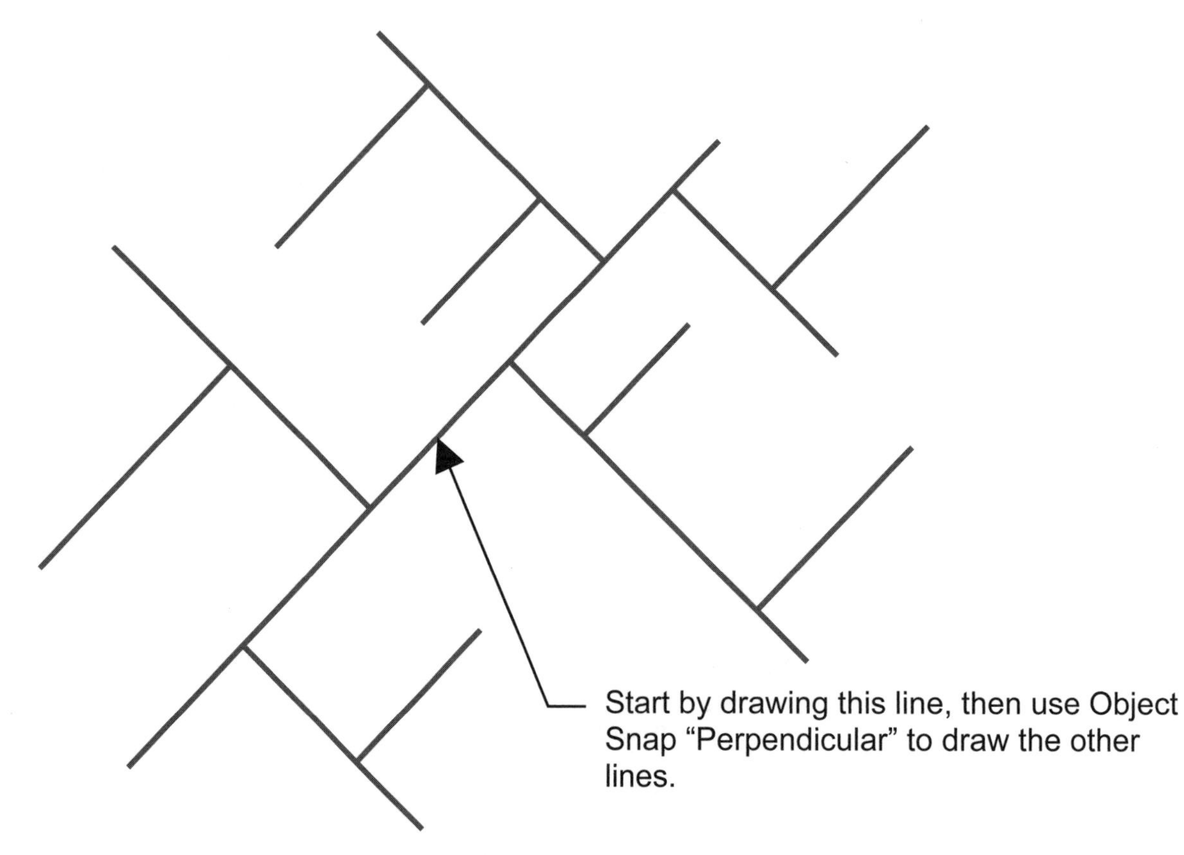

EXERCISE 4D

INSTRUCTIONS:

1. Start a **New** file and select **1workbook helper.dwt**.

2. Using **FORMAT / UNITS:**
   Set the units to **DECIMALS**
   Set the precision to **0.00**

3. Using **FORMAT / DRAWING LIMITS** set the drawing limits to:
   Lower left corner = **0,0**
   Upper right corner = **12, 9**

4. Use **VIEW / ZOOM / ALL** to make the screen adjust to the new limits.

5. Turn **OFF** the **GRIDS** (F7) **SNAP** (F9) and **ORTHO** (F8)
   (Your screen should be blank and your crosshair should move freely)

6. Draw the Lines below using:
   DRAW / LINE
   OBJECT SNAP = MIDPOINT

7. **Save** this drawing as: **EX4D**

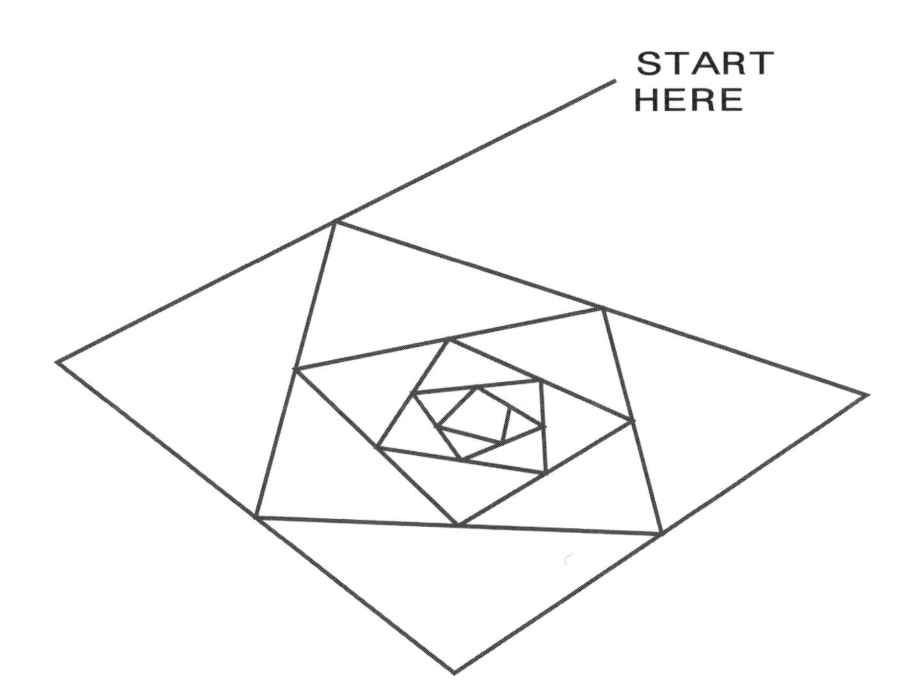

EXERCISE 4E

INSTRUCTIONS:

1. Start a **New** file and select **1workbook helper.dwt**

2. Draw the objects below using:
   DRAW / LINE
   ORTHO **ON** for Horizontal Lines
   OBJECT SNAP = ENDPOINT

3. **Save** this drawing as: **EX4E**

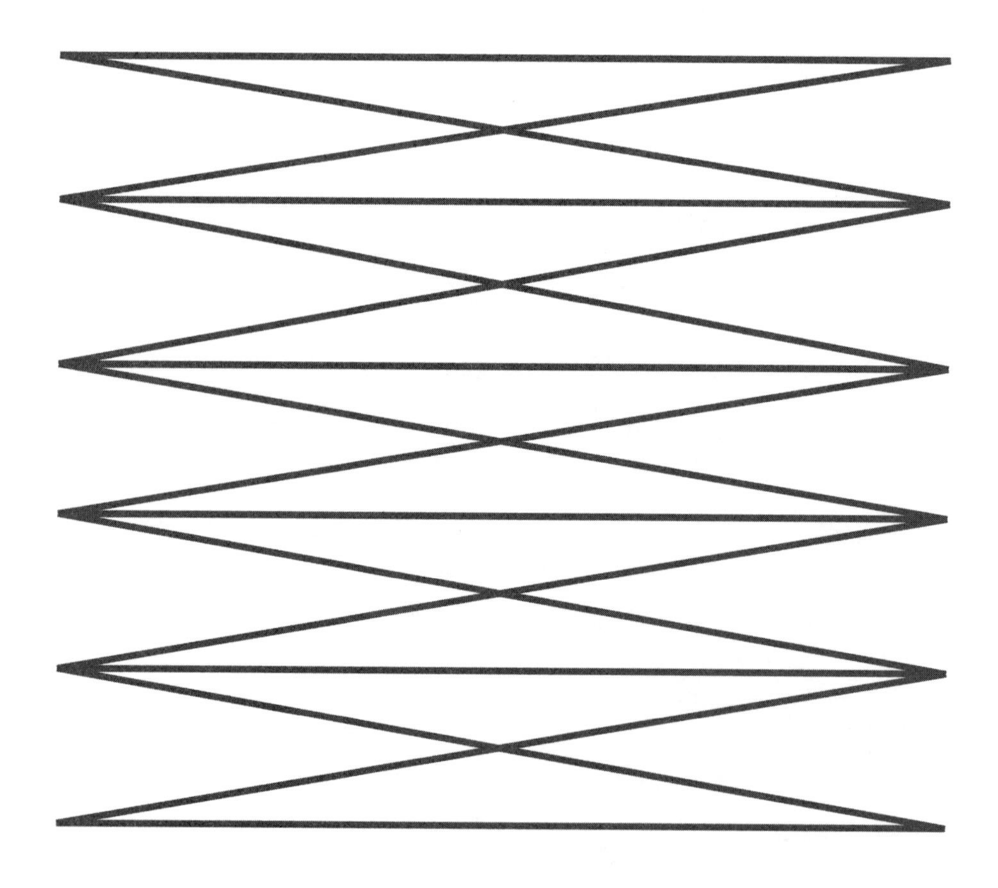

EXERCISE 4F

INSTRUCTIONS:

1. Start a **New** file and select **1workbook helper.dwt**

2. Draw the 2 vertical and 4 horizontal lines using:
   DRAW / LINE
   ORTHO (F8) = **ON**
   SNAP (F9) = **OFF**

3. Then draw the diagonal lines using:
   DRAW / LINE
   ORTHO & SNAP= **OFF**
   OBJECT SNAP = INTERSECTION

4. **Save** this drawing as: **EX4F**

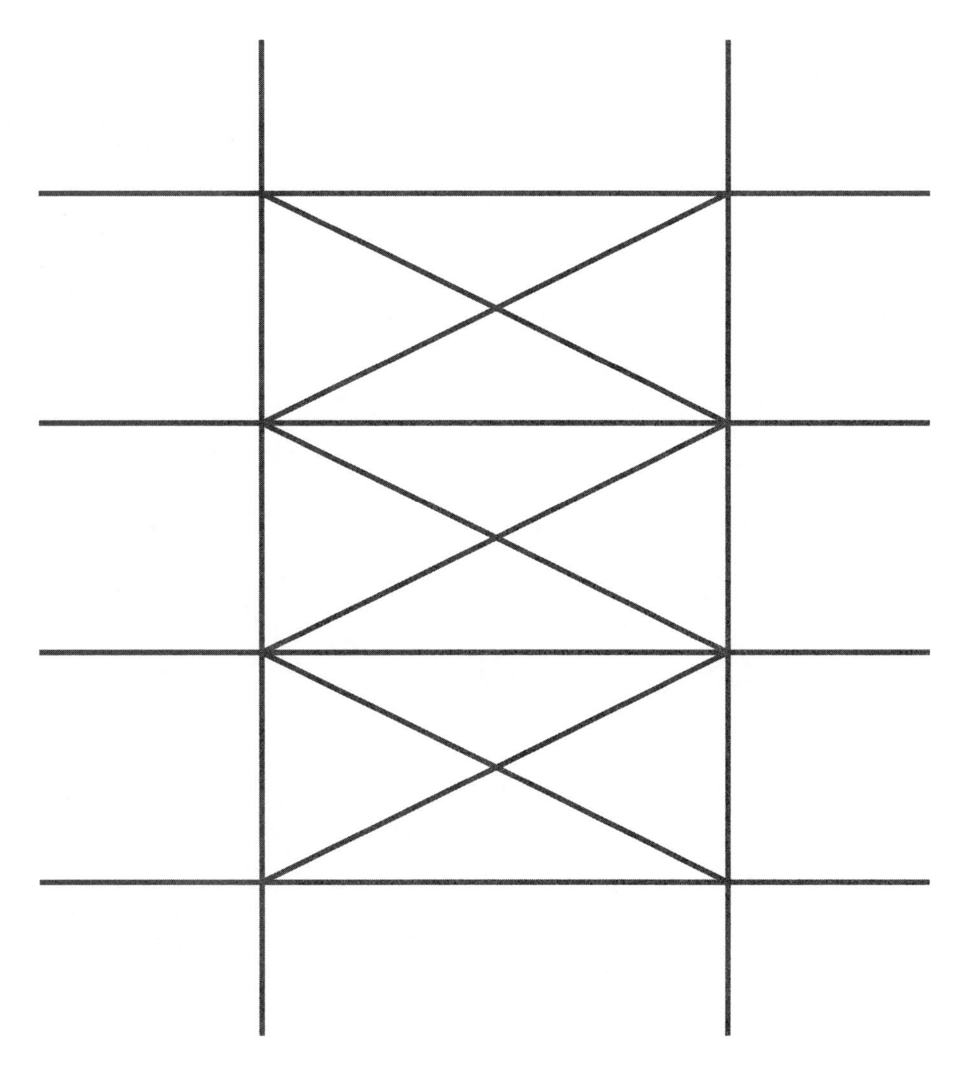

EXERCISE 4G

INSTRUCTIONS:

1. Start a **New** file and select **1workbook helper.dwt**.

2. Draw the 4 circles with the following Radii: 1, 2, 3, & 5
   (Use Object snap "Center" so all Circles have the same center)

3. Draw the LINES using:
   DRAW / LINE
   ORTHO and SNAP = **OFF**
   OBJECT SNAP = QUADRANT and TANGENT

4. Use Layers: Object and Center

5. **Save** this drawing as **EX4G**

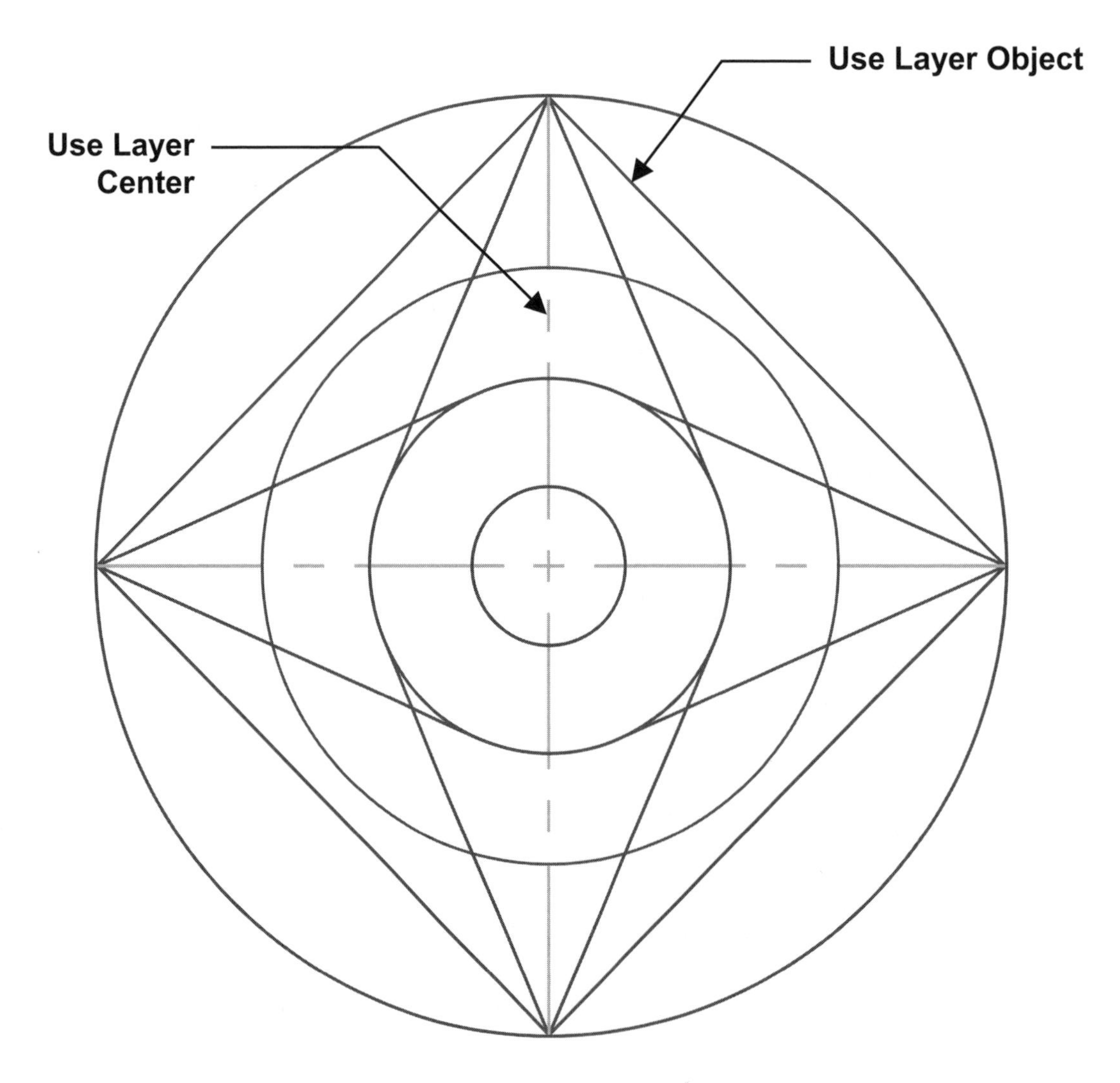

LEARNING OBJECTIVES

After completing this lesson, you will be able to:

1. Draw an Inscribed and Circumscribed Polygon.
2. Create an Ellipse using two different methods.
3. Draw an object called Donut.
4. Define a Point location.
5. Select various Point Styles.
6. Use 2 new Object Snap modes.

LESSON 5

POLYGON

A polygon is an object with multiple sides of equal length. You may specify from 3 to 1024 sides. A polygon appears to be multiple lines but in fact it is one object. You can specify the edge length or the center and a radius. The radius can be drawn inside (Inscribed) an imaginary circle or outside (circumscribed) an imaginary circle.

CENTER / RADIUS METHOD

1. Select the **Polygon** command using one of the following:

    TYPING = POL <enter>
    PULLDOWN = DRAW / POLYGON
    TOOLBAR = DRAW

2. The following prompts will appear on the command line:

    _polygon Enter number of sides <4>: ***type number of sides <enter>***
    Specify center of polygon or [Edge]: ***specify the center location***
    Enter an option [Inscribed in circle/Circumscribed about circle]<I>:***type I or C<enter>***
    Specify radius of circle: ***type radius or locate with cursor.***

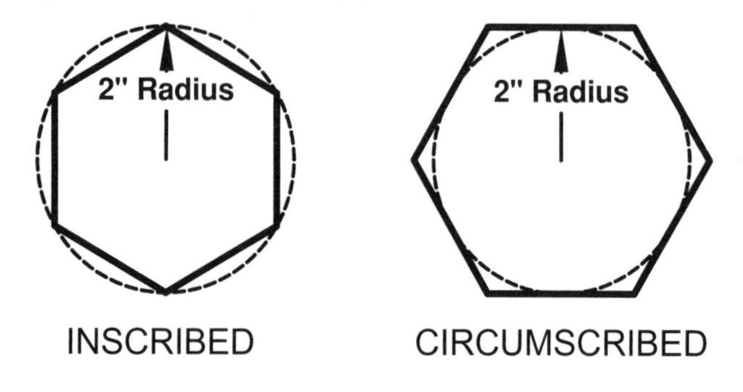

EDGE METHOD

1. Select the **Polygon** command using one of the following:

    TYPING = POL
    PULLDOWN = DRAW / POLYGON
    TOOLBAR = DRAW

2. The following prompts will appear on the command line:

    _polygon Enter number of sides <4>: ***type number of sides <enter>***
    Specify center of polygon or [Edge]: ***type E <enter>***
    Specify first endpoint of edge: ***place first endpoint of edge (P1)***
    Specify second endpoint of edge: ***place second endpoint of edge (P2)***

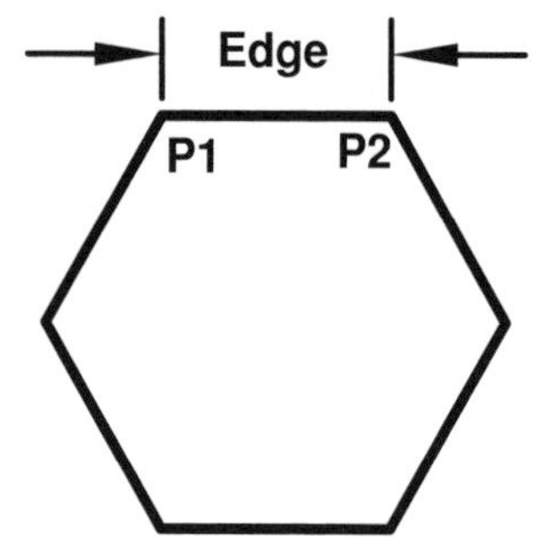

ELLIPSE

An Ellipse may be drawn by specifying the 3 points of the axes or by defining the center point and major / minor axis points.

AXIS END METHOD

1. Select the **ELLIPSE** command using one of the following:

 TYPING =EL <enter>
 PULLDOWN = DRAW / ELLIPSE
 TOOLBAR = DRAW

2. The following prompts will appear on the command line:

 Command: _ellipse
 Specify axis endpoint of ellipse or [Arc/Center]: ***place the first point of either the major or minor axis (P1).***
 Specify other endpoint of axis: ***place the other point of the first axis (P2)***
 Specify distance to other axis or [Rotation]: ***place the point perpendicular to the first axis (P3).***

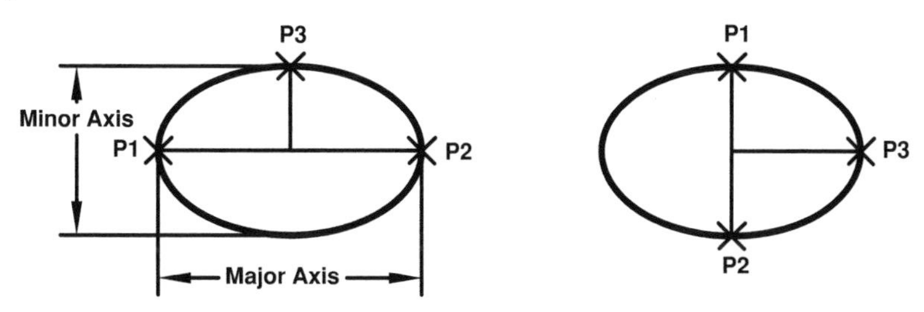

CENTER METHOD

1. Select the **ELLIPSE** command using one of the following:

 TYPING =EL
 PULLDOWN = DRAW / ELLIPSE
 TOOLBAR = DRAW

2. The following prompts will appear on the command line:

 Command: _ellipse
 Specify axis endpoint of ellipse or [Arc/Center]: ***type C <enter>***
 Specify center of ellipse: ***place center of ellipse (P1***
 Specify endpoint of axis: ***place first axis endpoint (either axis) (P2)***
 Specify distance to other axis or [Rotation]: ***place the endpoint perpendicular to the first axis (P3)***

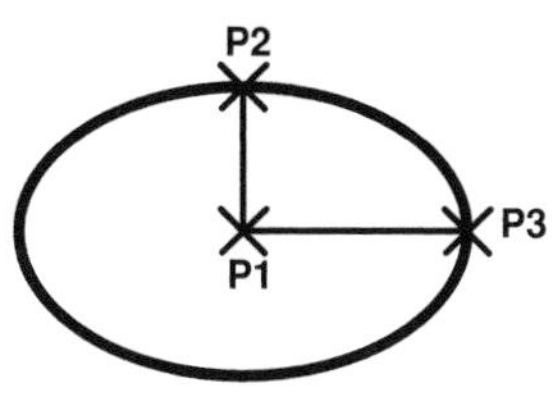

DONUT

A Donut is a circle with ***width.*** You will define the **Inside** and **Outside** diameters. The appearance is controlled by the **"Fill"** mode. If the **"Fill"** mode is **On**, the donut width will be solid. If the **"Fill"** mode is **Off**, the donut width will appear with lines somewhat like the spokes of a wheel. These lines enable you to visually differentiate between donuts and concentric circles.

1. Select the **DONUT** command using one of the following:

    TYPING =DO
    PULLDOWN = DRAW / DONUT
    TOOLBAR = DRAW (not a default icon) 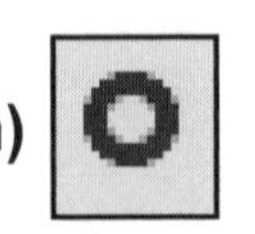

2. The following prompts will appear on the command line:

    Command: _donut
    Specify inside diameter of donut: ***type the inside diameter***
    Specify outside diameter of donut: ***type the outside diameter***
    Specify center of donut or <exit>: ***place the center of the first donut***
    Specify center of donut or <exit>: ***place the center of the second donut or <enter> to stop***

Controlling the "FILL MODE"

Command: ***type FILL <enter>***
Enter mode [ON/OFF] <OFF>: ***type ON or Off <enter>***

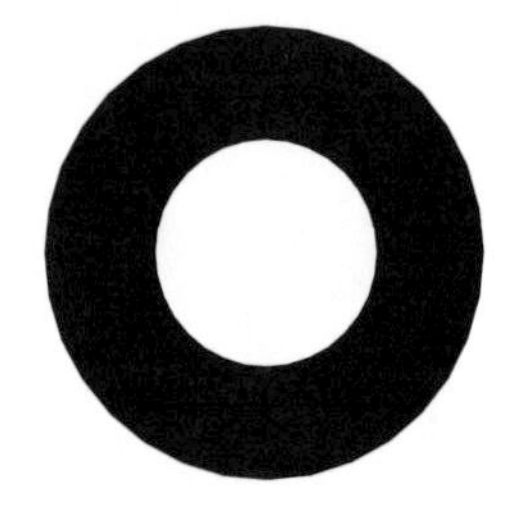

FILL = ON

FILL = OFF

Select **VIEW / REGEN** or type ***REGEN*** <enter> at the command line to update the drawing to the latest changes to the ***FILL*** mode.

POINT

A Point is an object that has no dimension and only has location. A Point may be represented by selecting one of many Point styles shown below. The default Point Style is a “dot”. Points are basically used to locate a point of reference.
You may select the “Single” or “Multiple” point option. The Single option creates one point. The Multiple option continues until you press the ESC key.
The only Object Snap mode that can be used with Point is “**Node**”. (See page 5-6)

1. Select the **POINT** command using one of the following:

   TYPING =PO
   PULLDOWN = DRAW / POINT (Single point or Multiple point)
   TOOLBAR = DRAW

2. The following prompts will appear on the command line:

   Command: _point
   Current point modes: PDMODE=3 PDSIZE=0.000
   Specify a point: ***place the point location***
   (If you selected “multiple point” you must press the “EXC” key to stop.)

TO SET THE “POINT STYLE”

1. Select one of the following:
   TYPE = DDPTYPE
   PULLDOWN = FORMAT / POINT STYLE
   TOOLBAR = NONE

2. This dialog box will appear.

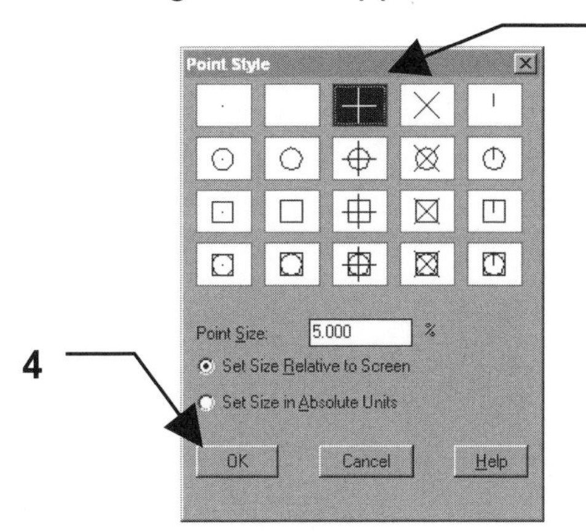

3. Select a tile.
4. Select the OK button.

MORE OBJECT SNAP

OBJECT SNAP OPTIONS:

NODe This option snaps to a POINT object.
Place the pick box on the POINT object.

NEArest Snaps to the nearest location on the nearest object.
Select the object anywhere on the object.

EXERCISE 5A

INSTRUCTIONS:

1. Start a **New** file and select **1workbook helper.dwt**

2. Draw the Circle first, Polygon second and Lines last.

3. Draw the objects below using:
   DRAW / CIRCLE (Center, Rad)
   DRAW / POLYGON (Circumscribed)
   Ortho (F8) = **ON**
   Snap (F9) = **OFF**
   Object Snap = Center, Midpoint, Endpoint and Quadrant

4. **Save** this drawing as: **EX5A**

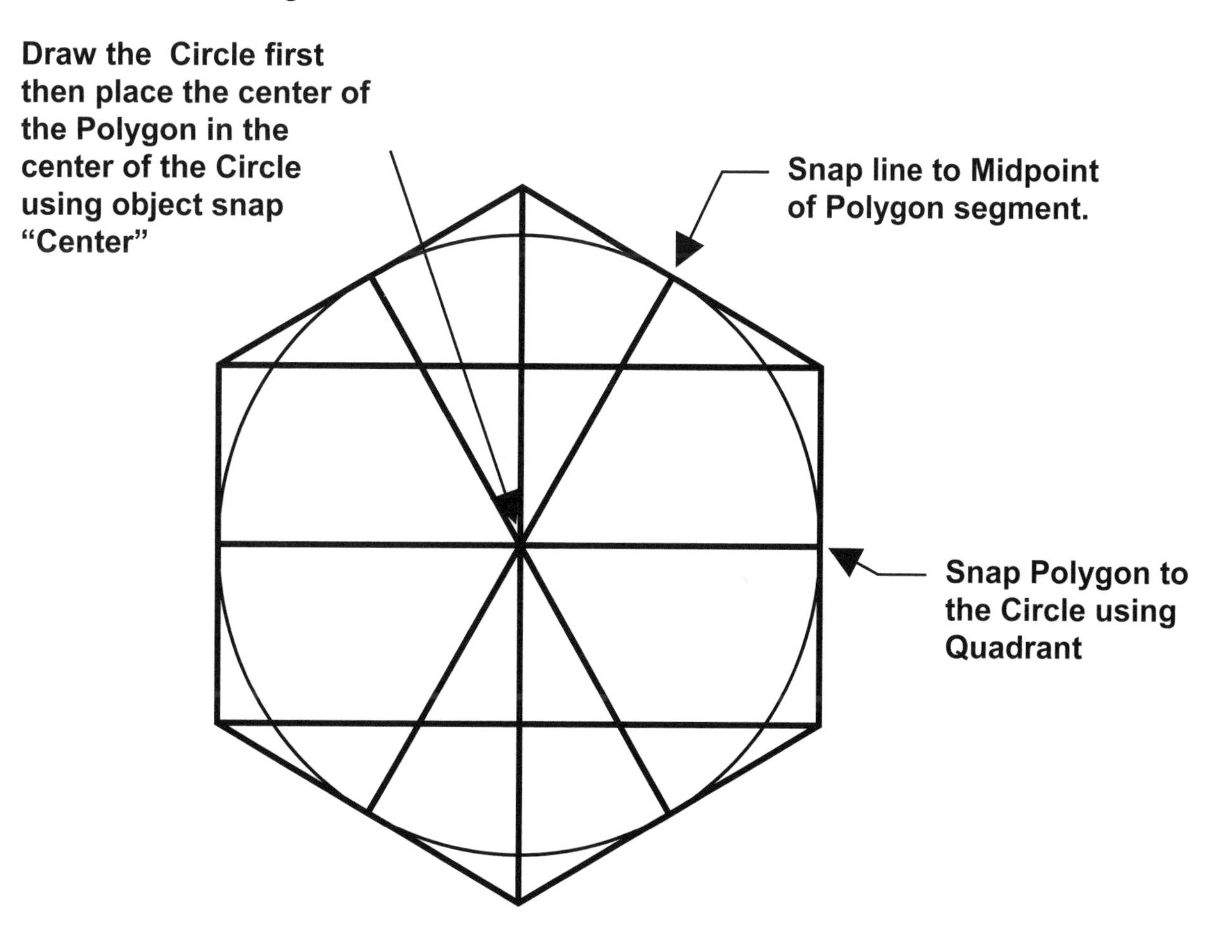

EXERCISE 5B

INSTRUCTIONS:

1. Start a **New** file and select **1workbook helper.dwt**.

2. Draw the Point first, then the Polygons

3. Draw the objects below using:
   DRAW / POINT and **POLYGON** (Inscribed)
   Ortho and Snap **ON**
   Object Snap = Node

4. Locate the center of the Polygon by snapping to the Point using Object snap: **Node**

5. **Save** this drawing as: **EX5B**

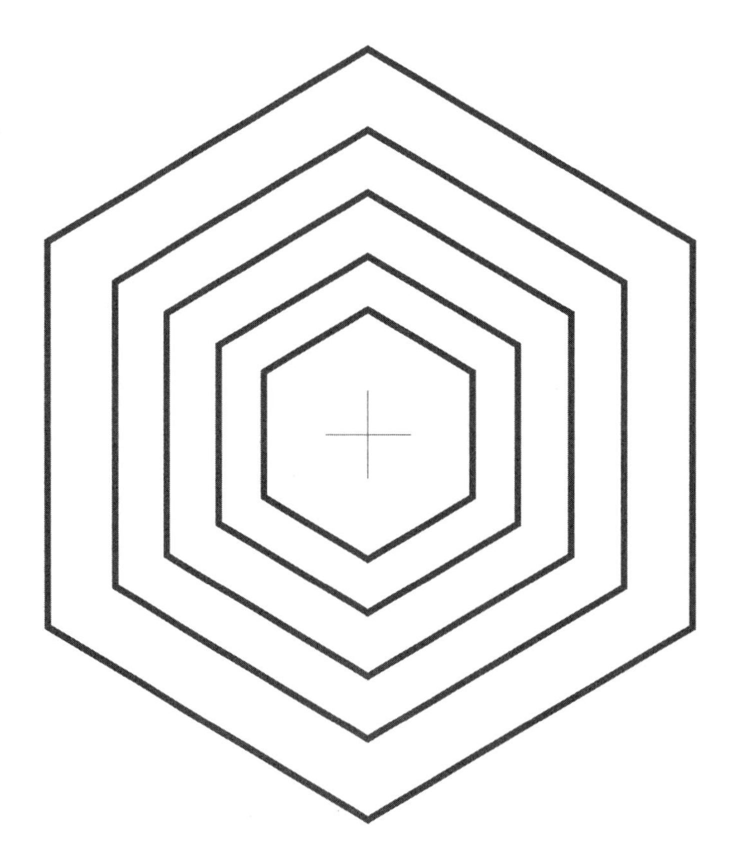

EXERCISE 5C

INSTRUCTIONS:

1. Start a **New** file and select **1workbook helper.dwt**.
2. Draw the objects below using:
   DRAW / ELLIPSE (Axis, End) **LINES** and **CIRCLE** (center, rad)
   Ortho **ON** and Snap **OFF**
   Object Snap = Perpendicular and Quadrant
3. Have fun with this one.
4. **Save** this drawing as **EX5C**

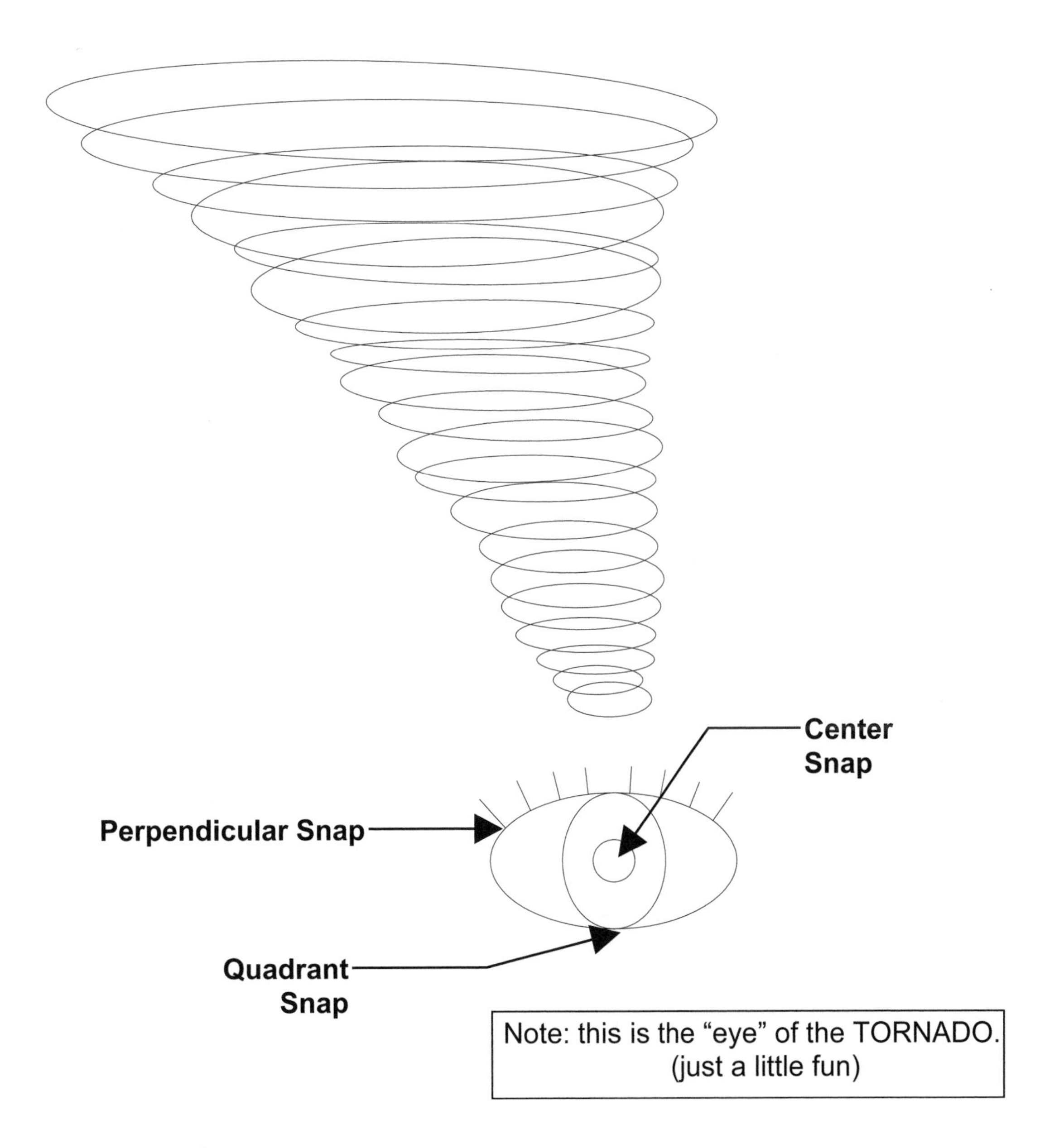

EXERCISE 5D

INSTRUCTIONS:

1. Start a **New** file and select **1workbook helper.dwt**.
2. Select the Point Style.
3. Draw the Point first, then the Ellipse
4. Draw the objects below using:
   DRAW / POINT
   DRAW / ELLIPSE (Use "Center" method)
   Ortho **ON** and Snap **OFF**
   Object Snap = Quadrant and Node
4. **Save** this drawing as: **EX5D**

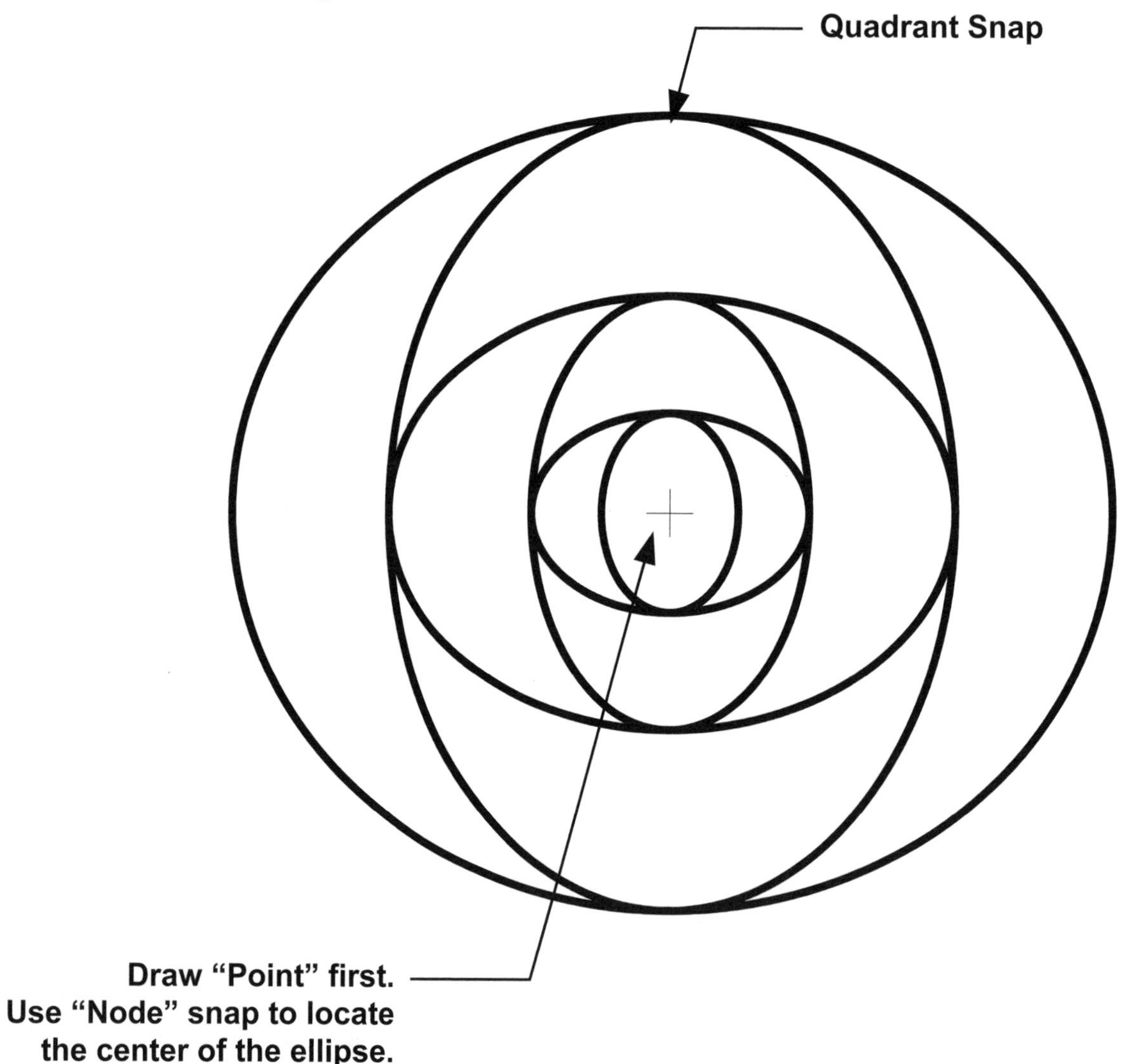

EXERCISE 5E

INSTRUCTIONS:

STEP 1.

1. Start a **New** file and select **1workbook helper.dwt**.
2. Draw the objects below using:
   DRAW / DONUT
   Ortho and Snap **OFF**
   Object Snap = Center

STEP 2.

3. Turn the **FILL** mode **OFF** (Type: Fill <enter> Off <enter>)
4. Select **View / Regen** or Type: **REGEN <enter>**

5. **Save** this drawing as **EX5E**

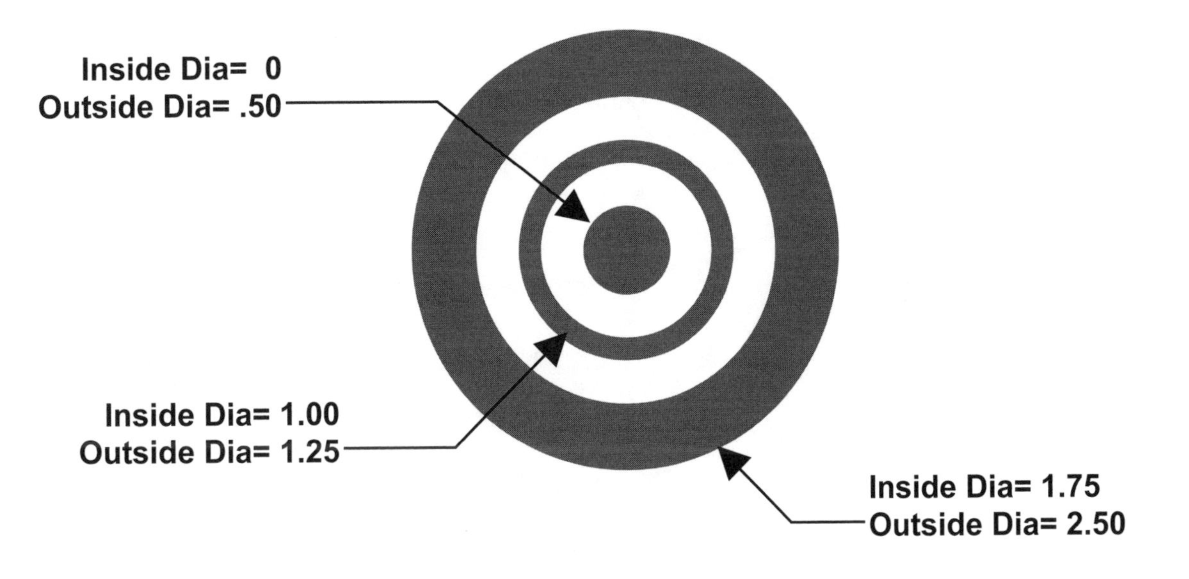

EXERCISE 5F

INSTRUCTIONS:

1. Start a **New** file and select **1workbook helper.dwt**.
2. Draw the objects below using:
   DRAW / POINT, POLYGON, ELLIPSE (center) and **DONUT**
   Ortho **ON** and Snap **OFF**
   Object Snap = Node, Endpoint and Midpoint
3. **Save** this drawing as: **EX5F**

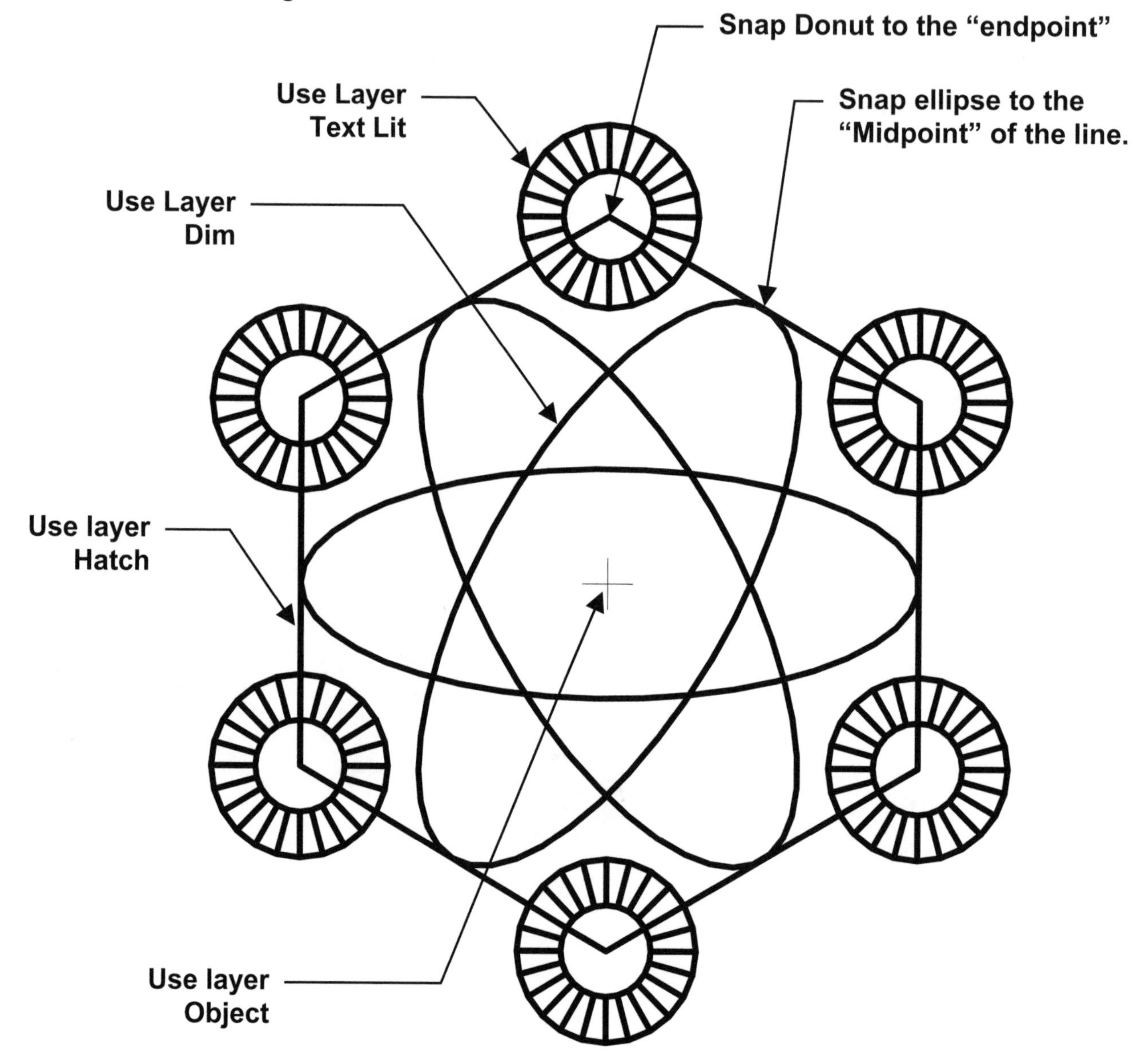

LEARNING OBJECTIVES

After completing this lesson, you will be able to:

1. Use the 4 Break command options.
2. Trim an object to a cutting edge.
3. Extend an object to a boundary.
4. Move an object(s) to a new location.
5. Explode objects into their primitive entities.

LESSON 6

BREAK

The ***BREAK*** command allows you to break a space in an object, break the end off an object or split a line in two. I think of it as taking a bite out of an object.
There are 4 break methods described below.

You may select the **BREAK** command by using one of the following:

TYPING = BR
PULLDOWN = MODIFY / BREAK
TOOLBAR = MODIFY

METHOD 1
How to break one object into two separate objects with no visible space in between. (Use this method if the location of the break is not important)

a. Draw a line.
b. Select the **BREAK** command using one of the methods listed above
c. _break Select object: ***pick the break location (P1) by clicking on it.***
d. Specify second break point or [First point]: ***type @ <enter>***
(this will duplicate the last point)
e. Now if you click on one end of the line you will see that there are 2 lines instead of just one.

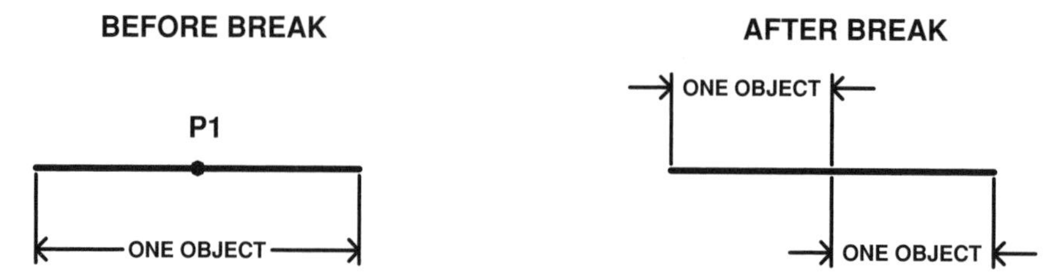

METHOD 2
This method is the same as method 1 except use this method if the location of the break is very specific.

a. Select the **BREAK** command using one of the methods listed above
b. _break Select objects: ***select the object to break (P1)***
c. Specify second break point or [First point]: ***type F <enter>***
d. Specify first break point: ***select break location (P2) accurately***
e. Specify second break point: ***type @ <enter>***

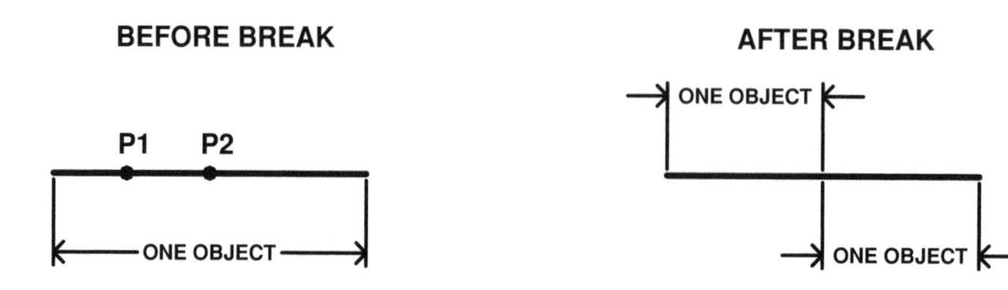

Method 2 can also be accomplished easily by selecting the "Break at point" icon. But I wanted you to understand how it works.

BREAK (continued)

<u>**METHOD 3**</u>

Take a bite out of an object. (Use this method if the location of the BREAK is not important.

(This is the Default option)

a. Select the **BREAK** command
b. _break Select object: ***pick the first break location (P1)***
c. Specify second break point or [First point]: ***pick the second break location (P2)***

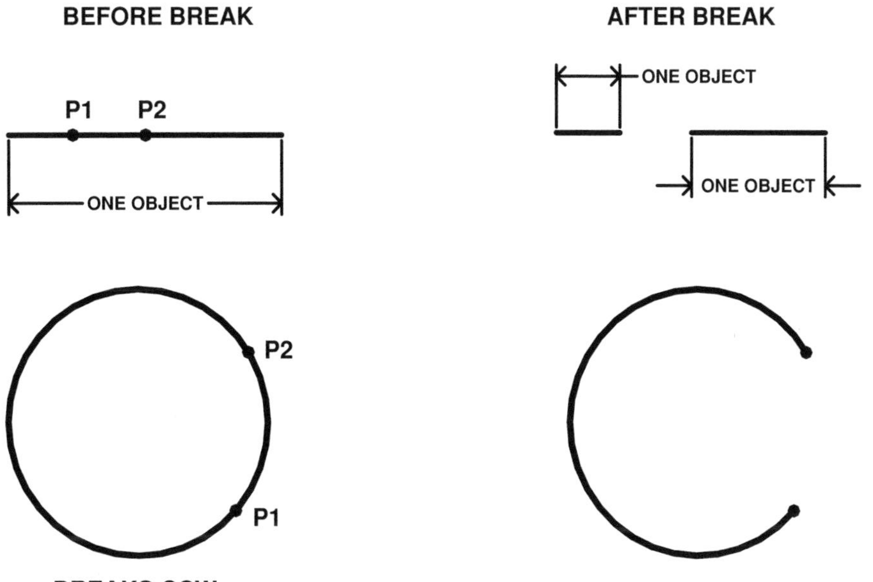

BREAKS CCW

Note: Circles can't be broken with "1 point". You must use 2 points.

<u>**METHOD 4**</u>

This method is the same as method 3 except <u>use this method if the location of the break is very specific.</u>

a. Select the **BREAK** command
b. _break Select objects: ***select the object to break (P1) anywhere on the object***
c. Specify second break point or [First point]: ***type F <enter>***
d. Specify first break point: ***select the first break location (P2) accurately***
e. Specify second break point: ***select the second break location (P3) accurately***

BEFORE BREAK **AFTER BREAK**

TRIM

The **TRIM** command is used to trim an object to a **cutting edge**. You first select the "Cutting Edge" and then select the part of the object you want to trim. The object to be trimmed must actually intersect the cutting edge or could intersect if the objects were infinite in length.

1. Select the Trim command using one of the following:

   TYPE = TR
   PULLDOWN = MODIFY / TRIM
   TOOLBAR = MODIFY

2. The following will appear on the command line:

Command: _trim
Current settings: Projection = UCS Edge = Extend
Select cutting edges ...
Select objects: ***select cutting edge(s) by clicking on the object (P1)***
Select objects: ***stop selecting cutting edges by pressing the <enter> key***
Select object to trim or shift-select to extend or [Project/Edge/Undo]: ***select the object that you want to trim. (P2) (Select the part of the object that you want to disappear, not the part you want to remain)***
Select object to trim or [Project/Edge/Undo]: ***press <enter> to stop***

> Note: You may toggle between Trim and Extend (page 6-5). Hold down the shift key and the Extend command is activated. Release the shift key and you return to Trim.

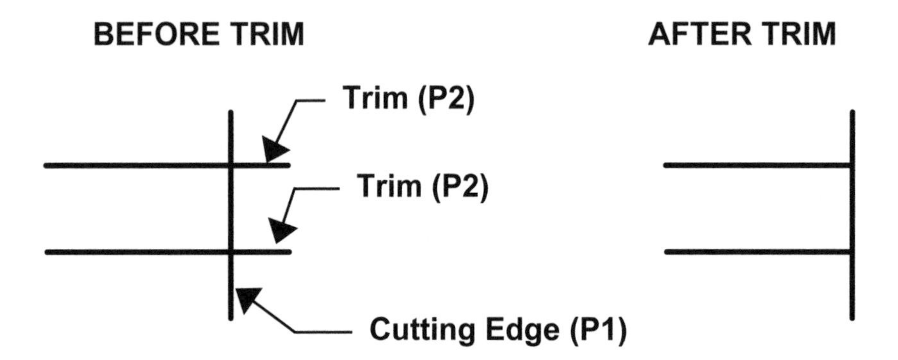

EDGE (Extend or No Extend)
In the **"Extend"** mode, (default mode) the cutting edges and the Objects to be trimmed need only apparently intersect if the objects were infinite in length.
In the **"No Extend"** mode the cutting edges and the objects to be trimmed must visibly intersect.

PROJECTION
Same as Edge except used only in "3D".

EXTEND

The **EXTEND** command is used to extend an object to a **boundary.** The object to be extended must actually or theoretically intersect the boundary.

1. Select the **EXTEND** command using one of the following:

   TYPE = EX
   PULLDOWN = MODIFY / EXTEND
   TOOLBAR = MODIFY

2. The following will appear on the command line:

Command: _extend
Current settings: Projection = UCS Edge = Extend
Select boundary edges ...
Select objects: ***select boundary (P1) by clicking on the object***
Select objects: ***stop selecting boundaries by selecting <enter>***
Select object to extend or shift-select to Trim or [Project/Edge/Undo]: ***select the object that you want to extend. (P2 and P3) (Select the end of the object that you want to extend.)***
Select object to extend or [Project/Edge/Undo]: ***stop selecting objects by selecting <enter>***

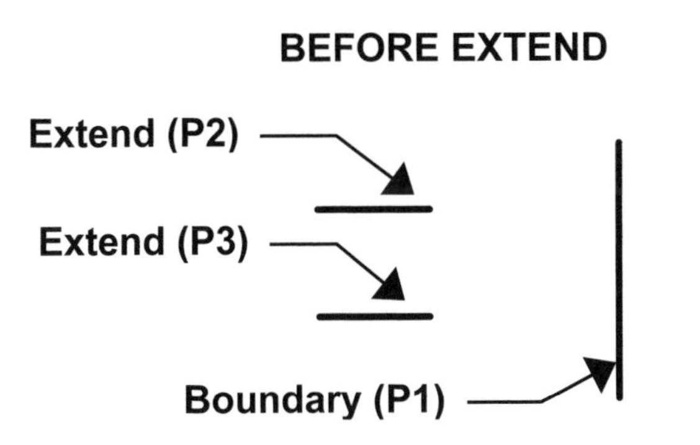

AFTER EXTEND

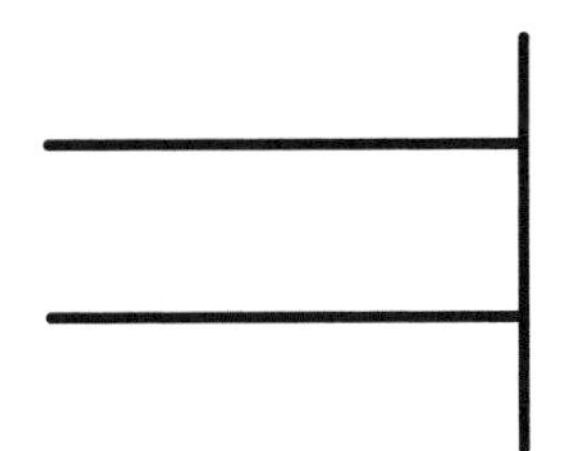

You may toggle between Extend and Trim (page 6-4). Hold down the shift key and the Trim command is activated. Release the shift key and you return to Extend.

EDGE (Extend or No Extend)
In the **"Extend"** mode, (default mode) the boundary and the Objects to be extended need only imaginarily intersect if the objects were infinite in length.
In the **"No Extend"** mode the boundary and the objects to be extended must visibly intersect.

PROJECTION Same as Edge except used only in "3D".

MOVE

The MOVE command is used to move object(s) from their current location (basepoint) to a new location (second displacement point)

1. Select the Move command using one of the following:

   TYPE = M
   PULLDOWN = MODIFY / MOVE
   TOOLBAR = MODIFY

2. The following will appear on the command line:

Command: _move
Select objects: ***select the object(s) you want to move (P1)***
Select objects: ***stop selecting object(s) by selecting <enter>***
Specify base point or displacement: ***select a location (P2) (usually on the object)***
Specify second point of displacement or <use first point as displacement>: ***move the object to it's new location (P3) and left click.***

Note: if you press <enter> instead of actually picking a new location (P3), Autocad will send it into Outer Space. If this happens, press U <enter> or select the " undo" icon and try again.

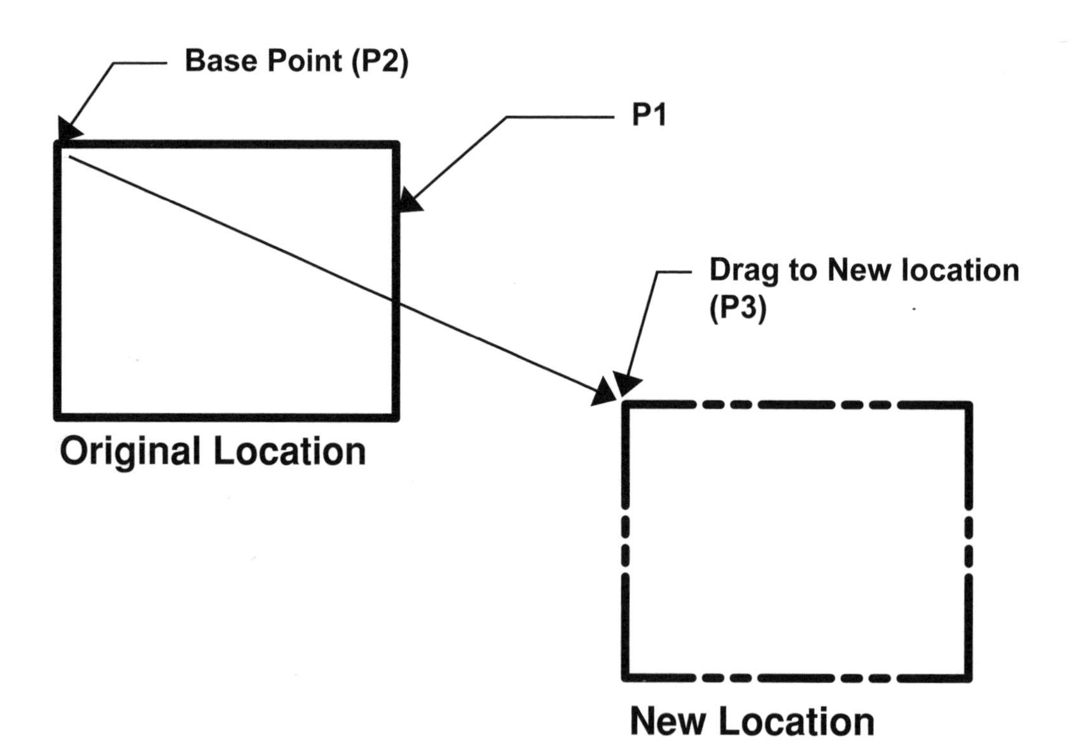

EXPLODE

The EXPLODE command changes (explodes) an object into its primitive objects. For example: a rectangle is originally one object, if you explode it, it changes into 4 lines. Visually you will not be able to see the change unless you select one of the lines.

1. Select the Explode command by using one of the following:

TYPE =X
PULLDOWN = MODIFY / EXPLODE
TOOLBAR = MODIFY

2. The following will appear on the command line:

Command: _explode
Select objects: ***select the object(s) you want to explode***
Select objects: ***select <enter>***

Before EXPLODE **After EXPLODE**

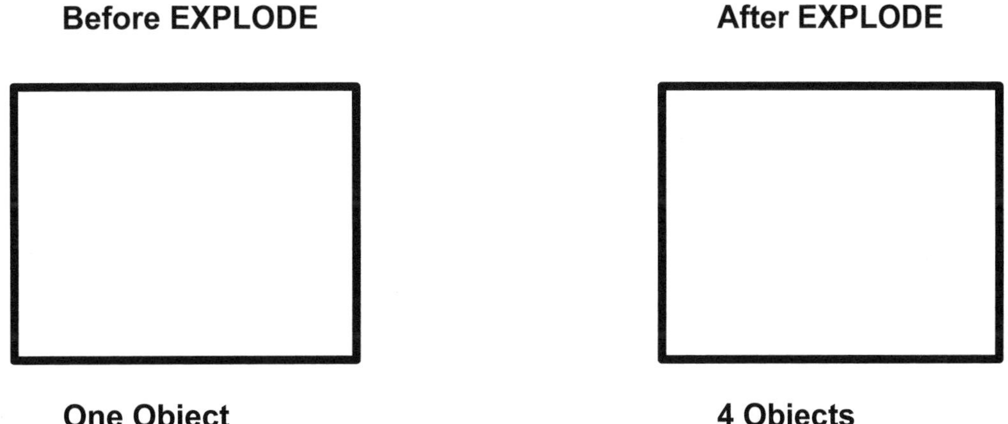

One Object (Rectangle) **4 Objects (4 Lines)**

(Notice there is no visible difference. But now you have 4 lines instead of 1 Rectangle)

Try this:
Draw a rectangle and then click on it. The entire object highlights.
Now explode the rectangle, then click on it again. Only the line you clicked on should be highlighted. Each line that forms the rectangular shape is now an individual object.

EXERCISE 6A

INSTRUCTIONS:

1. Start a **New** file and select **1workbook helper.dwt**.
2. Draw the objects below:

   Locate the center of the circles on the rectangle line
   using object snap = **Nearest and Midpoint**

   Ortho and Snap = **OFF**

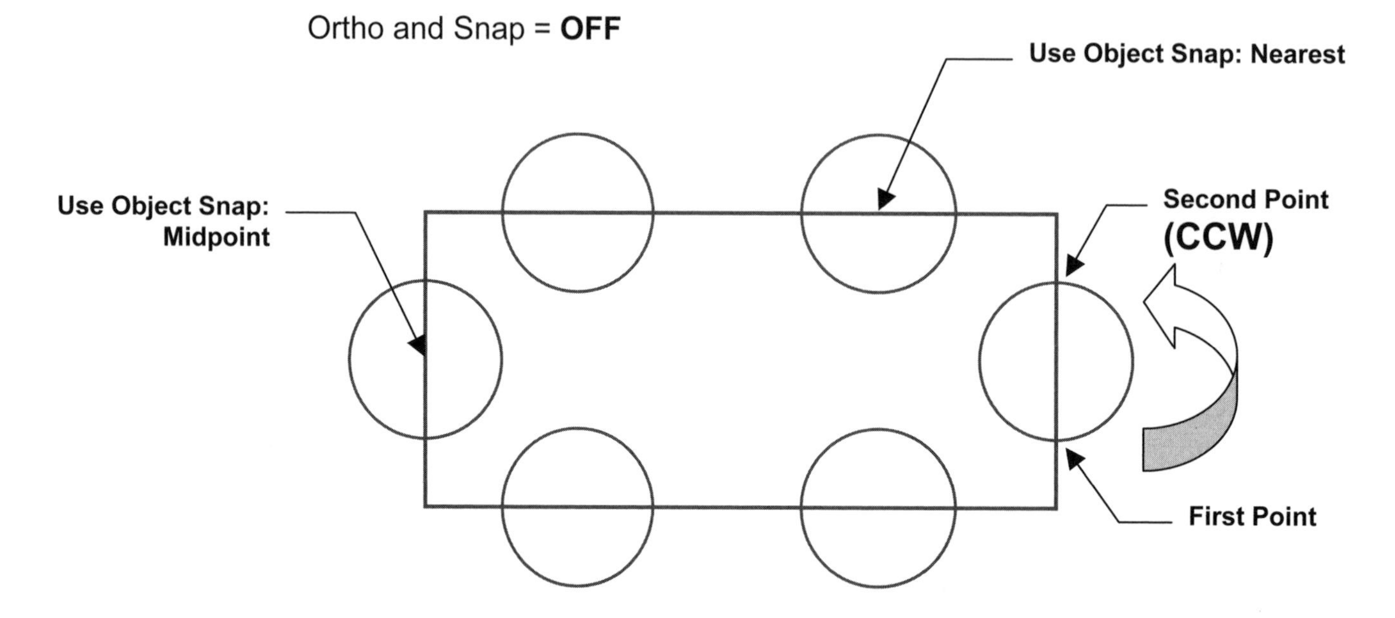

3. Modify the drawing above to look like the drawing below using:
   MODIFY / BREAK (Refer to 6-3, use Method 4)
   OBJECT SNAP = INTERSECTION

4. Remember the Circles break **CCW**.

5. **Save** this drawing as **EX6A.**

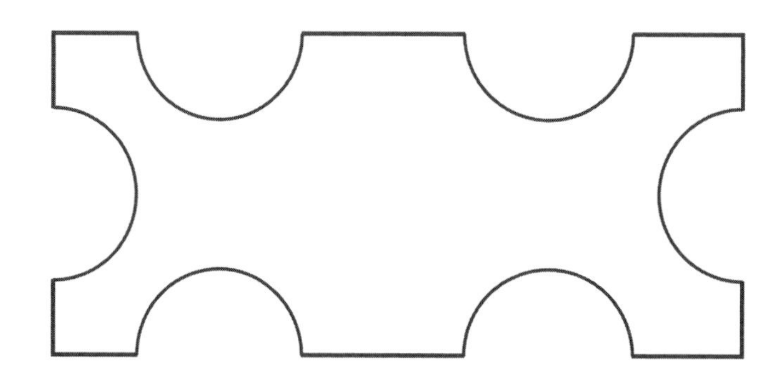

EXERCISE 6B

INSTRUCTIONS:

1. Start a **New** file and select **1workbook helper.dwt**.
2. Draw the objects below:
   OBJECT SNAP = CENTER
   ORTHO and SNAP **OFF**

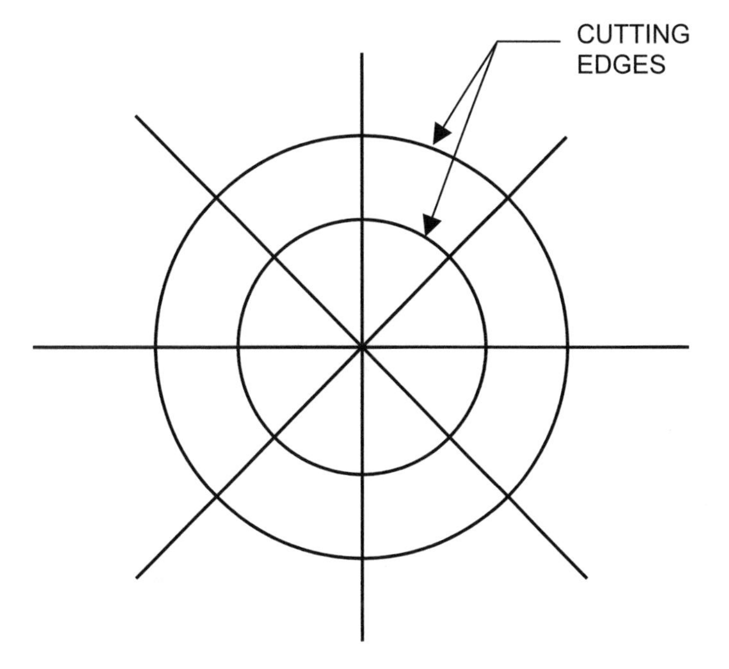

3. Modify the drawing above to look like the drawing below using:
   MODIFY / TRIM
4. **Save** this drawing as **EX6B.**

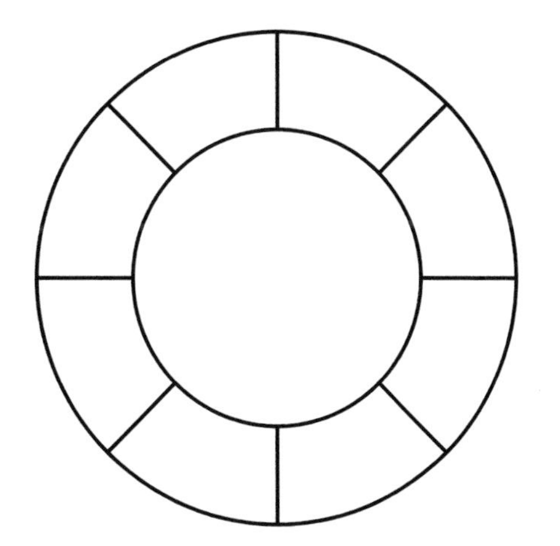

EXERCISE 6C

INSTRUCTIONS:

1. Start a **New** file and select **1workbook helper.dwt**.

2. Draw the LINES below exactly as shown.
   ORTHO and SNAP **ON**

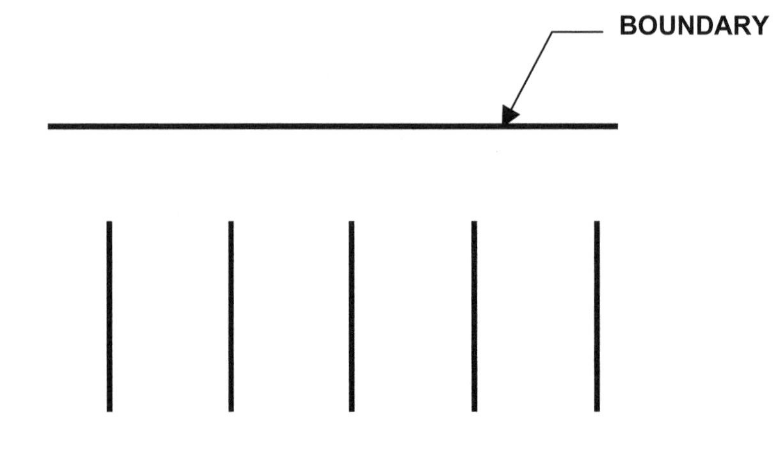

3. Modify the drawing above to look like the drawing below using:
   MODIFY / EXTEND

4. **Save** this drawing as **EX6C.**

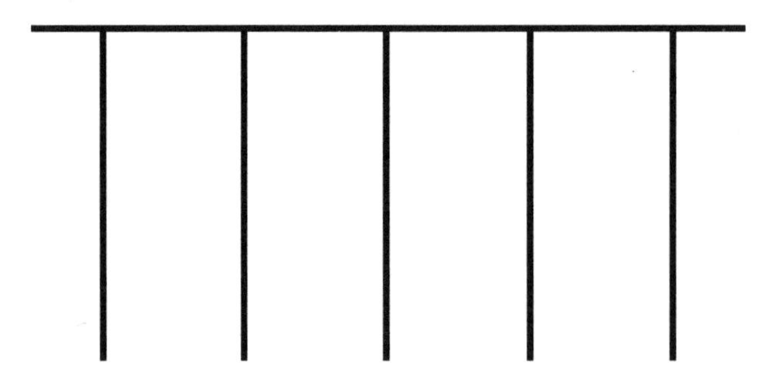

EXERCISE 6D

INSTRUCTIONS:

1. Start a **New** file and select **1workbook helper.dwt**.

2. Draw the drawing below using:
   RECTANGLE, CIRCLES and LINES (FOR THE X'S)

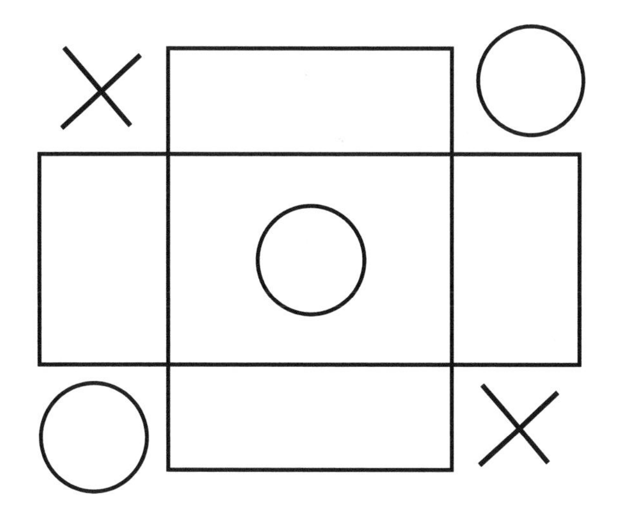

3. Modify the drawing above to look like the drawing below using:
   MODIFY / EXPLODE and ERASE

Note: You will not be able to erase the lines from the rectangle without Exploding the rectangle first. Now do you see why you need the Explode command?

4. **Save** this drawing as **EX6D.**

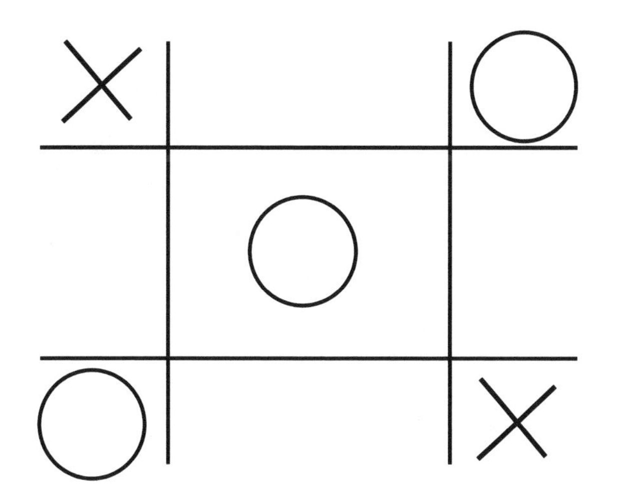

EXERCISE 6E

INSTRUCTIONS:

1. Start a **New** file and select **1workbook helper.dwt**.

2. Draw the drawing below using:
   CIRCLES, POINT and LINES

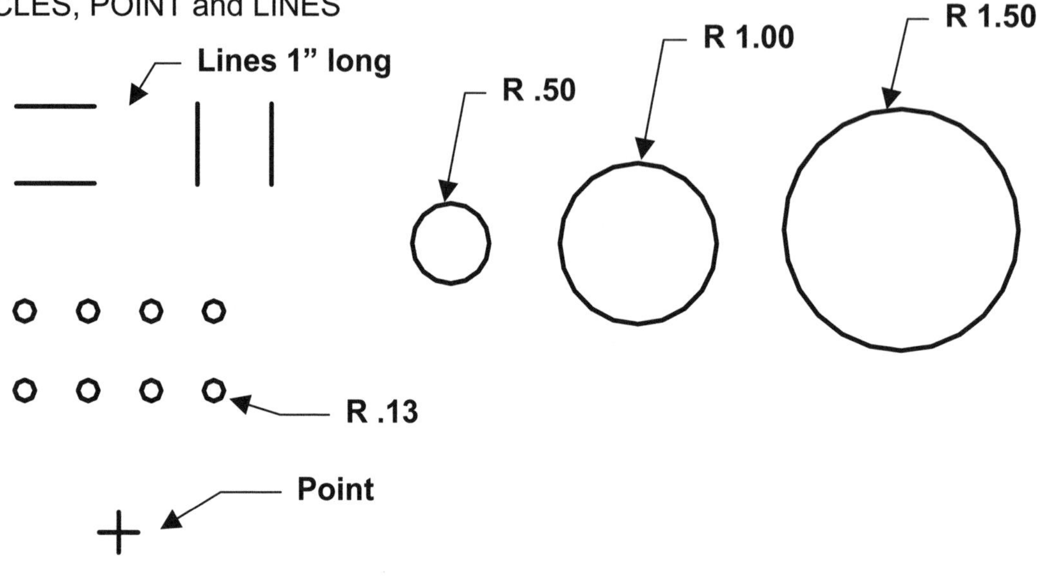

3. Assemble the objects as shown below using:
   MODIFY / MOVE
   OBJECT SNAP = CENTER, INTERSECTION, NODE, ENDPT and QUADRANT

Take your time an think about which object snap to use for each "basepoint" and "second point of displacement"

4. **Save** this drawing as **EX6E.**

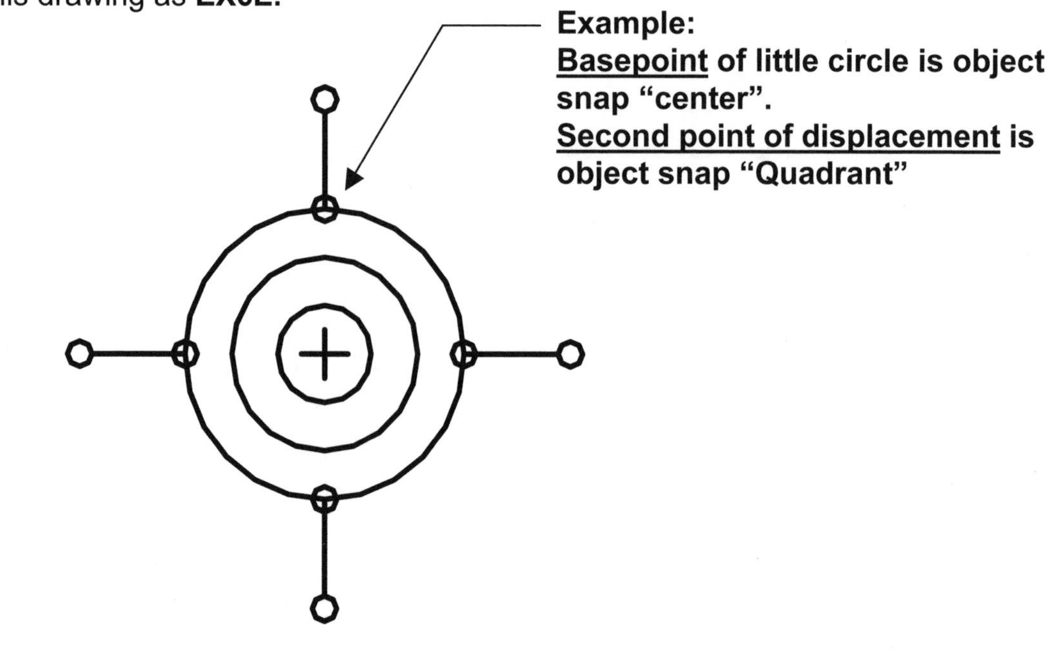

EXERCISE 6F

INSTRUCTIONS:

1. Start a **New** file and select **1workbook helper.dwt**.

2. Draw the drawing below using all the commands you have learned in the previous lessons. Use whatever layers you want and set snap and ortho on or off depending on the application.
   Be as creative as you like.

3. Save this drawing as **EX6F**

EXERCISE 6G

INSTRUCTIONS:

1. Start a **New** file and select **1workbook helper.dwt**.

The following exercise is exactly like exercise 6C but I want to show you a new method for selecting multiple objects without clicking on each object.

2. Draw the LINES below exactly as shown.
   ORTHO and SNAP **ON**

3. Select the "Extend" command.
4. Select the "Boundary"
5. Now instead of clicking on each vertical line, type F <enter>
6. Place your cursor approximately at location P1 and click.
7. Place your cursor approximately at location P2 and click.
8. Press <enter> <enter>

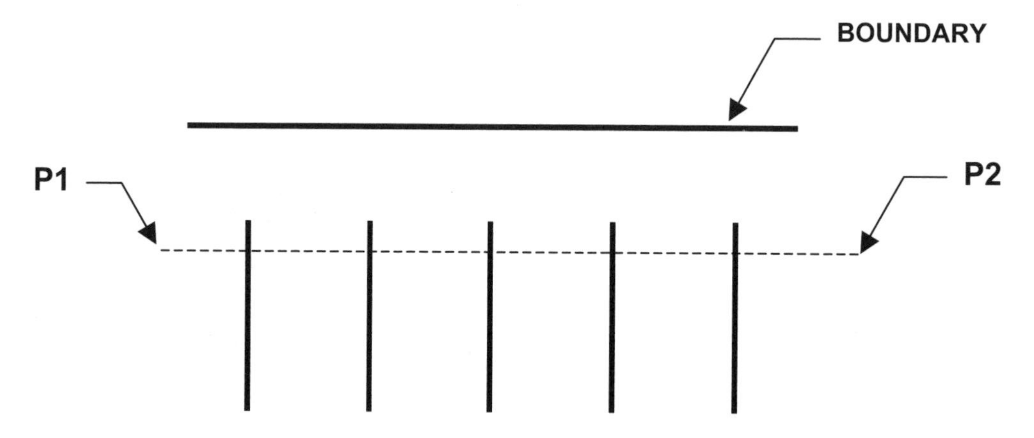

Note:
Be careful to place P1 and P2 above the midpoint of the vertical line. That tells AutoCAD that you want to extend the lines up. If you place P1 and P2 below the midpoint of the vertical line, AutoCAD will look for a boundary below. This will confuse AutoCAD because you did not select a boundary below.

FENCE
This method of selecting is called "Fence". The "F" you typed instead of clicking on each individual vertical line, told AutoCAD that you would like to use the "Fence" method for selecting objects.

4. **Save** this drawing as **EX6G.**

LEARNING OBJECTIVES

After completing this lesson, you will be able to:

1. Copy objects.
2. Make a mirrored image of one of more objects.
3. Add rounded corners to rectangular objects and lines.
4. Add angles to corners.

LESSON 7

COPY

The COPY command creates a duplicate set of the objects selected. The COPY command is similar to the MOVE command. You must select the objects to be copied, select a base point and a new location. The difference is, the Move command merely moves the objects to a new location. The Copy command makes a copy and you select the location for the new copy.

Select the Copy command using one of the following commands:

TYPE = CO
PULLDOWN = MODIFY / COPY
TOOLBAR = MODIFY

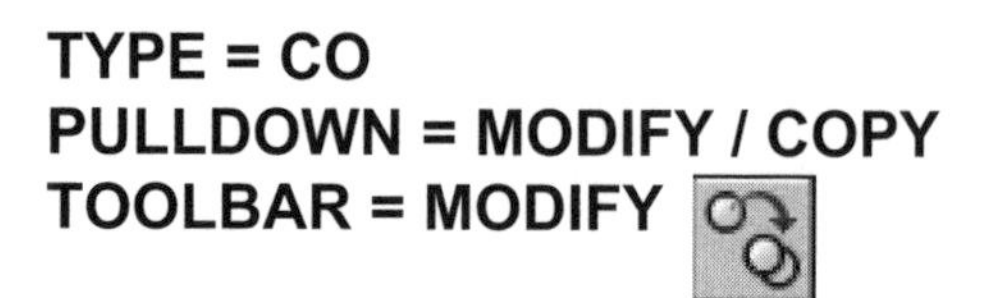

TO MAKE ONE COPY:
Command: _copy
Select objects: ***select the objects you want to copy***
Select objects: ***stop selecting objects by selecting <enter>***
Specify base point or displacement, or [Multiple]: ***select a base point (P1) (usually on the object)***
Specify second point of displacement or <use first point as displacement>: ***select the new location (P2) for the copy***

TO MAKE MULTIPLE COPIES:
Command: _copy
Select objects: ***select the objects you want to copy***
Select objects: ***stop selecting objects by selecting <enter>***
Specify base point or displacement, or [Multiple]: ***M <enter>***
Specify base point: ***select a point (P1) (usually on the object)***
Specify second point of displacement or <use first point as displacement>: ***select the new location for the first copy (P2)***
Specify second point of displacement or <use first point as displacement>: ***select the new location for the next copy***
Specify second point of displacement or <use first point as displacement>: ***select the new location for the next copy***
To Stop select <enter>

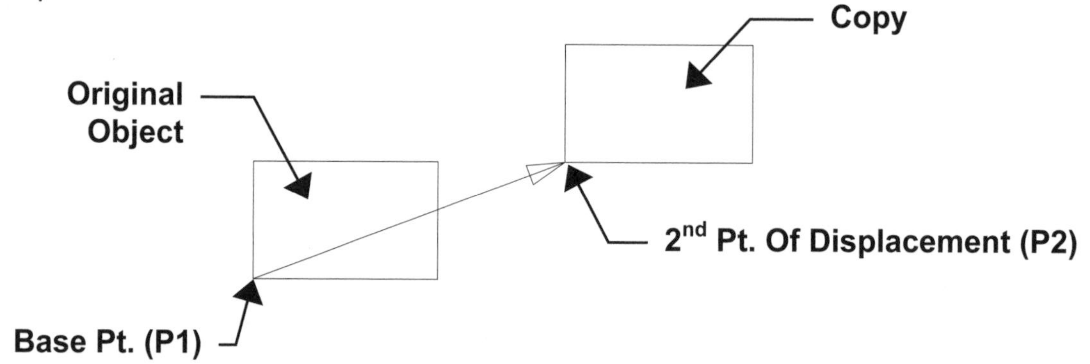

Note: Refer to Exercise Workbook for Advanced AutoCAD 2004 for additional Copy methods.

MIRROR

The MIRROR command allows you to make a mirrored image of any objects you select. You can use this command for creating right / left hand parts. You can draw a symmetrical object more efficiently by only drawing half of it.

Select the **MIRROR** command using one of the following:

TYPE = MI
PULLDOWN = MODIFY / MIRROR
TOOLBARS = MODIFY

The following will appear on the command line:

Command: _mirror
Select objects: ***select the objects to be mirrored***
Select objects: ***stop selecting objects by selecting <enter>***
Specify first point of mirror line: ***select the first end of the mirror line***
Specify second point of mirror line: ***select the second end of the mirror line***
Delete source objects? [Yes/No] <N>: ***select Y or N***

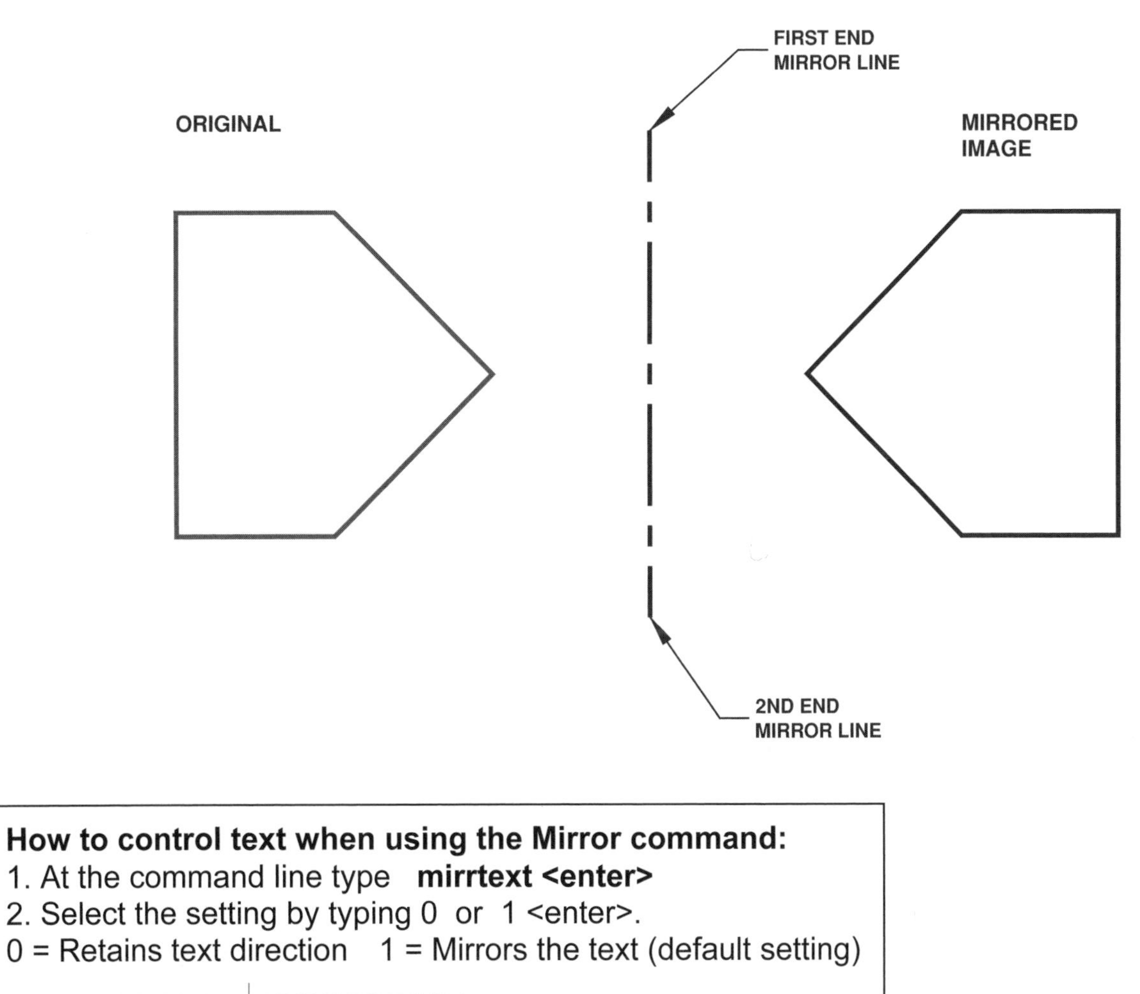

How to control text when using the Mirror command:
1. At the command line type **mirrtext <enter>**
2. Select the setting by typing 0 or 1 <enter>.
0 = Retains text direction   1 = Mirrors the text (default setting)

MIRRTEXT SETTING 0	MIRRTEXT SETTING 0
MIRRTEXT SETTING 1	MIRRTEXT SETTING 1

FILLET

The FILLET command will create a radius between two objects. The objects do not have to be touching. If two parallel lines are selected, it will construct a full radius.

1. Select the FILLET command using one of the following:

    TYPE = F
    PULLDOWN = MODIFY / FILLET
    TOOLBAR = MODIFY

The following will appear on the command line:

2. SET THE RADIUS OF THE FILLET
Command: _fillet
Current settings: Mode = TRIM, Radius = 0.000
Select first object or [Polyline/Radius/Trim/mUltiple]: ***type "R" <enter>***
Specify fillet radius <0.000>: ***type the radius <enter>***

3. NOW FILLET THE OBJECTS
Select first object or [Polyline/Radius/Trim/mUltiple]: ***select the first object to be filleted***
Select second object: ***select the second object to be filleted***

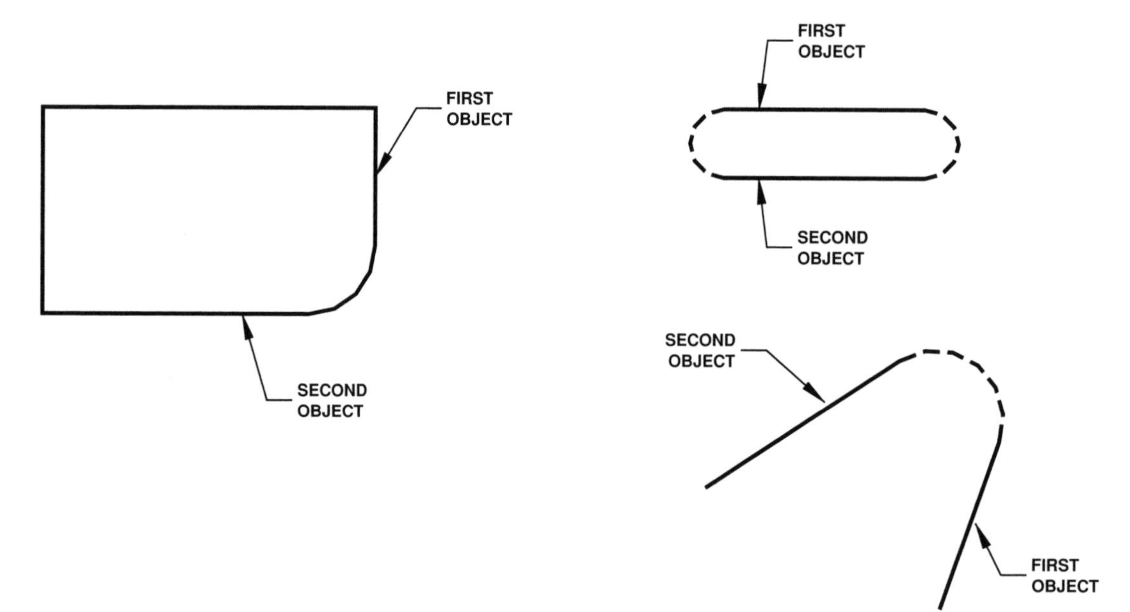

[Polyline/Radius/Trim/multiple] Options:

Polyline: This option allows you to fillet all intersections of a Polyline in one operation, such as all 4 corners of a rectangle.

Trim: This option controls whether the original lines are trimmed to the end of the Arc or remain the original length. (Set to Trim or No trim)

mUltiple: Repeats the fillet command until you press <enter> or esc key.

CHAMFER

The **CHAMFER** command allows you to create a chamfered corner on two lines. There are two methods: **Distance (page 7-5) and Angle (page 7-6)**.

DISTANCE METHOD

1. Select the CHAMFER command using one of the following:

   TYPE = CHA
   PULLDOWN = MODIFY / CHAMFER
   TOOLBAR = MODIFY

(Distance Method requires input of a distance for each side of the corner)

Command: _chamfer
(TRIM mode) Current chamfer Dist1 = 0.000, Dist2 = 0.000
Select first line or [Polyline/Distance/Angle/Trim/Method]: ***select "D"<enter>***
Specify first chamfer distance <0.000>: ***type the distance <enter>***
Specify second chamfer distance <1.000>: ***type the distance <enter>***

2. **NOW CHAMFER THE OBJECT**

Select first line or [Polyline/Distance/Angle/Trim/Method]: ***select the (First Line) to be chamfered. (dist 1 side)***
Select second line: ***select the (Second Line) to be chamfered. (dist 2 side)***

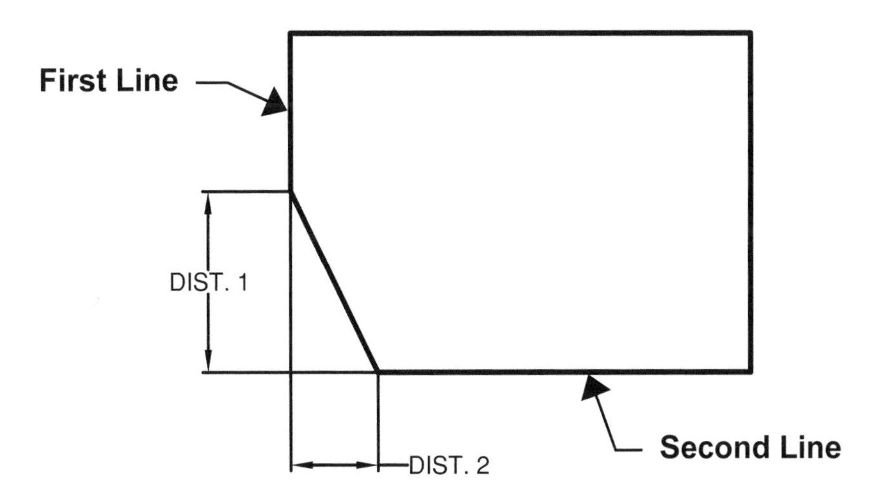

[Polyline/Distance/Angle/Trim/Method/mUltiple] options:

POLYLINE: This option allows you to Chamfer all intersections of a Polyline in one operation. Such as all 4 corners of a rectangle.

Trim: This option controls whether the original lines are trimmed or remain after the corners are chamfered. (Set to Trim or No trim)

Method: Allows you to switch between **Distance** and **Angle** method. The distance or angle must have been set previously.

mUltiple: Repeats the Chamfer command until you press <enter> or esc key.

CHAMFER (continued)

ANGLE METHOD

1. Select the CHAMFER command

(Angle method requires input for the length of the line and an angle)

Command: _chamfer
(TRIM mode) Current chamfer Dist1 = 1.000, Dist2 = 1.000
Select first line or [Polyline/Distance/Angle/Trim/Method]: ***select "A" <enter>***
Specify chamfer length on the first line <0.000>: ***type the chamfer length <enter> (dist 1)***
Specify chamfer angle from the first line <0>: ***type the angle <enter>***

2. **NOW CHAMFER THE OBJECT**

Select first line or [Polyline/Distance/Angle/Trim/Method/mUltiple]: ***select the (First Line) to be chamfered. (the length side)***
Select second line: ***select the (second line) to be chamfered. (the Angle side)***

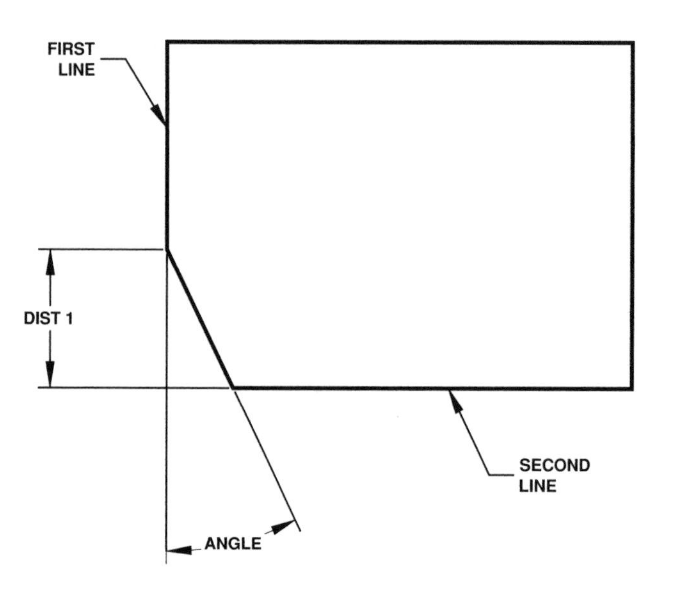

[Polyline/Distance/Angle/Trim/Method/mUltiple] options:

POLYLINE: This option allows you to Chamfer all intersections of a Polyline in one operation. Such as all 4 corners of a rectangle.

Trim: This option controls whether the original lines are trimmed or remain after the corners are chamfered. (Set to Trim or No trim)

Method: Allows you to switch between **Distance** and **Angle** method. The distance or angle must have been set previously.

mUltiple: Repeats the Chamfer command until you press <enter> or esc key.

EXERCISE 7A

INSTRUCTIONS:

1. Start a **New** file using **1workbook helper.dwt**.
2. **Draw** the rectangle below using:
   Draw / Rectangle
3. **Round** the corners using the FILLET command
4. **Save** this drawing as **EX7A**

BEFORE FILLET

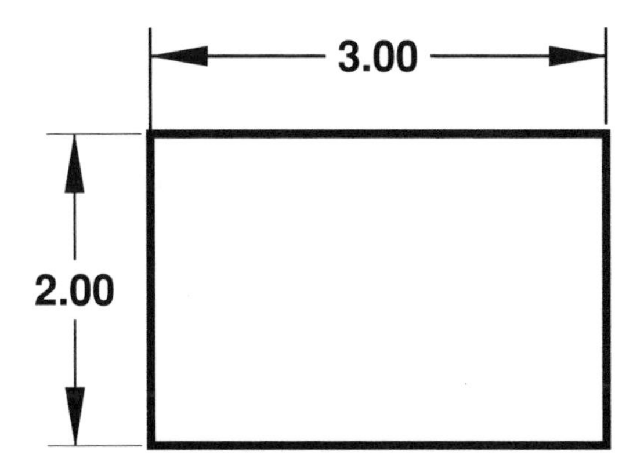

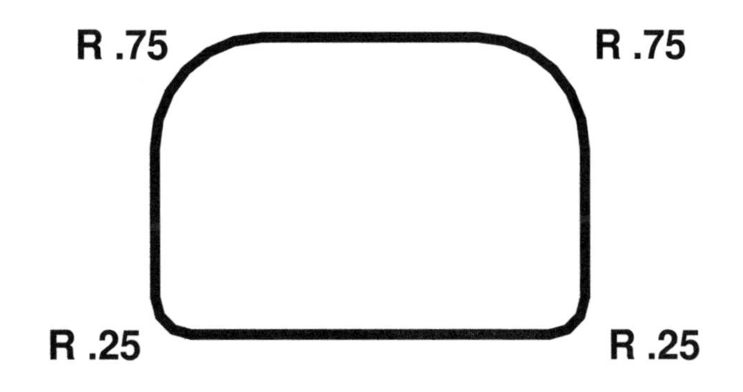

EXERCISE 7B

INSTRUCTIONS:

1. Start a **New** file **1 class template.dwt**.
2. **Draw** the rectangles below using:
   Draw / Rectangle
3. **CHAMFER** the corners using the CHAMFER command
4. **Save** this drawing as **EX7B**

Before Chamfer

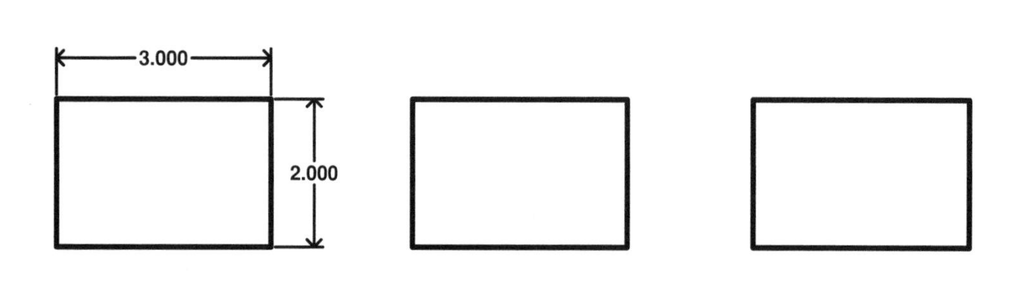

After Chamfer

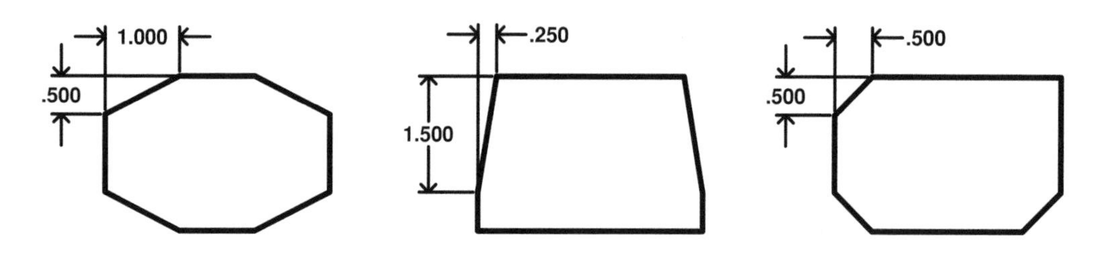

EXERCISE 7C

INSTRUCTIONS:

1. Start a **New** file and select **1workbook helper.dwt**.
2. **Draw** the rectangle below using:
   Draw / Rectangle
3. Chamfer the corners using the **CHAMFER** command
4. **Save** this drawing as **EX7C**

BEFORE CHAMFER

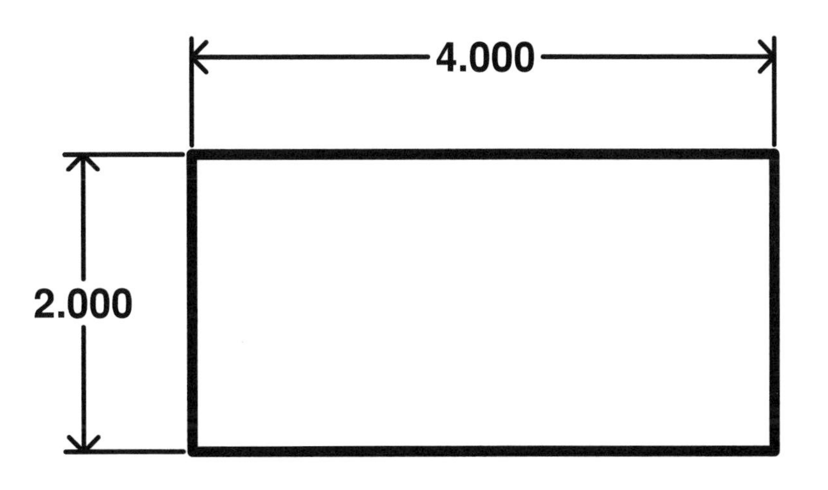

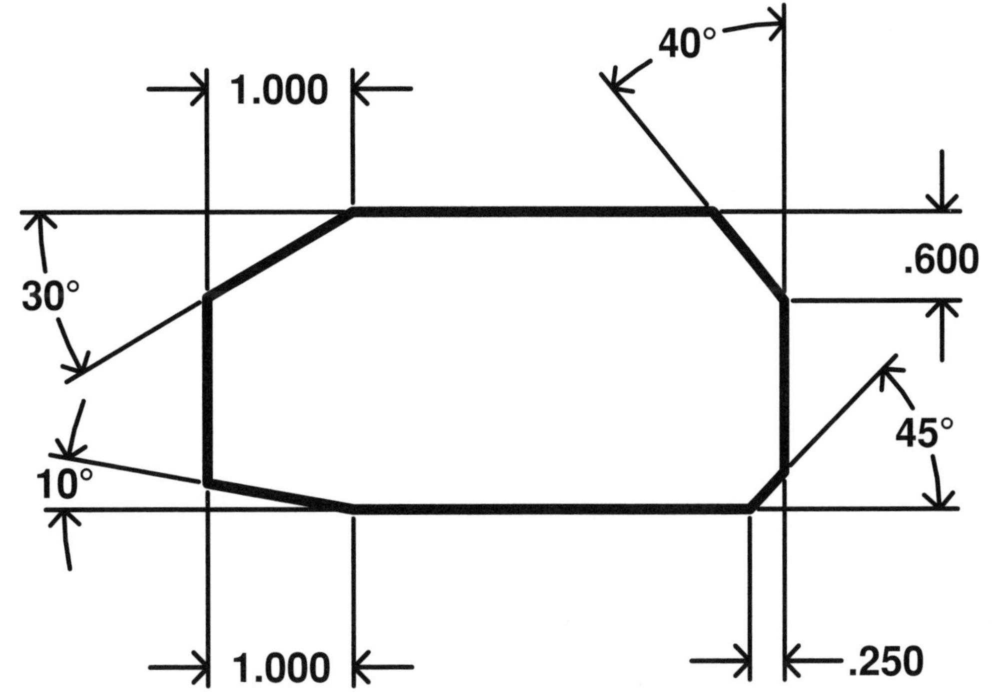

EXERCISE 7D

INSTRUCTIONS:

1. Start a **New** file and select **1workbook helper.dwt**.
2. **Draw** the Lines and one Circle as shown below

BEFORE COPY

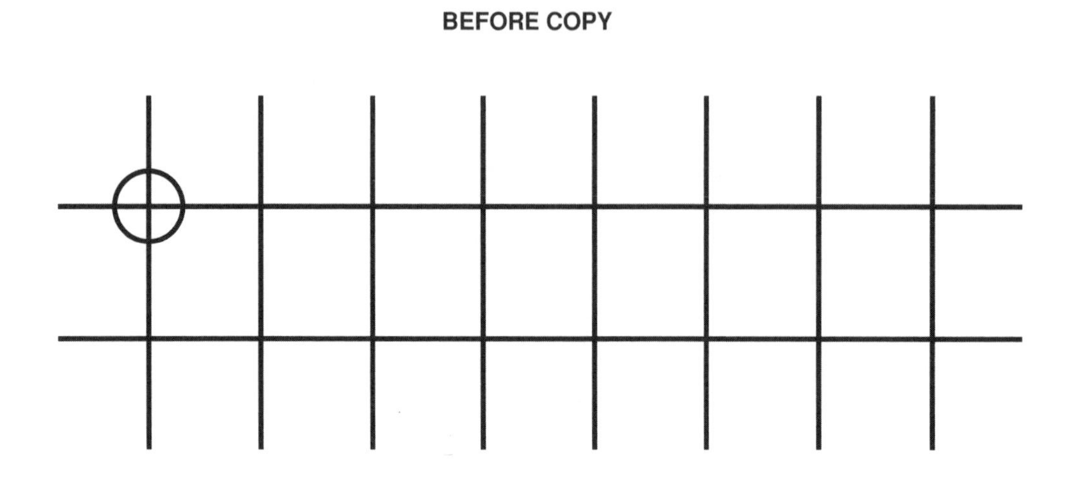

3. Complete the drawing as shown below using:
   a. Copy
   b. Select the Circle
   c. Select Multiple
   d. Select the basepoint on the Original circle. (Notice the basepoints are not all the same)
   e. Select the New location (2nd point of displacement)
4. Save this drawing as **EX7D**

AFTER COPY

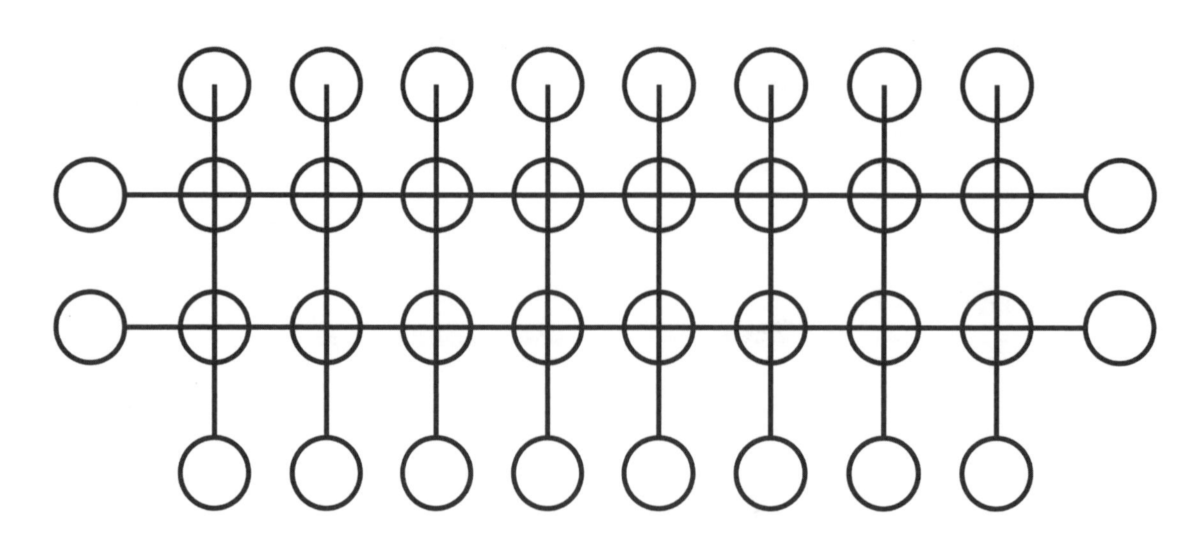

EXERCISE 7E

INSTRUCTIONS:

1. Start a **New** file and select **1workbook helper.dwt**.
2. Draw the half house below using:
   Lines, Circles and Rectangles
3. Use at least 4 different layers.

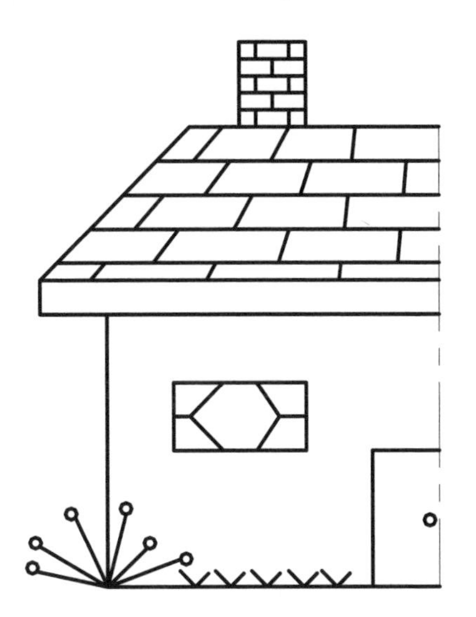

4. Now create a MIRRORED IMAGE of the half house using:
   a. Modify / Mirror
   b. Select the objects to be mirrored
   c. Select the 1st endpoint of the mirror line
   d. Select the 2nd endpoint of the mirror line
   e. Answer **No** to "Delete old objects?"

5. Save this drawing as **EX7E**

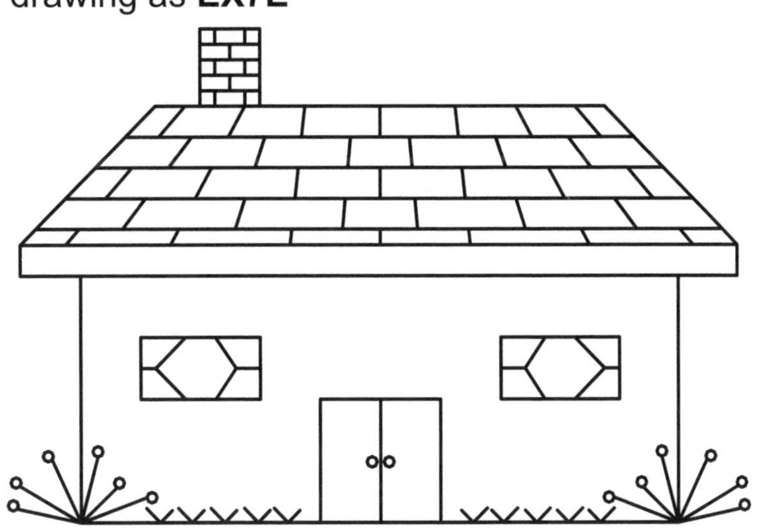

NOTES:

LEARNING OBJECTIVES

After completing this lesson, you will be able to:

1. Add a “Single Line” of text to your drawing.
2. Add a paragraph, using “Multiline Text”.
3. Control tabs, indents and line spacing.
4. Edit text already in the drawing.
5. Scale Text

LESSON 8

SINGLE LINE TEXT

SINGLE LINE TEXT allows you to draw one or more lines of text. The text is visible as you type. To place the text in the drawing, you may use the default **START POINT** (the lower left corner of the text), or use one of the many styles of justification described on the next page.

USING THE DEFAULT START POINT

1. Select the **SINGLE LINE TEXT** command using one of the following:

TYPE = DT or TEXT
PULLDOWN = DRAW / TEXT / SINGLE LINE TEXT
TOOLBAR = DRAW

Command: _dtext
Current text style: "STANDARD" Text height: 0.250

2. Specify start point of text or [Justify/Style]: ***Place the cursor where the text should start and left click.***
3. Specify height <0.250>: ***type the height of your text***
4. Specify rotation angle of text <0>: ***type the rotation angle then <enter>***
5. Enter text: ***type the text string; press enter at the end of the sentence***
6. Enter text: ***type the text string; press enter at the end of the sentence***
7. Enter text: ***type the next sentence or press <enter> to stop***

USING JUSTIFICATION

If you need to be very specific, where your text is located, you must use the Justification option. For example if you want your text in the middle of a rectangular box, you would use the justification option "Middle".

<u>The following is an example of Middle justification</u>.

1. Draw a Rectangle 6" wide and 3" high.
2. Draw a Diagonal line from one corner to the diagonal corner.
3. Select the SINGLE LINE TEXT command
   Command: _dtext
   Current text style: "STANDARD" Text height: 0.250
4. Specify start point of text or[Justify/Style]: ***type "J"***
5. Enter an option [Align/Fit/Center/Middle/Right/TL/TC/TR/ML/MC/MR/BL/BC/BR]: ***type M***
6. Specify middle point of text: ***snap to the midpoint of the diagonal line***
7. Specify height <0.250>: ***1 <enter>***
8. Specify rotation angle of text <0>: ***0 <enter>***
9. Enter text: ***type: HHHH <enter>***
10. Enter text: ***press <enter> to stop***

OTHER JUSTIFICATION OPTIONS:

ALIGN
Aligns the line of text between two points specified.
The height is adjusted automatically.

FIT
Fits the text between two points specified.
The height is specified by you and does not change.

CENTER HyyHHyyHHyy
This is a tricky one. Center is located at the bottom center of Upper Case letters.

MIDDLE HHHHHHHHH HHyyHHyyHHyy
If only uppercase letters are used: located in the middle, horizontally and vertically.
If both uppercase and lowercase letters are used: located in the middle, horizontally and vertically, of the lowercase letters.

RIGHT HyyHHyyHHyy
Bottom right of upper case text.

TL, TC, TR HyyHHyyHHyy
Top left, Top center and Top right of upper and lower case text

ML, MC, MR HyyHHyyHHyy
Middle left, Middle center and Middle right of upper case text.
(Notice the difference between "Middle" and "MC"

BL, BC, BR HyyHHyyHHyy
Bottom left, Bottom center and Bottom right of lower case text.

MULTILINE TEXT or MText

MULTILINE TEXT command allows you to easily add a sentence, paragraph or tables. The Mtext editor has most of the text editing features of a word processing program. You can underline, bold, italic, add tabs for indenting, change the font, line spacing, and width of the paragraph.

When using MText you must first define a text boundary box. The text boundary box is defined by entering where you wish to start the text (first corner) and approximately where you want to end the text (opposite corner) It is very similar to drawing a rectangle. The paragraph is considered one object rather than several individual sentences.

USING MULTILINE TEXT

1. Select the MULTILINE TEXT command using one of the following:

   TYPE = MT
   PULLDOWN = DRAW / TEXT / MULTILINE TEXT
   TOOLBAR = DRAW

   The command line will lists the current style and text height. The cursor will then appear as crosshairs with the letters "abc" attached. These letters show you how the text will appear using the current font and text height.

   Mtext Current text style: "STANDARD" Text height: .250

2. Specify first corner: ***Place the cursor at the location of the upper left corner of the new text boundary box and left click.***

3. Specify opposite corner or [Height / Justify / Line Spacing / Rotation / Style / Width]: ***Place the cursor at the location of the lower right corner of the new text boundary box and left click.***

   The (2 piece) Mtext Editor will appear.

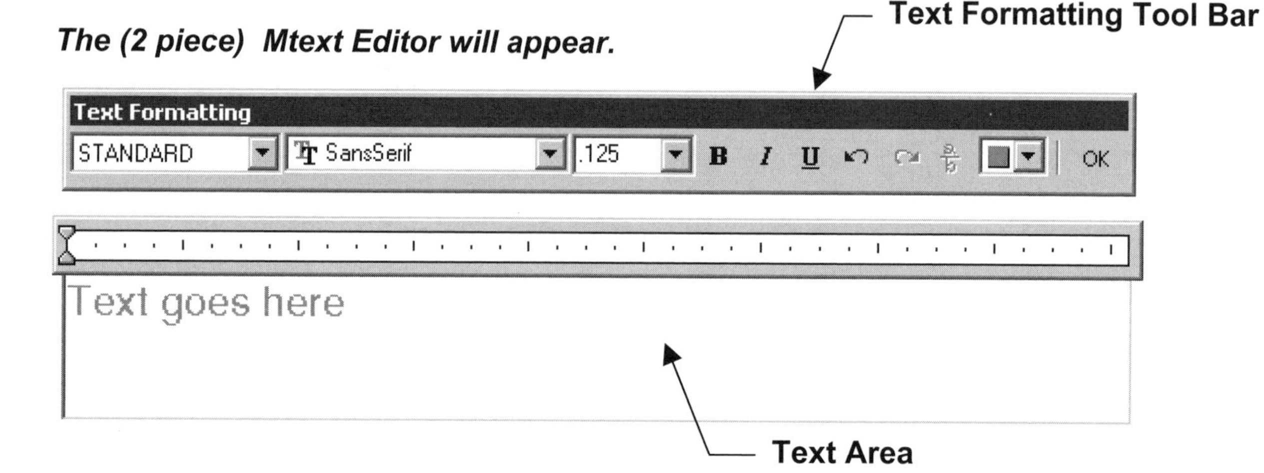

Notice the **Mtext Editor** is in 2 pieces. The **Text Formatting toolbar** and the **Text Area**. They can be moved independently so you can see the drawing underneath.

The **Text Formatting toolbar** allows you to select the Text Style, Font, and Height. You can add features such as bold, italics, underline and color. There is even an UNDO button.

The **Text Area box** allows you to enter the text, add tabs, adjust left hand margins and change the width of the paragraph.

4. After you have entered the text in the Text Area box, do one of the following to add the new text to the drawing and close the Mtext Editor:
   a. Select the OK button
   b. Press Ctrl + <enter>
   c. Left Click anywhere outside the Mtext Editor, but within the drawing area.

To adjust the width of the Mtext paragraph

Click and drag the right edge of the ruler (P1)

To change the width of the Text Area box itself

Click and drag the right edge of the Text Area box. (P2)
(This <u>will not</u> affect the actual width of the text that is added to the drawing)

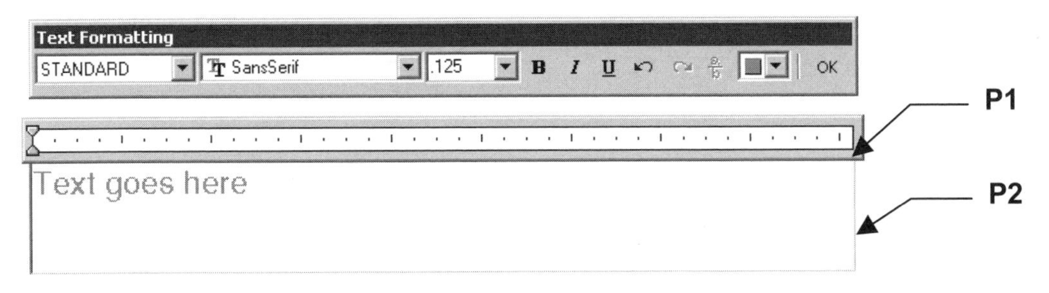

How to change the "abc", on the crosshairs, to other letters.

You can personalize the letters that appear attached to the crosshairs using the **MTJIGSTRING** system variable. (10 characters max).
Type MTJIGSTRING <enter> on the command line. Type the new letters <enter>.
The letters will be saved to the computer not the drawing. They will appear anytime you use Mtext and will remain until you change them again.

Note: Depending on whether you have zoomed in or out AutoCAD will resize these letters for easy viewing. If you would like to stop the resizing you may use the **MTEXTFIXED** system variable. Type MTEXTFIXED <enter> on the command line. Type 1 <enter>.

How to set tabs and indent

Refer to page 8-6 and 8-7

How to set Line Spacing

Refer to page 8-8

TABS

Setting and removing Tabs is very easy.

The default setting for tabs is 1”. (You may set as many tabs as you need.)
Set or change the stop positions, at anytime, using one of the following methods.

Method 1.
Place the cursor on the “Ruler” where you want the tab and left click. A little dark “L” will appear. The tab is set.
If you would like to remove a tab, just click and drag it off the ruler and it will disappear.

Method 2.
1. Right click on the Ruler.
2. Select “Indents and Tabs” from the short cut menu.
3. Set a tab position by typing the position in the upper box then select the “Set” button. Clear a tab position by highlighting the position in the lower box then select the “Clear” button.
4. Select the OK button.

Tab Stop Position
Sets tab positions for the current paragraph or selected paragraphs. The list below the text box shows the current tab stops.

Set
Copies the value in the Tab Stop Position box to the list below the box.

Clear
Removes the selected tab stop from the list.

Indents and Tabs
Indentation
First line:
.000
Paragraph:
.000
Tab stop position
1.000
1.000
3.500
6.000
Set
Clear
OK
Cancel

INDENTS

Sliders on the ruler show indention relative to the left side of the text boundary box. The top slider indents the first line of the paragraph, and the bottom slider indents the other lines of the paragraph.

You may change their positions, at anytime, using one of the following methods.

Method 1.
Place the cursor on the "Slider" and click and drag it to the new location.

Method 2.
1. Right click on the Ruler.
2. Select "Indents and Tabs" from the short cut menu.
3. Type the indent position in the First line and or Paragraph box.
4. Select OK button.

First Line:
Sets indentation for the first line of the current paragraph or selected paragraphs.

Paragraph:
Sets indentation for the current paragraph or selected paragraphs.

Indents and Tabs
Indentation
First line:
.000
Paragraph:
.000
Tab stop position
1.000
1.000
3.500
6.000
Set
Clear
OK
Cancel

MText – LINE SPACING

The Multiline text command allows you to set the spacing between the bottom of the first line of text to the bottom of the following lines of text. This is accomplished using the **Line spacing** option within the MText command.

You may set the Line spacing to a **Factor** of the "original line spacing", or enter the **Specific** distance desired.

If you choose to use the Factor method, you may use a factor up to 4X the "original line spacing". AutoCAD has established the "original line spacing" as 1.66 of the text height. For example, the line spacing for 1" text is 1.66 from the bottom of the first line to the bottom of the second line.

Line spacing
(Original line spacing distance is 1.66 X Text ht.)
Line spacing

If you choose to enter a specific distance, the original line spacing is ignored.

Here is how it works.

1. Select the MText command.

   Command: _mtext Current text style: "STANDARD" Text height: 1.00
2. Specify first corner: ***Place the cursor, and left click, to locate the first corner of the text boundary.***

3. Specify opposite corner or [Height/Justify/Line spacing/Rotation/Style/Width]: ***Select the Line spacing option by typing L <enter> or right click and select Line spacing from the short cut menu.***

4. Enter line spacing type [At least/Exactly] <At least>: ***Select " Exactly" option by typing E <enter> or right click and select "Exactly" from the short cut menu.*** ("At least" is the default option that merely insures that the text does not overlap)

5. Enter line spacing factor or distance <1x>: ***Enter a factor or a specific distance. (Note: you must include the "x" when entering the factor number.)***

6. Specify opposite corner or [Height/Justify/Line spacing/Rotation/Style/Width]: ***Place the cursor, and click, to locate the opposite corner of the text paragraph.***

   The MText editor will appear.

7. Type the text and select OK.

EDITING TEXT

SINGLE LINE TEXT

Editing **Single Line Text** is somewhat limited compared to Multiline Text. In the example below you will learn how to edit the text within a Single Line Text sentence. (In Lesson 12 you will learn additional options for editing Single Line text by using the Properties command.)

1. Double click on the Single Line text you want to edit.

The following "Edit Text" box will appear:

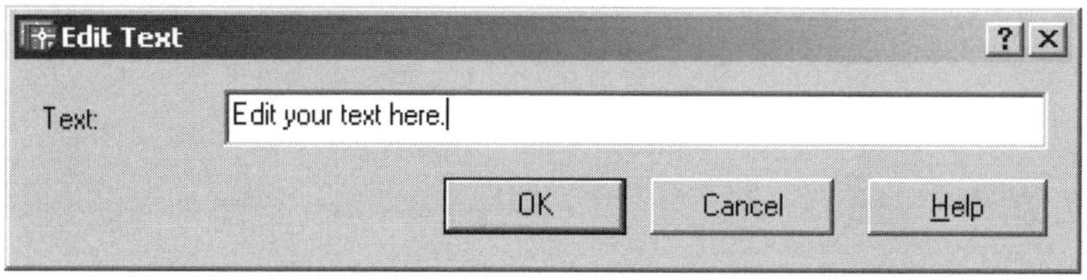

2. Make the changes then select the OK button.
3. Select the next line of text to be edited or press <enter> to stop

MULTILINE TEXT

Multiline Text is as easy to edit as it is to input originally. You may change the style, font, height, color, indent and add text features such as bold, italic and underline.

1. Double click on the Multiline text you want to edit.

The following "Mtext Editor" will appear:

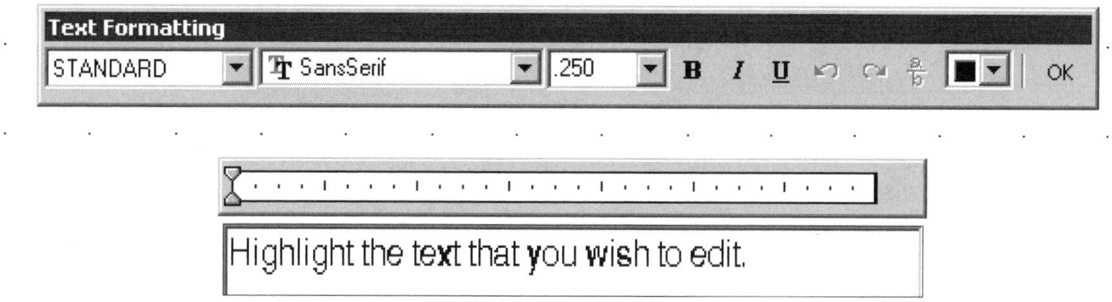

2. Highlight the text, that you want to change, using click and drag.
3. Make the changes then select the OK button.

Note: If you right click in the Text Area you may also change the Justification, Change Case and add a Symbol.

You may also type "**DDEDIT**" on the command line, then select the text you want to edit. This method works with Single or Multiline Text.

SCALING TEXT

The **Scaletext** command allows you to scale Single or Multiline text using **height**, **factor** or **match** a previously drawn text. The text will be scaled proportionately in the X and Y axis. In other words, it gets larger or smaller all over. You must select a "base point" from which the text will enlarge or reduce. The base point is a justification option. If you would like it to scale from the middle, you will select "middle" as the base point. If you would like the text to scale from the top, you will select "TC". Most of the time you will simple use "existing". Existing is the justification used when the text was originally created.

Select the **SCALETEXT** command using one of the following:

TYPE = scaletext
PULLDOWN = MODIFY / OBJECT / TEXT / SCALE
TOOLBAR = DRAW

HEIGHT

1. Select the line of text you want to edit.
2. Select objects: ***select more text or <enter> to stop selecting***
3. Enter a base point option for scaling [Existing/Left/Center/Middle/Right/TL/TC/TR/ML/MC/MR/BL/BC/BR] <Existing>:***select the base point***
4. Specify new height or [Match object/Scale factor] <.250>:***type new height <enter>***

MATCH OBJECT

1. Select the line of text you want to edit.
2. Select objects: ***select more text or <enter> to stop selecting***
3. Enter a base point option for scaling [Existing/Left/Center/Middle/Right/TL/TC/TR/ML/MC/MR/BL/BC/BR] <Existing>:***select the base point***
4. Specify new height or [Match object/Scale factor] <.250>:***select Match***
5. Select a text object with the desired height: ***select the text to match***
   Height = ***the new height will be shown here***

SCALE FACTOR

1. Select the line of text you want to edit.
2. Select objects: ***select more text or <enter> to stop selecting***
3. Enter a base point option for scaling [Existing/Left/Center/Middle/Right/TL/TC/TR/ML/MC/MR/BL/BC/BR] <Existing>:***select the base point***
4. Specify new height or [Match object/Scale factor] <.250>:***select Scale factor***
5. Specify scale factor or [Reference] <2.000>: ***type the factor***

EXERCISE 8A

INSTRUCTIONS:

1. Start a **New** file and select **1workbook helper.dwt**
2. Duplicate the text shown below using SINGLE LINE text.
3. Use Layer TXT-HVY.
4. Select **DRAW / TEXT / SINGLE LINE**
5. Follow the instructions in each block of text. To start the text in the correct location, stated in each example, move your cursor while watching the coordinate display.
6. Save this drawing as EX8A.

DO NOT DUPLICATE NOTES WITH ARROWS

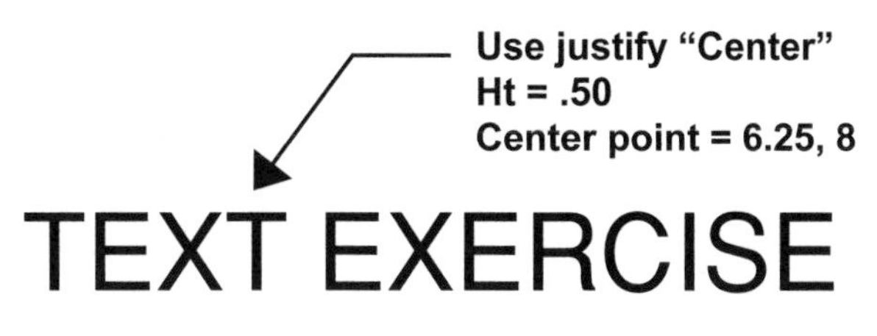

THIS TEXT'S START POINT IS .75, 7.50
AND THE HEIGHT IS .13

THIS TEXT IS JUSTIFIED, RIGHT.
THE ENDPOINT IS 11.50, 7.50.
THE HEIGHT IS .13.
THE TEXT WILL LOOK A LITTLE STRANGE
BECAUSE IT SEEMS TO BE JUSTIFIED LEFT.
BUT WHEN YOU FINISH TYPING AND PRESS
ENTER TWICE, THE TEXT WILL MOVE TO
THE RIGHT JUSTIFIED.

USE JUSTIFY "ALIGN" FOR THIS TEXT. FIRST ENDPT IS .75, 5.5. SECOND ENDPT IS 11.5, 5.5.

USE JUSTIFY "ALIGN" AGAIN. FIRST ENDPT IS .75, 4.75. SECOND ENDPT IS 11.75, 5.25.

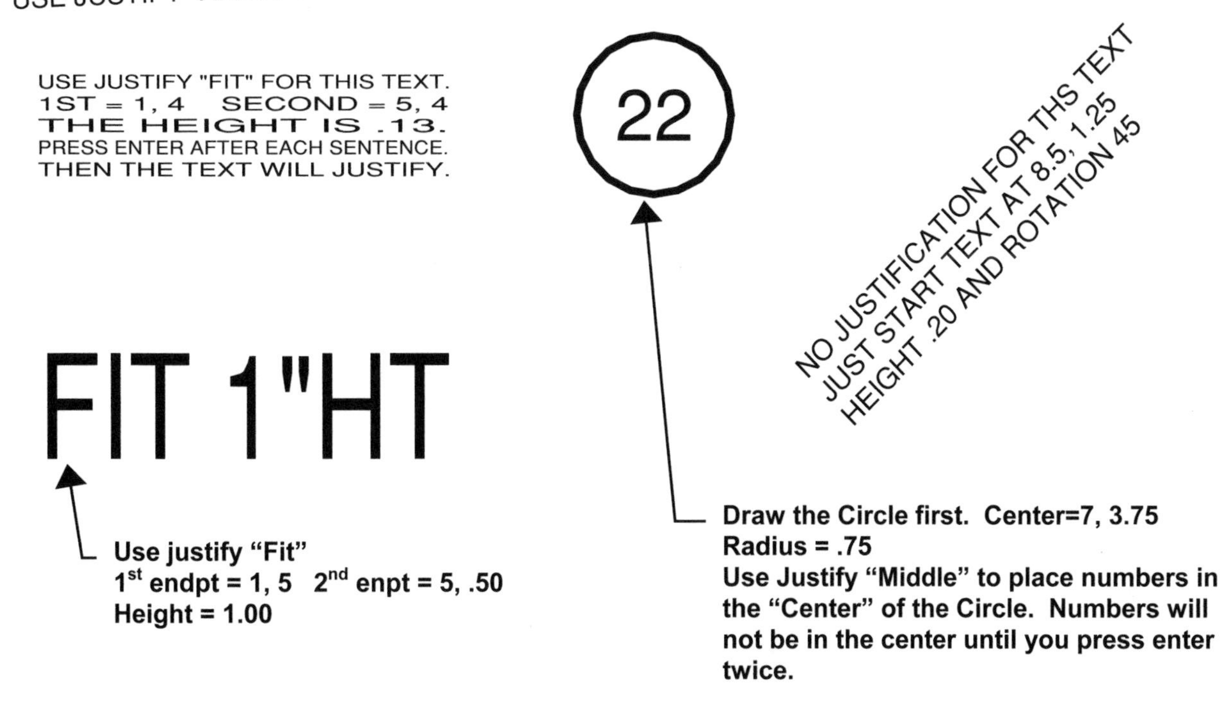

EXERCISE 8B

INSTRUCTIONS:

1. Open **Workbook Helper.dwg**
2. Duplicate the text shown below using **MULTILINE** text.
3. Select **DRAW / TEXT / MULTILINE**
4. Use Text Style: **Standard**
5. Use font: **SansSerif**
6. Text Height: **.250**
7. Set tabs to: **4.50** and **8.00**
8. Header Line is **bold** and **underlined**.
9. Enter all text shown below.
10. Save as: **EX8B**

The following is a practice exercise for tabs, indent and the features bold and underline.

    1. This sentence should be indented 1 inch.

        a. This sentence should be indented 1/2 inch more.

And now back to the left margin.

Isn't this fun!

STUDENT NAME	**STUDENT ID**	**GRADE**
Susie Que	1234567	A
John Smith	8910116	B

EXERCISE 8C

INSTRUCTIONS:

1. Open Workbook Helper.dwg

The following exercise is designed to teach you how to insert text into the exact middle of a rectangular area. You will use both Single Line Text and Multiline Text.

2. Follow the instructions for each text box below.
3. Save as: **EX8C**

SINGLE LINE TEXT

1. Draw a 6” wide by 3” high rectangle.
2. Draw a diagonal line from corner to corner, as shown.
   (Use endpoint object snap to be accurate.)
3. Use the following settings:
   Justify: Middle
   Text Ht: 1inch

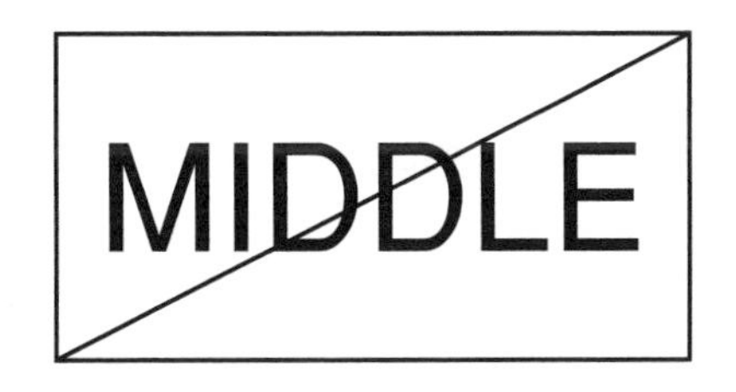

MULTILINE TEXT

1. Draw a 13” wide by 2” high rectangle.
2. Start the Text boundary box at the upper left corner of the rectangle.
   Place the opposite corner at the lower right corner of the rectangle.
   (Use endpoint object snap to be accurate.)
3. Use the following settings:
   Justify: Middle Center
   Text Ht: 1inch

MIDDLE CENTER

EXERCISE 8D

INSTRUCTIONS:

1. Open Workbook Helper.dwg
2. Draw two 6" long lines as shown.
3. Select Draw / Text / Single Line Text
   a. Select **Justify - Center**.
   b. Use Midpoint snap to place the justification point at the midpoint of the line.
   c. Use text height 1" and rotation angle 0.
   d. Type the word "Happy" <enter> <enter>

4. Select Draw / Text / Single Line Text again.
   a. This time select **Justify - BC**. (bottom center)
   b. Use Midpoint snap to place the justification point at the midpoint of the line.
   c. Use text height 1" and rotation angle 0.
   d. Type the word "Happy" <enter> <enter>

Notice the difference between <u>Center</u> and <u>Bottom Center</u>?

"Center" only considers the Upper Case letters when justifying.
"Bottom Center" is concerned about those Lower Case letters.
Can you see how you could accidentally place your text too high or too low? Think about the difference between Center and Bottom Center.

EXERCISE 8E

INSTRUCTIONS:

1. Open **Workbook Helper.dwg**
2. Select Draw / Text / Multiline Text
3. Specify the size of the multiline paragraph by entering the coordinates for the first corner and then the coordinates for the opposite corner:
   First corner Type: 0,0<enter> Opposite Corner Type: 7, 3 <enter>
3. Use the following:
   Text style: Standard Font: SansSerif Height: .250
4. Type the following as **one continuous line**. ***Do not press <enter>.***
5. When you are finished typing, select OK.

Today is the day that I will learn how to use
AutoCAD's text editing features. I will be a
good student and practice **all** the features
because I **really** want to learn!

Hopefully your text appears the same as the text shown above.

Now you are going to change the width of this paragraph.

5. Select the multiline editor by double clicking on the paragraph.
6. Reduce the "ruler" to 4 inches .
7. Select OK.

Does your text look like the example below?

Today is the day that I will
learn how to use
AutoCAD's text editing
features. I will be a good
student and practice **all**
the features because I
really want to learn!

See how easily you can manipulate the Multiline text.

EXERCISE 8F

INSTRUCTIONS:

1. Open **Workbook Helper.dwg**.
2. Draw the Text on the left using "Single Line" text option for the first sentence and "Multiline" text option for the second sentence. (Layer = Txt-Hvy)
3. Type **Mirrtext** at the command line.
4. Select setting 0
5. Using the **Mirror** command, mirror both sentences using a vertical mirror line, P1 to P2, approximately as shown.

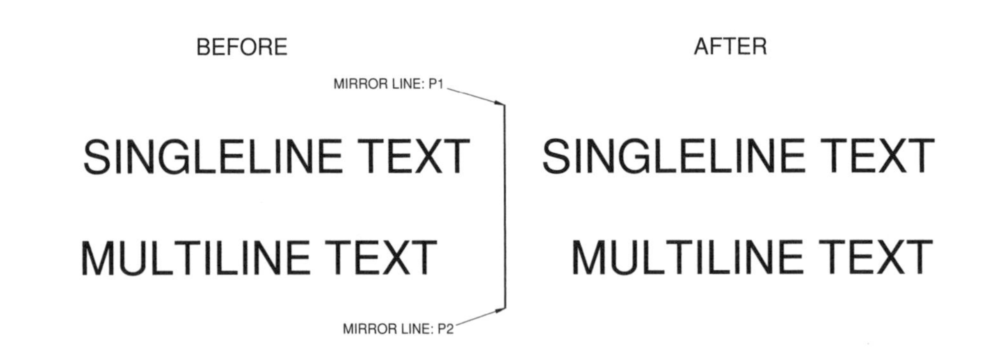

6. Now change the **Mirrtext** setting to 1
7. Try the Mirror command again.

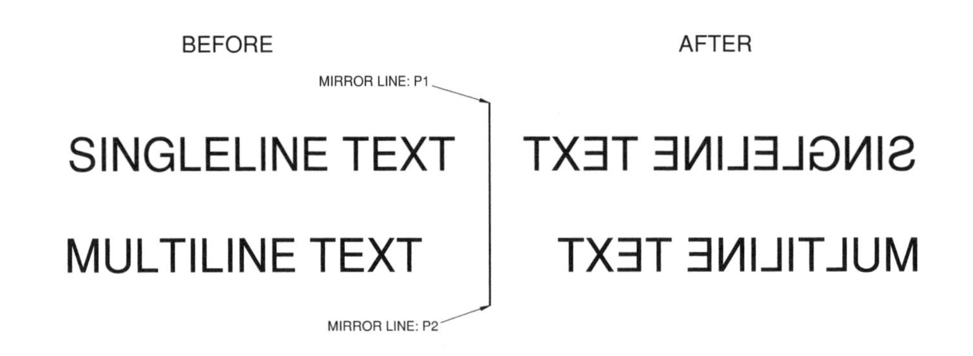

8. Save this drawing as **EX8F**.

LEARNING OBJECTIVES

After completing this lesson, you will be able to:

1. Understand the ORIGIN.
2. Draw Objects using Coordinate Input.
3. Input Absolute and Relative coordinates.
4. Use Direct Distance Entry.
5. LIST information about objects.
6. Determine the distance between two points.
7. Identify a location within the drawing.
8. Create your own 11 x 17 Master Border.
9. Print 11 x 17 drawings in Model Space.

LESSON 9

COORDINATE INPUT

In the previous lessons you have been using the cursor to place objects.
In this lesson you will learn how to place objects in specific locations by entering coordinates. This process is called **Coordinate Input**.

This is not difficult, so do not start to worry.

Autocad uses the ***Cartesian Coordinate System.***

The Cartesian Coordinate System has 3 axes, X, Y and Z.

The **X** is the Horizontal axis. (*Right and Left*)
The **Y** is the Vertical axis. (*Up and Down*)
The **Z** is Perpendicular to the X and Y plane.
*(The **Z** axis, which is not discussed in this workbook, is used for 3D.)*

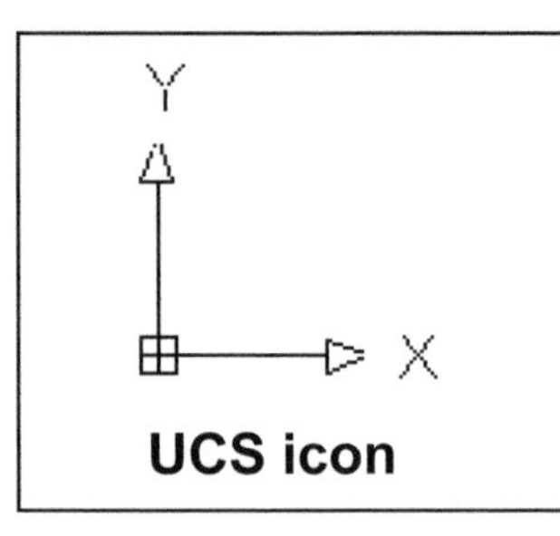

UCS icon

Look at the User Coordinate System (UCS) icon in the lower left corner of your screen. The arrows are pointing in the positive direction.

The location where the X , Y and Z axes intersect is called the **ORIGIN**. (0,0,0)
Currently the Origin is located in the lower left corner of the screen.
When you move the cursor away from the Origin, in the direction of the arrows, the X and Y coordinates are positive.
When you move the cursor in the opposite direction, the X and Y coordinates are negative.

Using this system, every point on the screen can be specified using positive and negative X and Y coordinates.

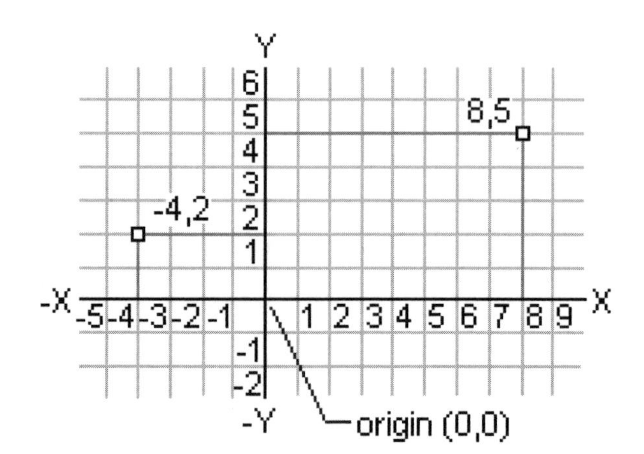

There are 3 types of Coordinate input, **Absolute, Relative** and **Polar**.
(Polar will be discussed in Lesson 11)

ABSOLUTE COORDINATES (The input format is: **X, Y)**

Absolute coordinates come ***from the ORIGIN*** and are typed as follows: **8, 5 .**
The first number (8) represents the **X-axis** (horizontal) distance from the Origin and the second number (5) represents the **Y-axis** (vertical) distance from the Origin.
The two numbers must be separated by a **comma**.

An absolute coordinate of **4, 2** will be **4** units to the right (horizontal) and **2** units up (vertical) from the current location of the Origin.

An absolute coordinate of **-4, -2** will be **4** units to the left (horizontal) and **2** units down (vertical) from the current location of the Origin.

The following are examples of Absolute Coordinate input.
Notice where the Origin is located in each example.

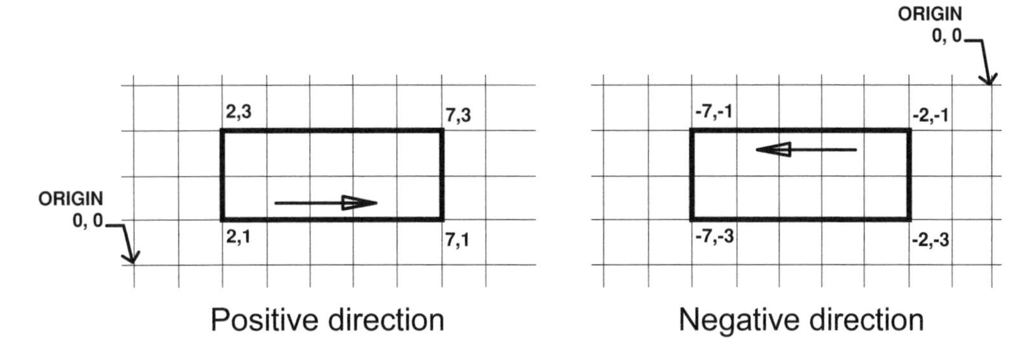

Positive direction   Negative direction

RELATIVE COORDINATES (The input format is: **@X, Y**)

Relative coordinates come ***from the last point entered***. The first number represents the **X-axis** (horizontal) and the second number represents the **Y-axis** (vertical).
The two numbers must be separated by a **comma**. To distinguish between Absolute and Relative, use the **@** symbol and then the X and Y coordinates.

A Relative coordinate of **@5, 2** will go to the **right** 5 units and **up** 2 units from the last point entered.

A Relative coordinate of **@-5, -2** will go to the **left** 5 units and **down** 2 units from the last point entered.

The following is an example of Relative Coordinate input.

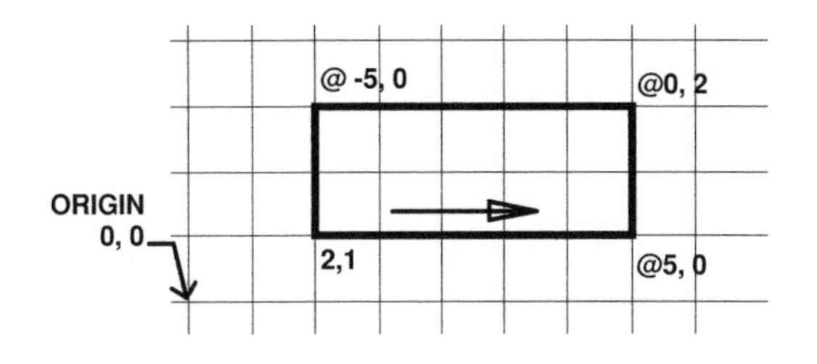

EXAMPLES OF COORDINATE INPUT

Scenario 1.
You want to draw a line with the first endpoint "at the Origin" and the second endpoint 3 units in the positive X direction.

1. Select the Line command.
2. You are prompted for the first endpoint: **Type 0, 0 <enter>**
3. You are then prompted for the second endpoint: **Type 3, 0 <enter>**

What did we do?
The first endpoint coordinate input, 0,0 means that you do not want to move away from the Origin. You want to start "on" the Origin.

The second endpoint coordinate input, 3, 0 means that you want to move 3 units in the positive X axis. The "0" means you do not want to move in the Y axis. So the line will be exactly horizontal.

Scenario 2.
You want to start a line 8 units directly above the origin and it will be 4 units in length, perfectly vertical.

1. Select the Line command.
2. You are prompted for the first endpoint: **Type 0, 8 <enter>**
3. You are prompted for the second endpoint: **Type @0, 4 <enter>**

What did we do?
The first endpoint coordinate input, 0, 8 means you do not want to move in the X axis direction but you do want to move in the Y axis direction.

The second endpoint coordinate input @0, 4 means the you do not want to move in the X axis "from the last point entered" but you do want to move in the Y axis "from the last point entered. (Remember the @ symbol tells AutoCAD that the coordinates typed are relative to the "last point entered" not the Origin.

Scenario 3.
Now try drawing 5 connecting lines sements.

1. Select the Line command.
2. First endpoint: 2, 4 <enter>
3. Second endpoint: @ 2, -3 <enter>
4. Second endpoint: @ 0, -1 <enter>
5. Second endpoint: @ -1, 0 <enter>
6. Second endpoint: @ -2, 2 <enter>
7. Second endpoint: @ 0, 2 <enter> <enter>

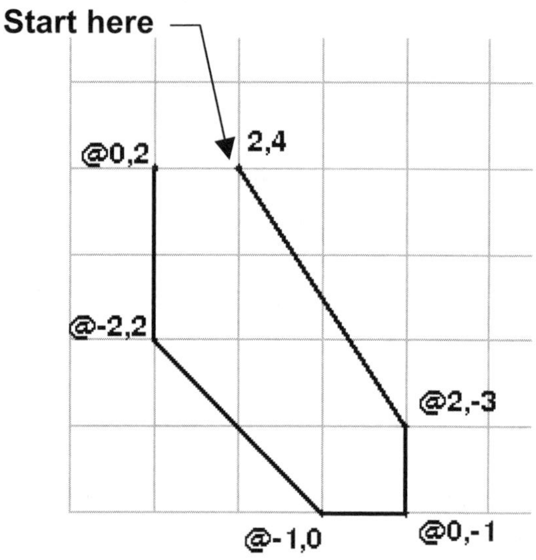

Notice the "@" symbol for relative coordinates.

Note: If you enter an incorrect coordinate, just type "U" <enter> and the last segment will disappear and you will have another chance at entering the correct coordinate.

DIRECT DISTANCE ENTRY (DDE)

DIRECT DISTANCE ENTRY is a combination of keyboard entry and cursor movement. **DDE** is used to specify distances in the horizontal or vertical axes from the last point entered. **DDE is a *Relative Input*.** Since it is used for Horizontal and Vertical movements, **Ortho** must be **ON**.

(Note: to specify distances on an angle, refer to Lesson 11)

Using DDE is simple. Just move the cursor and type the distance.
Negative and positive is understood automatically by moving the cursor up (positive), down (negative), right (positive) or left (negative) from the last point entered. No minus sign necessary.

Moving the cursor to the right and typing 5 and <enter> tells AutoCAD that the 5 is positive and Horizontal.
Moving the cursor to the left and typing 5 and <enter> tells AutoCAD that the 5 is negative and Horizontal.
Moving the cursor up and typing 5 and <enter> tells AutoCAD that the 5 is positive and Vertical.
Moving the cursor down and typing 5 and <enter> tells AutoCAD that the 5 is negative and Vertical.

EXAMPLE:

1. Ortho must be ON.
2. Select the Line command.
3. Type: 1, 2 <enter> to enter the first endpoint using Absolute coordinates.
4. Now move your cursor to the right and type: 5 <enter>
5. Now move your cursor up and type: 4 <enter>
6. Now move your cursor to the left and type: 5 <enter> <enter> to stop

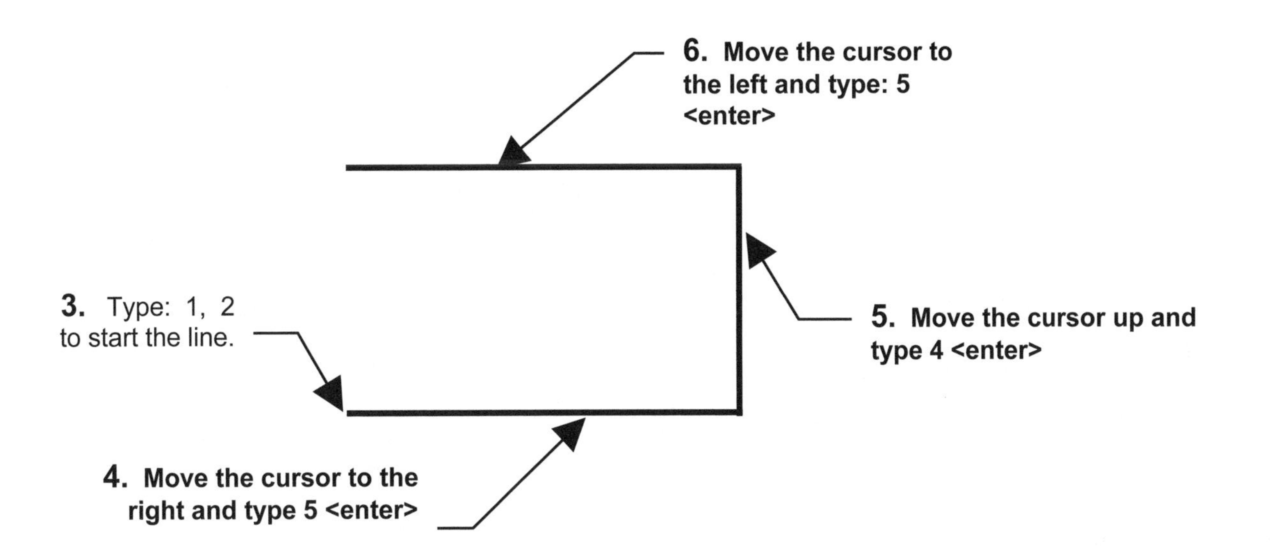

INQUIRY

The INQUIRY command allows you to Inquire about objects on the screen. There are 8 commands in the inquiry menu but we will only discuss 3 of those commands at this time. The remaining 5 commands will be discussed in the "Advanced" workbook.

LIST

The LIST command will list the type of object you have selected, coordinate location and properties that apply to the object.

1. Select the LIST command using one of the following:

TYPE = LIST
PULLDOWN = TOOLS / INQUIRY / List
TOOLBAR = STANDARD

2. Select the object: ***select the object***
3. Select the object: ***press <enter> to stop***

(The information will be listed in the "Text Screen". Press F2 to close the Text Screen.)

DISTANCE

The DISTANCE command will list the distance between two points that you select.

1. Select the DISTANCE command using one of the following:

TYPE = DI
PULLDOWN = TOOLS / INQUIRY / Distance
TOOLBAR = STANDARD

2. First point: ***select the first point***
3. Second point: ***select the second point***
   Distance = ***distance between the two points will be listed here***

ID POINT

The ID POINT or LOCATE POINT command will list the X and Y coordinates of the point that you selected. The coordinates will be from the ORIGIN.

1. Select the ID POINT command

TYPE = ID
PULLDOWN = TOOLS / INQUIRY / ID Point
TOOLBAR = TOOLS

2. Point: ***select a point anywhere on the drawing***
   X =***coordinate listed here*** Y = ***coordinate listed here*** Z = ***coordinate listed here***

Note: the **ID POINT** can also be used to create a ***LAST POINT***. This enables you to use Relative coordinates (the @ symbol) for the location of the next object.

LINEWEIGHTS

You learned in Lesson 3 (page 3-6) that it is “good drawing management” to draw related objects on the same layer. It is also “good drawing management” to establish a contrast in line weights between layers. For example, objects, such as a house or a paper clip, should be drawn with the “Object” layer and should have a greater line weight than the dimension layer or text layer.

The following are instructions for assigning Lineweights to Layers.

FIRST YOU NEED TO CHANGE THE LINEWEIGHT SETTINGS BOX.

1. Select **Format / Lineweight...**

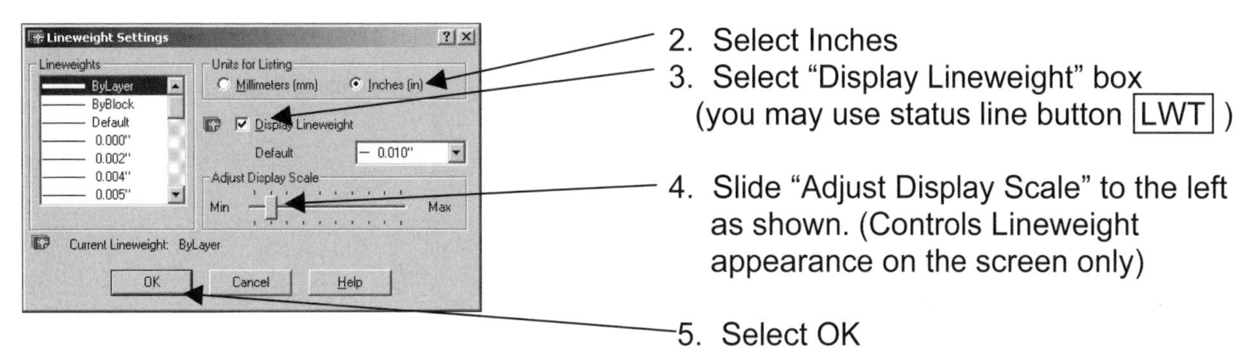

2. Select Inches
3. Select “Display Lineweight” box (you may use status line button LWT)
4. Slide “Adjust Display Scale” to the left as shown. (Controls Lineweight appearance on the screen only)
5. Select OK

(These settings will be saved to the computer not the drawing and will remain until you change them.)

ASSIGNING LINEWEIGHTS TO LAYERS

1. Select **Format / Layer**

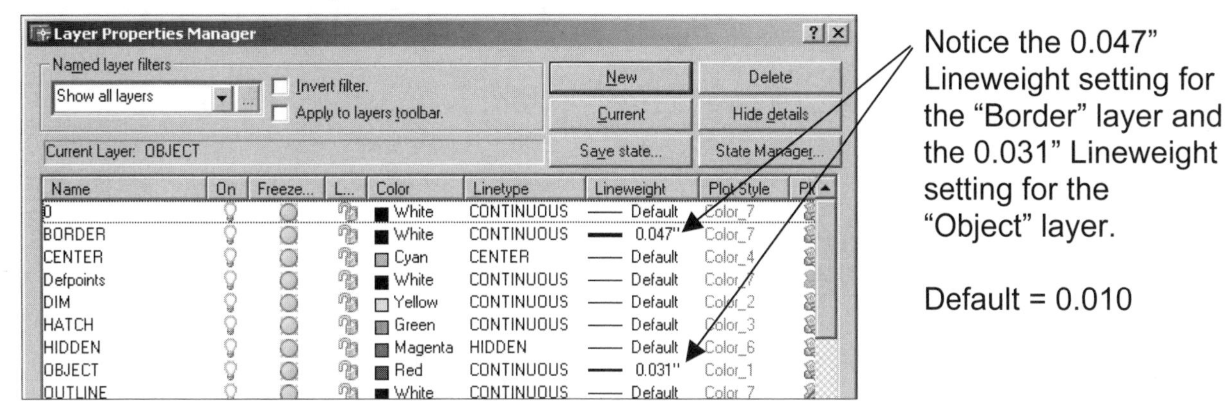

Notice the 0.047” Lineweight setting for the “Border” layer and the 0.031” Lineweight setting for the “Object” layer.

Default = 0.010

2. Select the “Border” layer. (Click on the name “Border”)

3. Select the Lineweight down arrow (▼) to select a Lineweight.

Note: If the “Details” area does not appear at the bottom of the “Layer Properties Manager” dialog box, click on the “Show details” button located at the upper right corner.

*(Lineweight changes will be saved **only to the current drawing** and will not affect any other drawing)*

PRINTING IN COLOR OR BLACK ONLY

Even though you draw with colors, you may print the drawing in color or black.
To accomplish this, you must create a "color-dependent plot style table". (**.ctb** file.)

HOW TO CREATE A "COLOR-DEPENDENT PLOT STYLE TABLE"

1. Select **File / Plot**

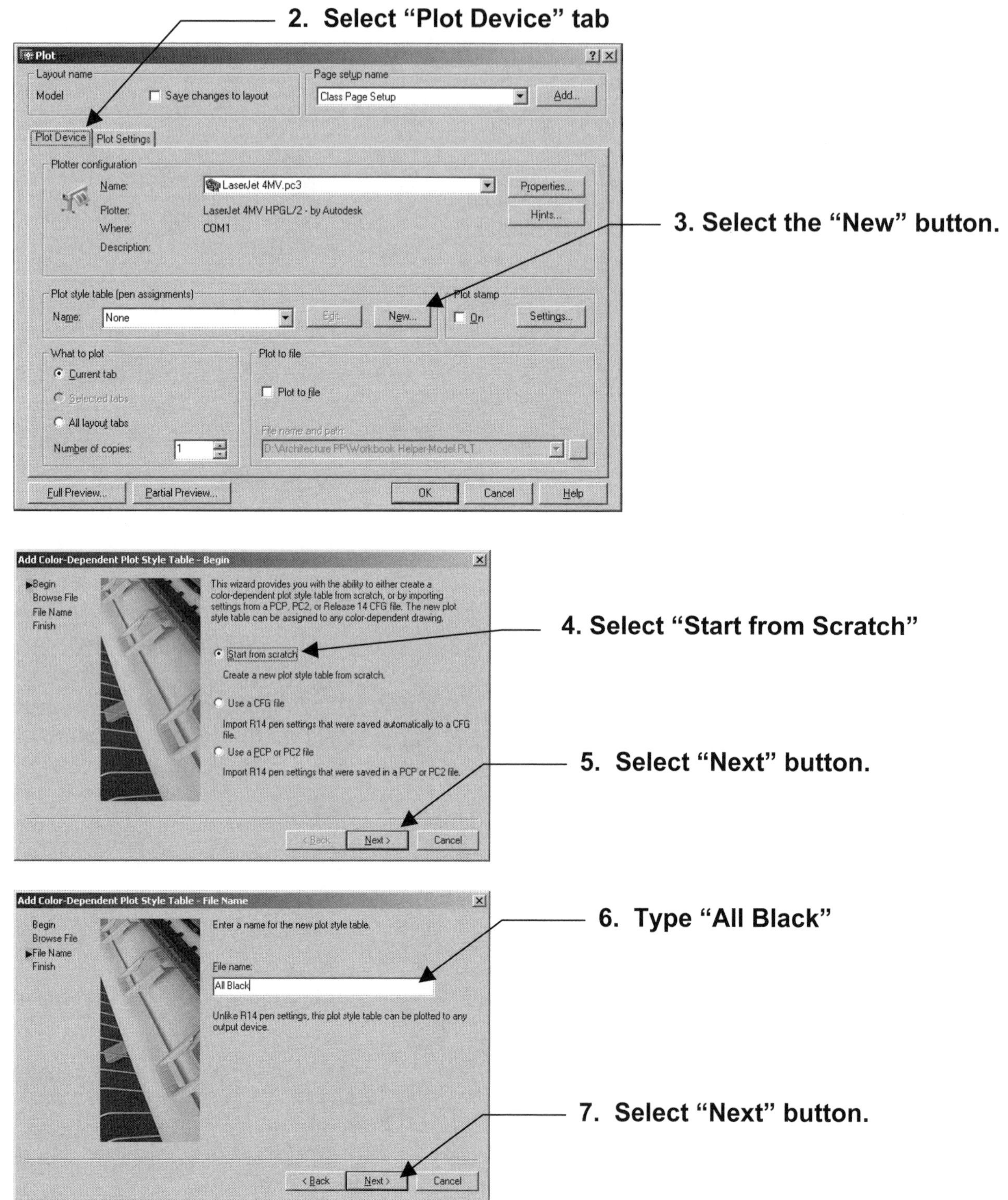

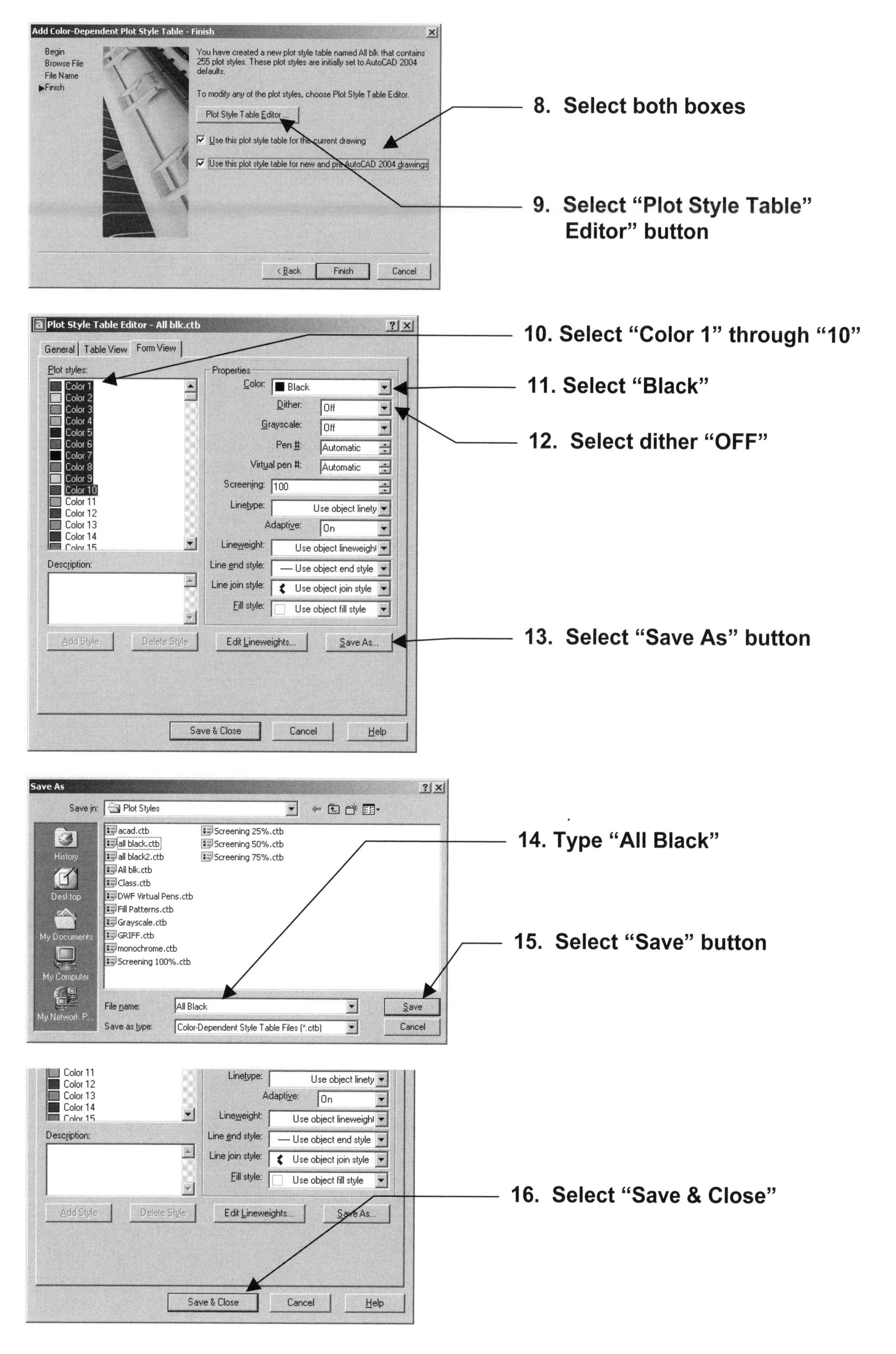
Add Color-Dependent Plot Style Table - Finish
Begin
Browse File
File Name
Finish
You have created a new plot style table named All blk that contains 255 plot styles. These plot styles are initially set to AutoCAD 2004 defaults.
To modify any of the plot styles, choose Plot Style Table Editor.
Plot Style Table Editor...
Use this plot style table for the current drawing
Use this plot style table for new and pre AutoCAD 2004 drawings
< Back
Finish
Cancel
8. Select both boxes
9. Select "Plot Style Table" Editor" button
Plot Style Table Editor - All blk.ctb
General
Table View
Form View
Plot styles:
Color 1
Color 2
Color 3
Color 4
Color 5
Color 6
Color 7
Color 8
Color 9
Color 10
Color 11
Color 12
Color 13
Color 14
Properties
Color:
Black
Dither:
Off
Grayscale:
Off
Pen #:
Automatic
Virtual pen #:
Automatic
Screening:
100
Linetype:
Use object linety
Adaptive:
On
Lineweight:
Use object lineweight
Line end style:
Use object end style
Line join style:
Use object join style
Fill style:
Use object fill style
Description:
Add Style
Delete Style
Edit Lineweights...
Save As...
Save & Close
Cancel
Help
10. Select "Color 1" through "10"
11. Select "Black"
12. Select dither "OFF"
13. Select "Save As" button
Save As
Save in:
Plot Styles
acad.ctb
all black.ctb
all black2.ctb
All blk.ctb
Class.ctb
DWF Virtual Pens.ctb
Fill Patterns.ctb
Grayscale.ctb
GRIFF.ctb
monochrome.ctb
Screening 100%.ctb
Screening 25%.ctb
Screening 50%.ctb
Screening 75%.ctb
History
Desktop
My Documents
My Computer
File name:
All Black
Save as type:
Color-Dependent Style Table Files (*.ctb)
Save
Cancel
14. Type "All Black"
15. Select "Save" button
16. Select "Save & Close"

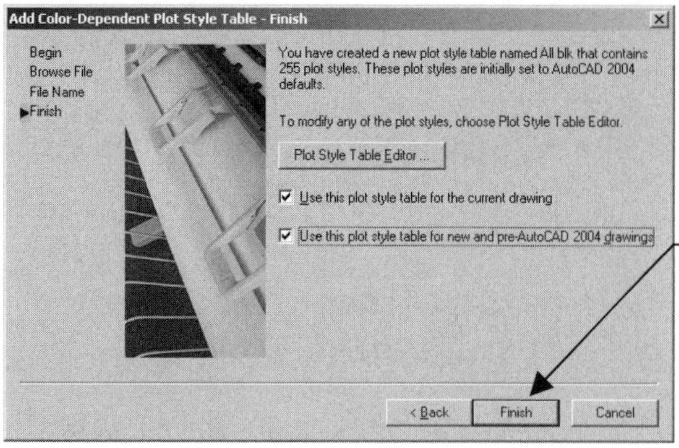

17. Select "Finish" button.

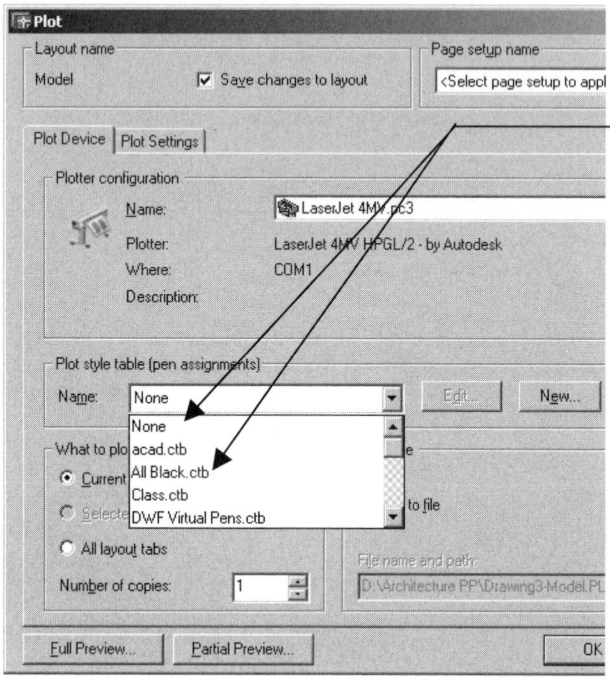

Now you have 2 .ctb files:

"None" to be used when you want to print with color.

"All Black" to be used when you want to print Black only.

I know you are not too clear how this will be used yet, but you will understand after you have completed Exercises 9A and 9B.

HOW TO CREATE A NEW "COLOR-DEPENDENT PLOT STYLE TABLE" TO BE USED WITHOUT LINEWEIGHTS

Each color, used within a drawing, can be assigned different plotting properties. Plotting properties are line weight, line type, line end style, joint style, fill pattern, gray scale, screen percentage etc. When you plot your drawing using a specific Plot Style Table, the plotting properties you have assigned will apply to the color of the objects in your drawing.

For example: You could create a Plot Style Table called "XYZ" with the line weight of color red set to .0831 inches. When you plot the drawing you will select the "XYZ" Plot Style Table. All the objects that are red, in the drawing, will have a line weight of .0831 inches on the paper when plotted.

A. Select **FILE / PLOT STYLE MANAGER**

B. Select "**Add-A-Plot Style Table" Wizard**

The following dialog boxes will appear.

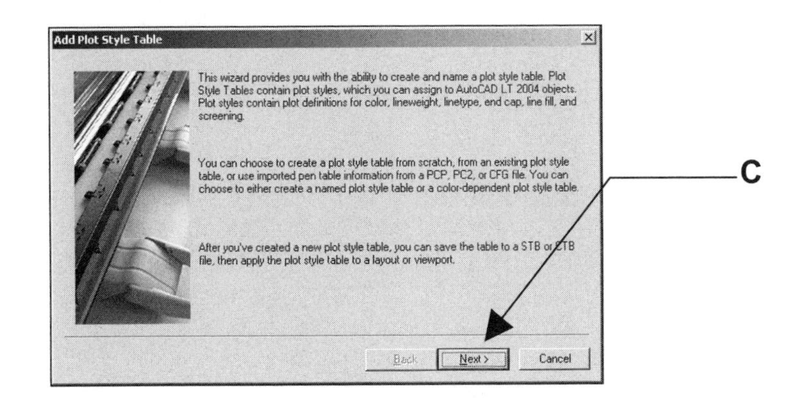

C. Select the **Next** button.

D. Select "**Start from Scratch"** then the **Next** button.

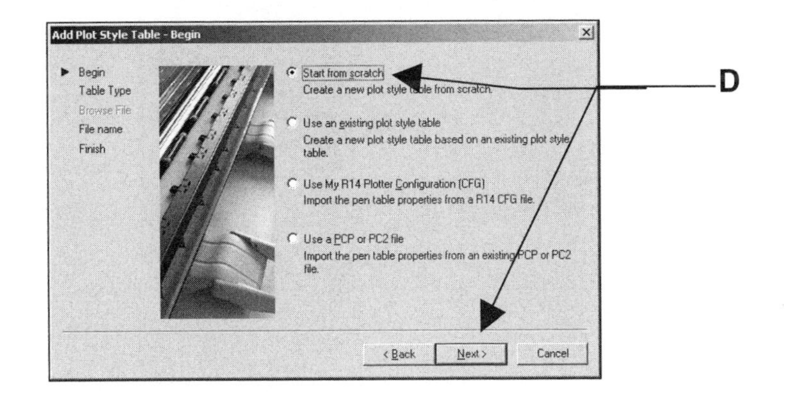

E. Select **"Color-Dependent Plot Style Table"** then the **Next** button.

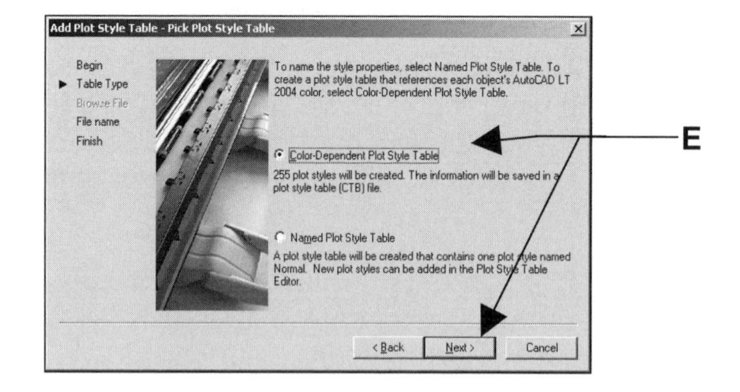

F. Type the new Plot Style Table **name** then select the **Next** button.

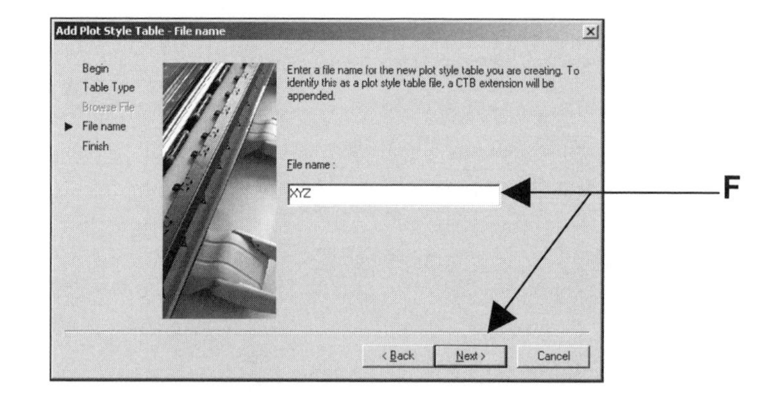

G. Select the **"Plot Style Table Editor"** button.

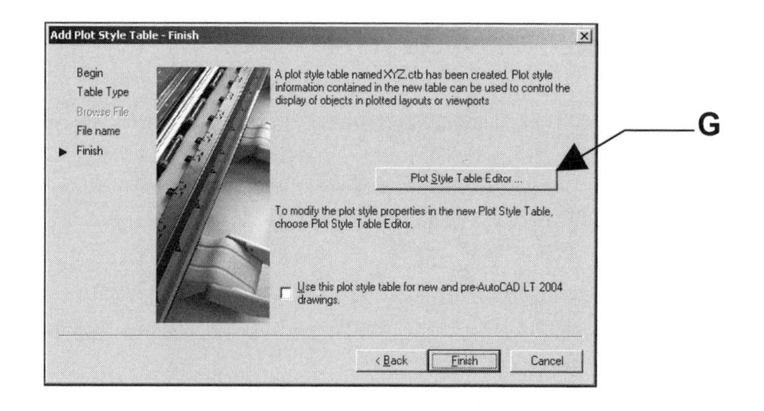

H. Make changes to the **"PROPERTIES"** then select the **Save & Close** button.

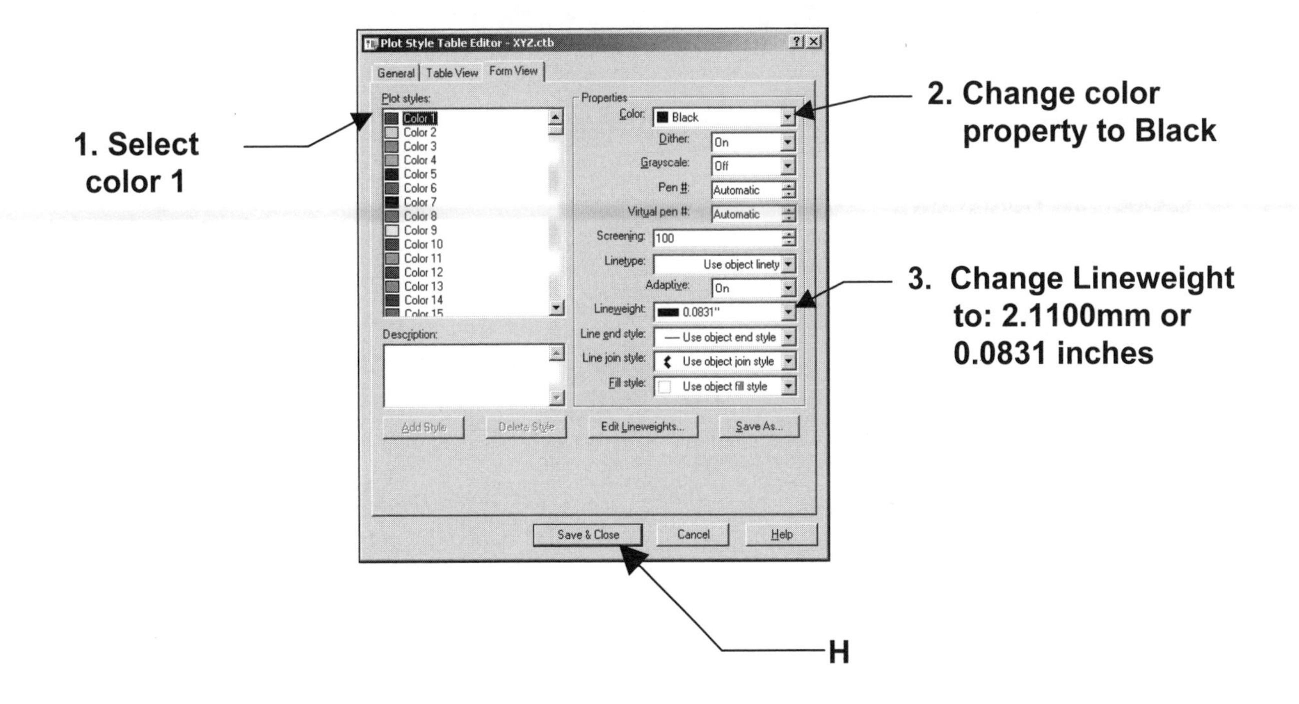

I. Select the **FINISH** button.

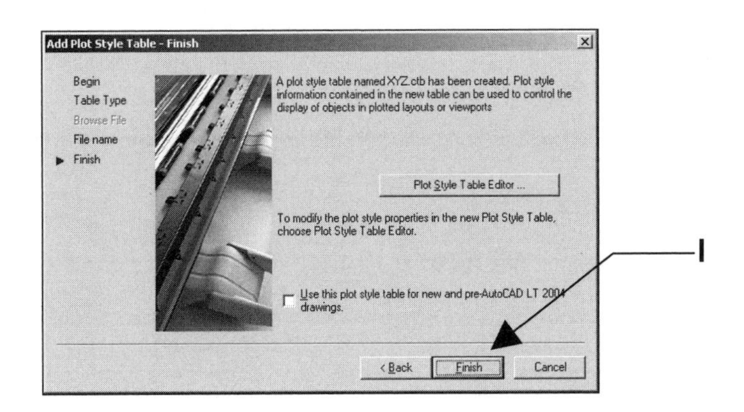

Now select File / Plot.
Check the list of Plot Style tables to see if your new style is listed.

EXERCISE 9A

INSTRUCTIONS:

1. Open **9A Helper.dwg**
(If you do not have this drawing refer page 2-2)

2. Set **UNITS** using:
Format / Units
Units = Decimal
Precision = 0.000

3. Set **DRAWING LIMITS** using:
Format / Drawing Limits
Lower Left Corner = 0,0
Upper Rt. Corner = 17, 11

4. Show the new limits using:
View / Zoom / All

5. Set **GRIDS** and **SNAP** using:
Tools / Drafting Settings
Snap = ON = .250
Grids = ON = .500

6. **Draw the border below** to scale using the dimensions shown.
Use layer **BORDER**.

7. Place the **TEXT** as shown.
Use "**Single Line Text**"
Use Layer = **Text-Hvy**
Text Ht. = .25

8. **Save** this border as **BSIZE**.

9. **PLOT** using:
File / Plot
Refer to "Basic Plotting from Model Space" on the next page.

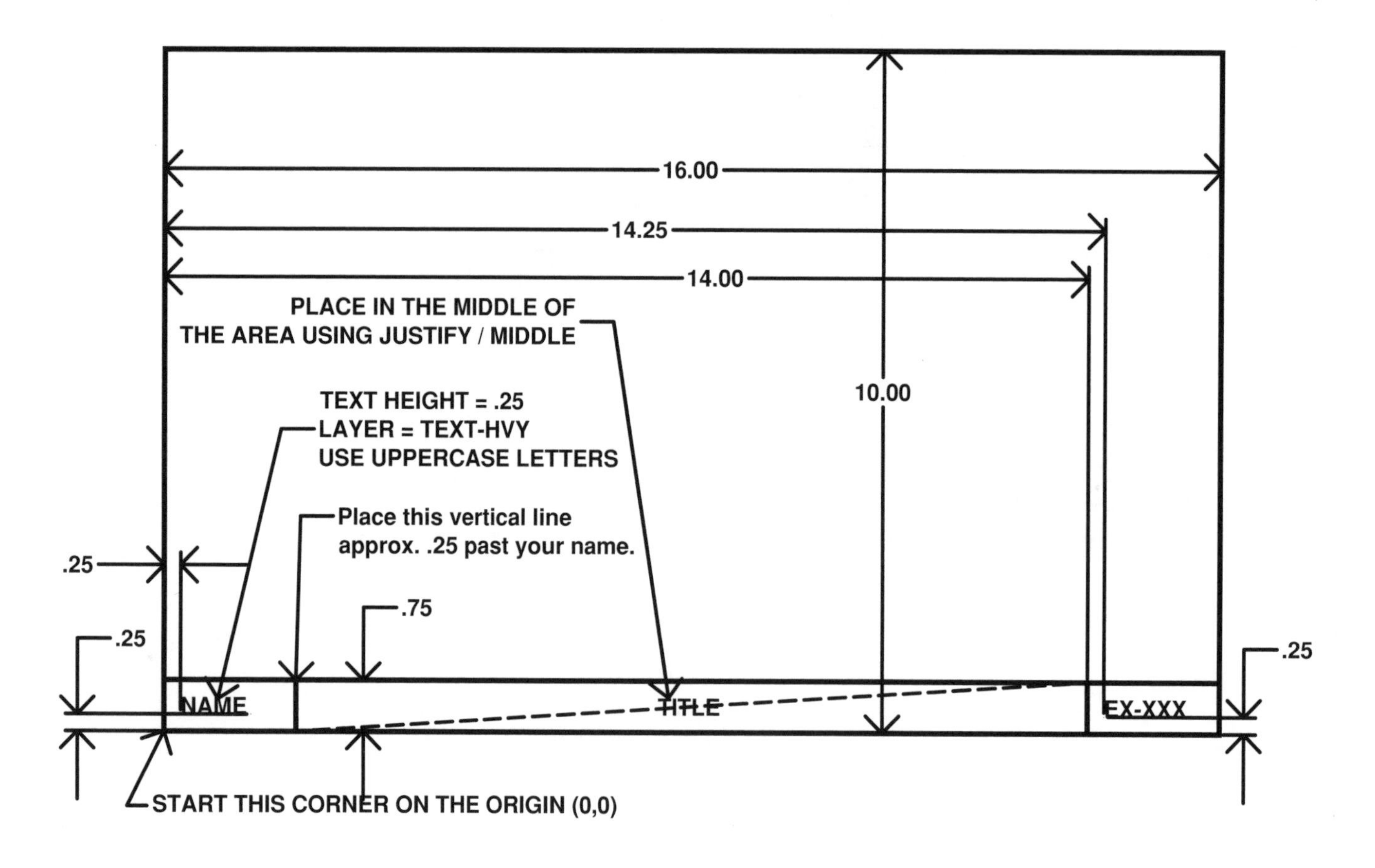

BASIC PLOTTING FROM MODEL SPACE

Note: More Advanced plotting methods will be explained in Lessons 26 and 27.

1. **Important:** Open the drawing you want to plot.
2. Make sure that the Model tab is selected. (Model tab)
3. Select the **Plot** command by "right clicking" on the "Model" tab or using one of the following methods listed below:

   Type = Print or Plot
   Pulldown = File / Plot
   Tool bar = Standard

The Plot dialog box below should appear.

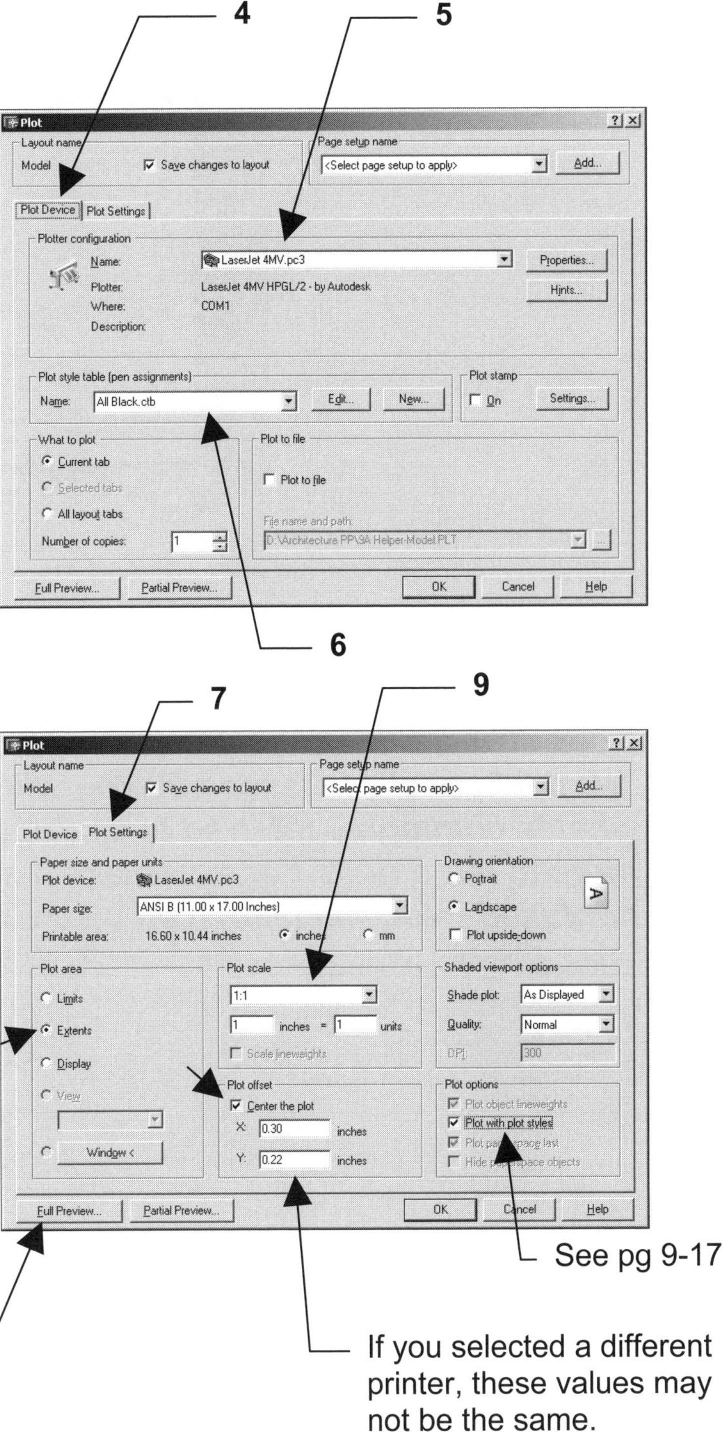

4. Select the **"Plot Device"** tab.

5. Select the Printer (Plot device)
   Note: This printer represents a size 17 x 11 for the exercises in this workbook. If it is not in the list, it needs to be configured.
   Refer to Appendix A

   Notice: You may configure a printer even though your computer is not attached to it.

6. Select the Plot Style Table
   Select "**All Black.ctb**" to print black.
   (If this .ctb file is not in the list, refer to page 9-8.)

   Select "**None**" to print in color.

7. Select the "**Plot Settings**" tab.

8. Check all the settings to make sure they match the workbook example.

9. If you would like to print your drawing on a 8-1/2 X 11 printer, select the printer and change the scale to "Scaled to Fit".

10. Select **Full Preview** button.

See pg 9-17

If you selected a different printer, these values may not be the same. That's OK.

11. If your drawing appears approximately like the display shown below, press <enter>.

12. Select the **OK** button to send the drawing to the printer or select **Cancel** if you do not want to print the drawing at this time or if you are not attached to this printer.

If your drawing did not appear approximately the same as shown below, recheck your settings.
Do you have the correct printer configured, paper size and plot area?
Remember, you do not have to be attached to the printer configured to preview the plot.
Refer to Appendix A, Add a Printer / Plotter.
Did you forget to open the drawing before attempting to print?

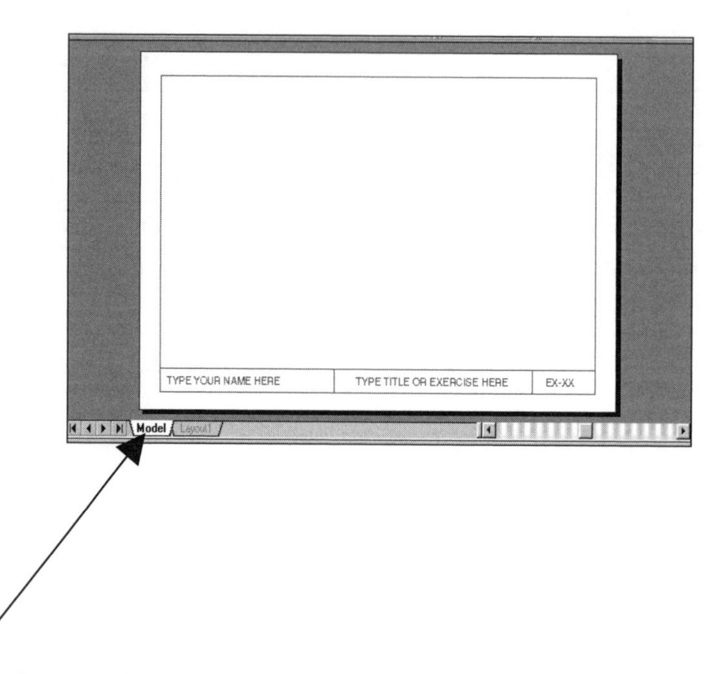

Note: The instructions above are to be used when plotting while in "Model Space". In lesson 26 you will learn how to plot in "Layout" or what is commonly referred to as Paperspace. But you have much more to learn first. Let's just take it step by step and not get confused. You will learn it all by the end of this workbook.

PLOT OPTIONS

The following describes the two plot options Plot Object lineweights and Plot with Plot Styles. The remaining two options will be discussed in lesson 26.

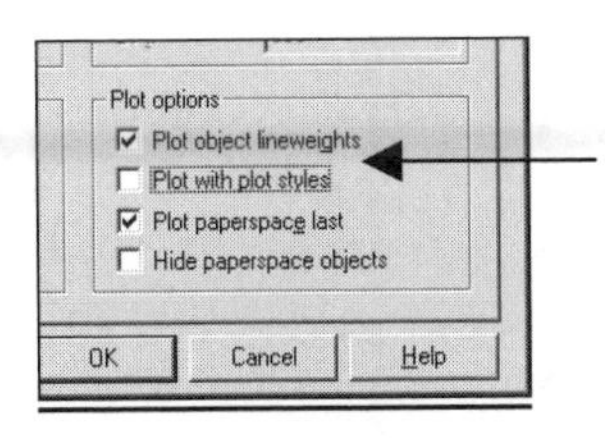

☐ PLOT OBJECT LINEWEIGHTS

Lineweight
The printer/plotter will use the Lineweights set within the drawing.

Plot Style Table
The Plot Style Table selected will be completely disregarded. (Overridden)

Color or Black
If you are using a color printer/plotter, it will print the colors you see on the screen.

☐ PLOT WITH PLOT STYLES (This is the default setting and gives you the most control)

Plot Style Table
The Printer/plotter will use the settings within the plot style table selected.

Color
If you are using a color printer/plotter and the "Color" properties box, within the Plot Style table, is set to "Use object Color".

Black
If the "Color" properties box, within the Plot Style table, is set to "Black".
Or
You are using a "black ink only" printer/plotter.

Lineweight
The Printer/Plotter will use the setting in the "Lineweight" properties box within the Plot Style table. (Refer to page 9-7)

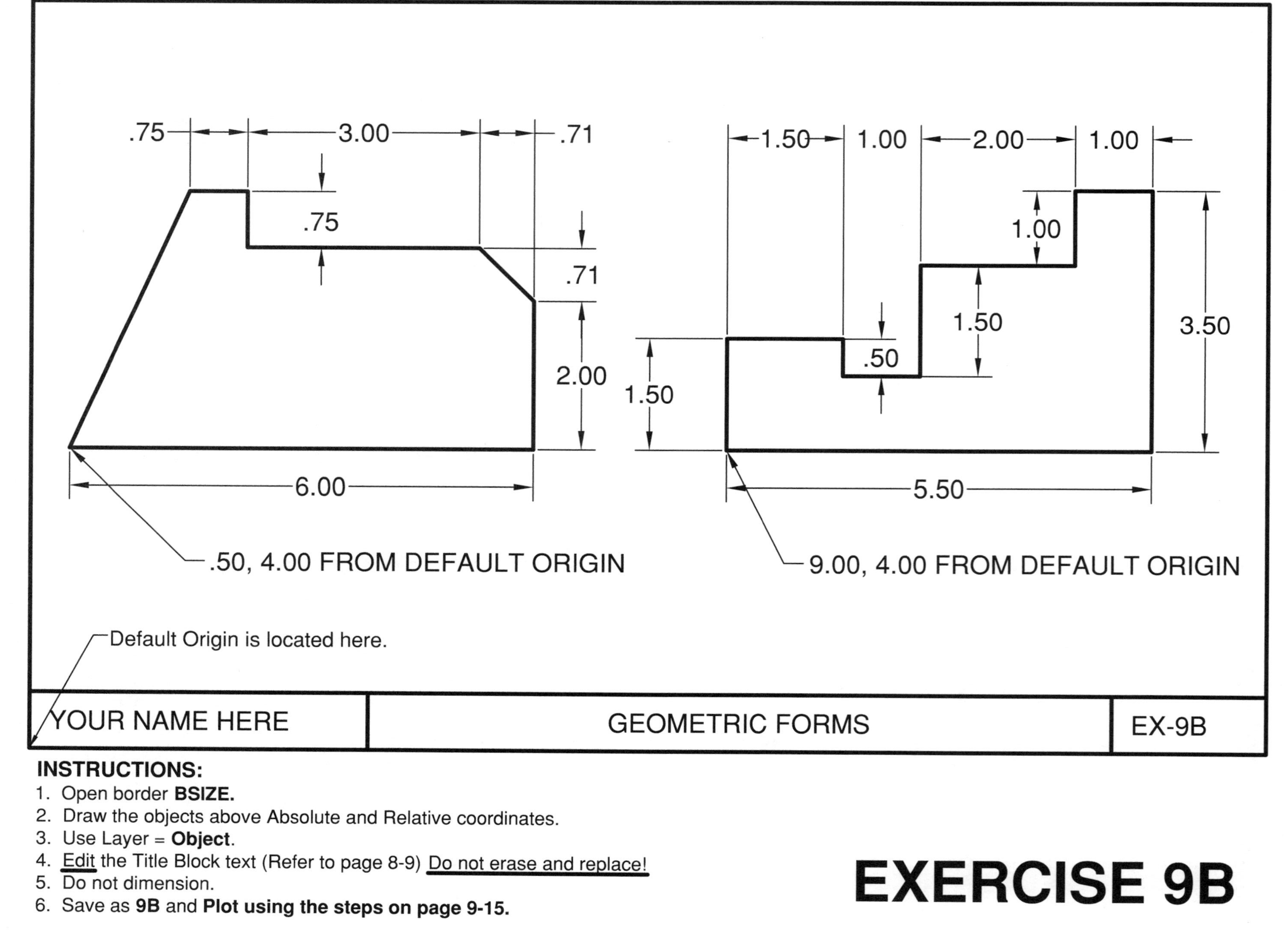

INSTRUCTIONS:

1. Open border **BSIZE.**
2. Draw the objects above Absolute and Relative coordinates.
3. Use Layer = **Object**.
4. Edit the Title Block text (Refer to page 8-9) Do not erase and replace!
5. Do not dimension.
6. Save as **9B** and **Plot using the steps on page 9-15.**

EXERCISE 9B

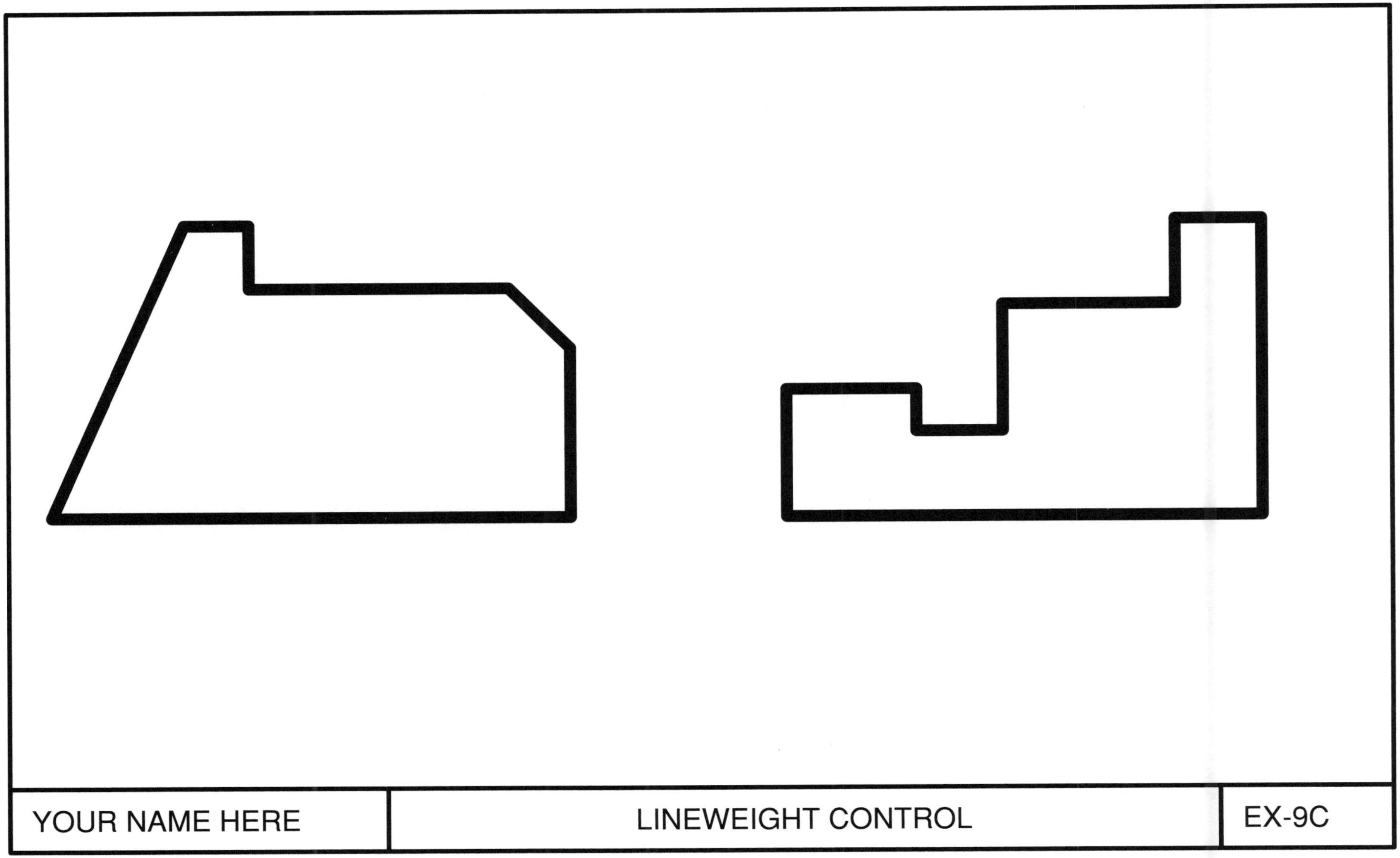

YOUR NAME HERE	LINEWEIGHT CONTROL	EX-9C

INSTRUCTIONS:

1. Open **9B.dwg.**
2. Plot this drawing again using the steps on page 9-15 **EXCEPT**:
3. Select Plot Style table XYZ.ctb instead of "All Black" (If this .ctb file is not in the list refer to page 9-11)

Do not save -- plotting exercise only. Are the object lines heavier than the previous plot?

EXERCISE 9C

NOTES:

LEARNING OBJECTIVES

After completing this lesson, you will be able to:

1. Move the Origin.
2. Turn the UCS Icon On and Off.
3. Command the UCS Icon to move with the Origin.

LESSON 10

MOVING THE ORIGIN

As previously stated, in Lesson 9, the **ORIGIN** is where the X, Y, and Z axes intersect. The Origin's (0,0,0) default location is in the lower left-hand corner of the drawing. But you can move the Origin anywhere on the screen using the UCS command.
(The default location is designated as the "**World**" option.)

You may move the Origin many times while creating a drawing. This is not difficult and it will make it much easier to draw objects in specific locations. You will understand this better after completing 10A, 10B and 10C.

To MOVE the Origin

1. Select one of following:

    TYPING = UCS <enter> or M <enter>
    PULLDOWNS = TOOLS / Move UCS
    UCS II TOOLBAR =

Command: _ucs
Current ucs name: *World*
Enter an option [New/Move/orthoGraphic/Prev/Restore/Save/Del/Apply/?/World]
<World>: ***m <enter>***

2. Specify new origin point or [Zdepth]<0,0,0>: ***type coordinate or use the cursor to place.***

To RETURN the Origin to the default "World" location (the lower left corner):

1. Select one of the following:

    TYPING = UCS <enter> W <enter>
    PULLDOWNS = TOOLS / New UCS / World
    UCS TOOLBAR =

DISPLAYING THE UCS ICON

The UCS icon is merely a drawing aid. You control how it is displayed.
It can be visible (on) or invisible (off). It can move with the Origin or stay in the default location. You can even change it's appearance.

Select the following pull-down menu:

View / Display / UCS Icon

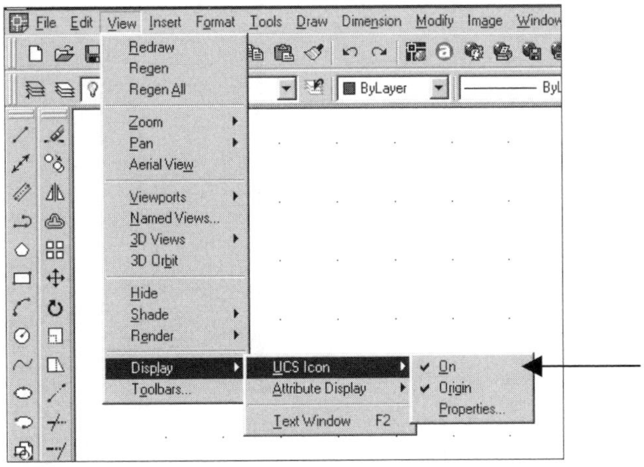

ON: A check mark beside the word ON means the Origin icon will be Visible. Remove the check mark and the Origin icon will disappear.
It will be very helpful to have the Origin Icon "On" most of the time. You will need it as you learn more about AutoCAD. (Set to ON for the workbook exercises)

ORIGIN: A check mark beside the word Origin will move the UCS Icon with the Origin each time you move the Origin. Remove the check mark and the icon will not move. *I find it very helpful to know where the Origin is at all times by merely looking for the UCS icon.* (Set to ON for the workbook exercises)

PROPERTIES: This setting allows you to change the appearance of the UCS icon. When you select this option the dialog box, shown below, will appear.
You may change the Style, Size and Color at any time.

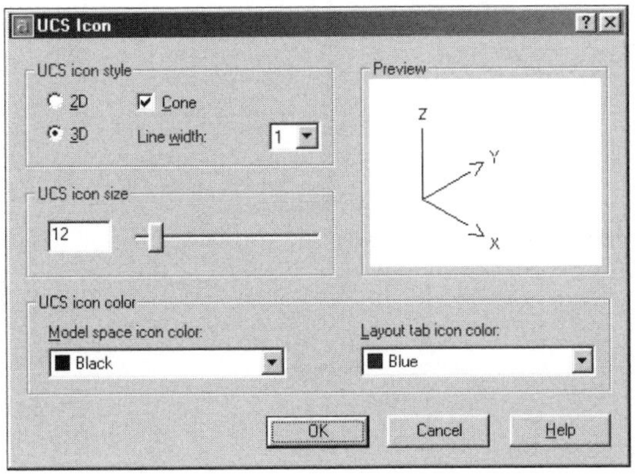

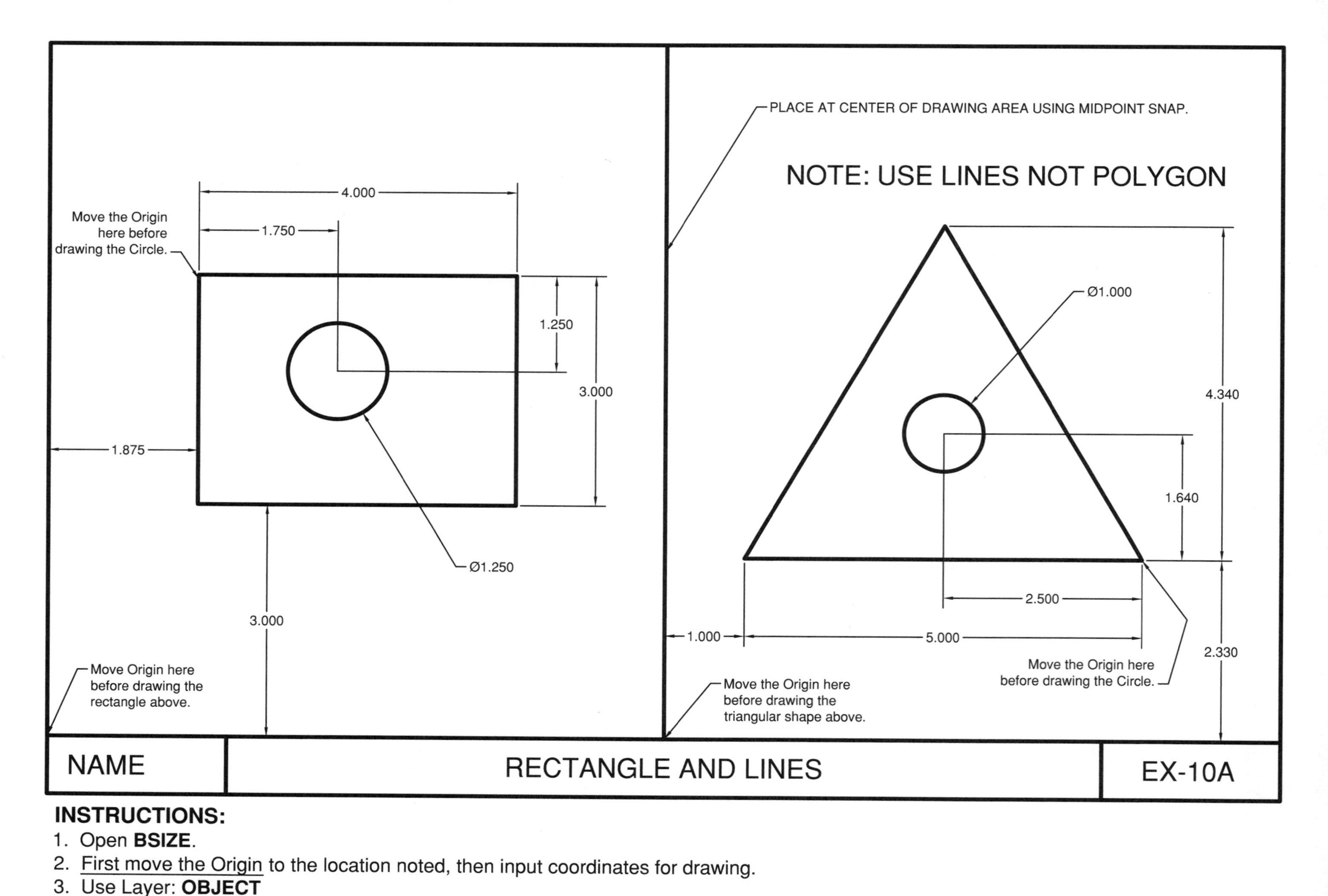

INSTRUCTIONS:

1. Open **BSIZE**.
2. First move the Origin to the location noted, then input coordinates for drawing.
3. Use Layer: **OBJECT**
4. Do not dimension.
5. Edit the Title Block text
6. Save as **10A** and **Plot**.

EXERCISE 10A

EXERCISE 10B

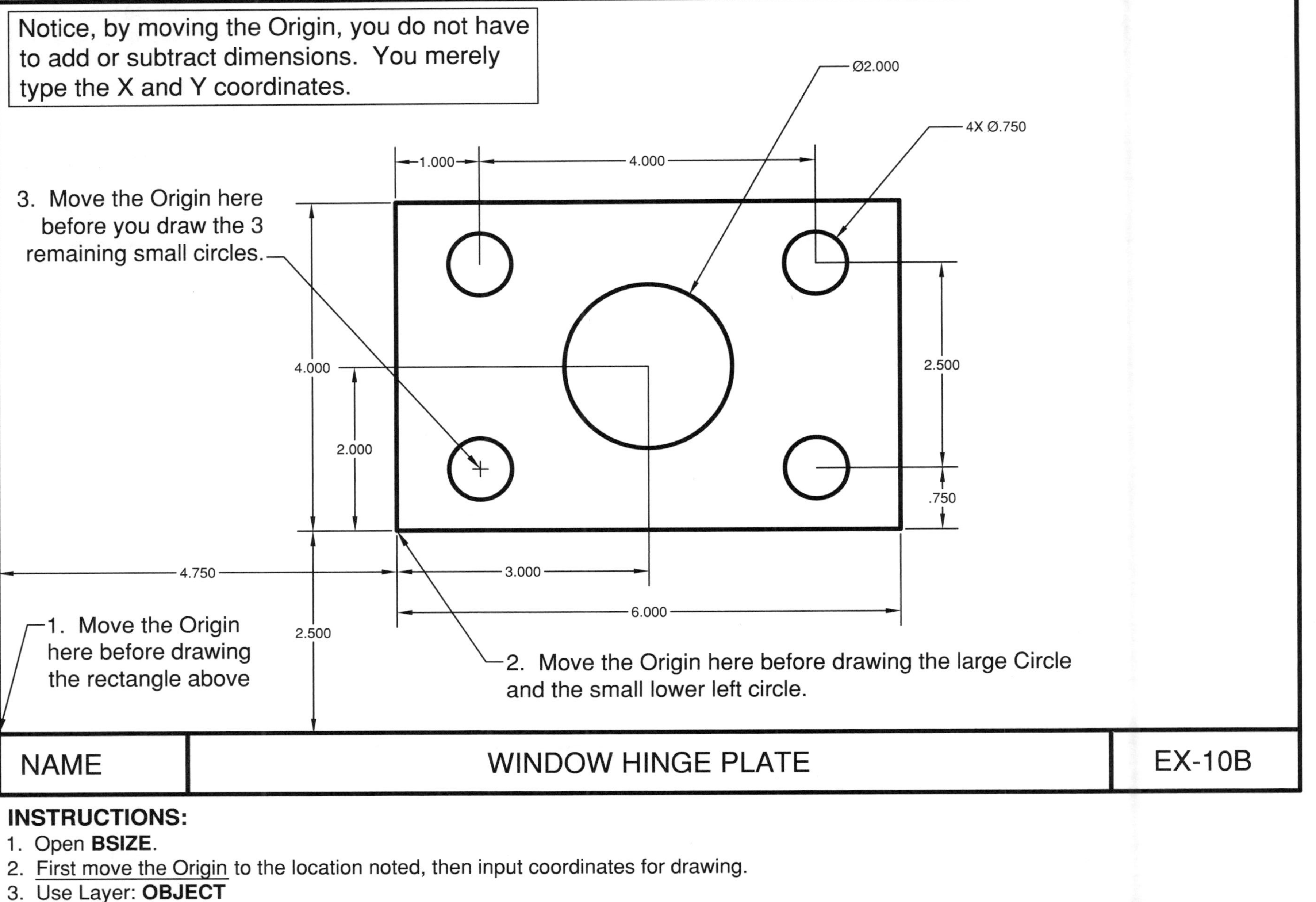

INSTRUCTIONS:

1. Open **BSIZE**.
2. First move the Origin to the location noted, then input coordinates for drawing.
3. Use Layer: **OBJECT**
4. Do not dimension.
5. Edit the Title Block text.
6. Save as **10B** and **Plot**.

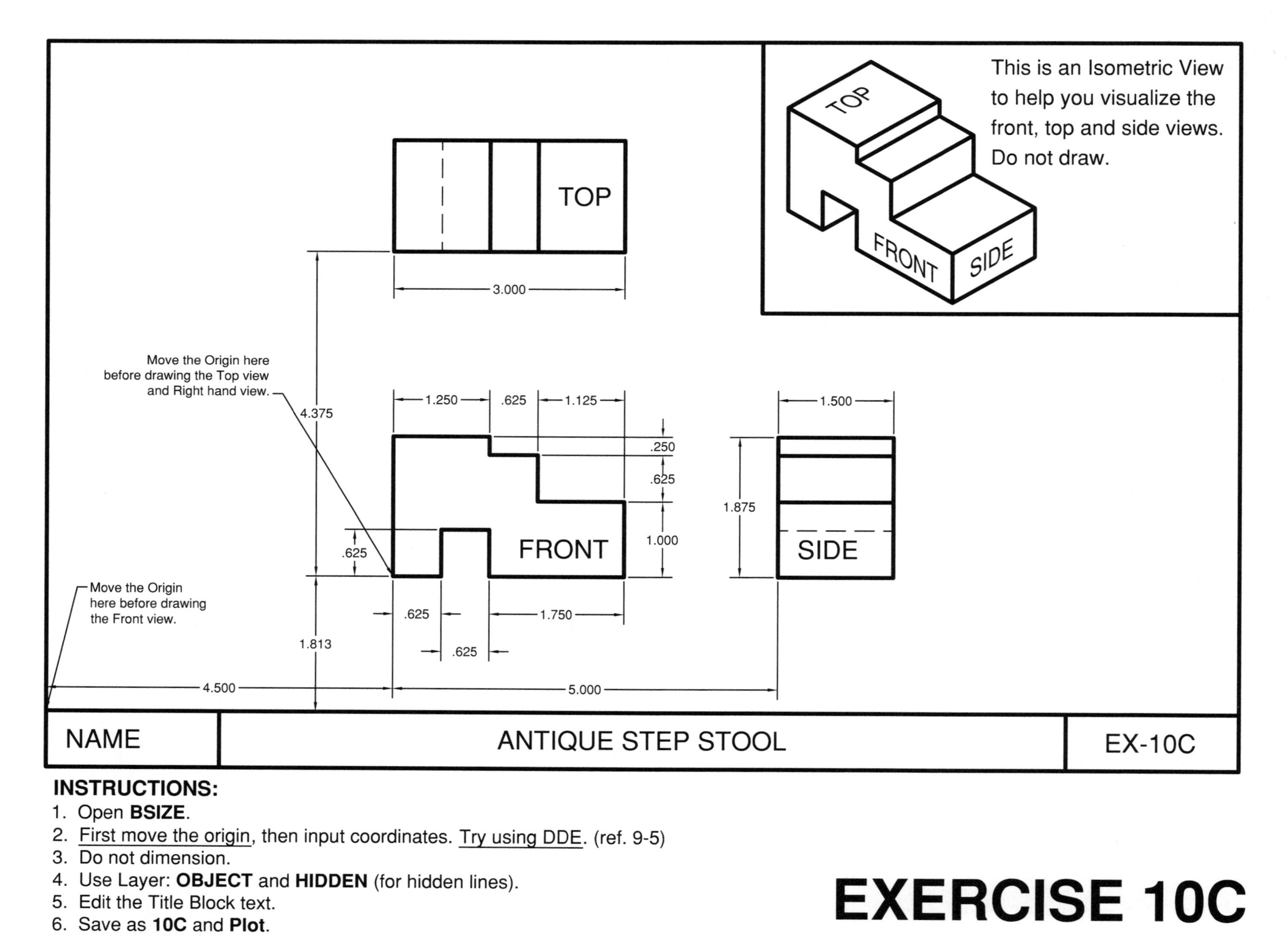

INSTRUCTIONS:

1. Open **BSIZE**.
2. First move the origin, then input coordinates. Try using DDE. (ref. 9-5)
3. Do not dimension.
4. Use Layer: **OBJECT** and **HIDDEN** (for hidden lines).
5. Edit the Title Block text.
6. Save as **10C** and **Plot**.

EXERCISE 10C

LEARNING OBJECTIVES

After completing this lesson, you will be able to:

1. Understand the Polar Degree Clock.
2. Draw Lines to a specific length and angle
3. Draw Objects using Polar Coordinate Input.
4. Use Polar Tracking.
5. Construct an Isometric view.

LESSON 11

POLAR COORDINATE INPUT

In Lesson 9 we learned to control the length and direction of horizontal and vertical lines using Relative Input and Direct Distance Entry. Now we will learn how to control the length and **ANGLE** of a line using **POLAR COORDINATE INPUT**.

UNDERSTANDING THE *"POLAR DEGREE CLOCK"*

Previously when drawing Horizontal and Vertical lines you controlled the direction using a Positive or Negative input. ***Polar Input is different***. The Angle of the line will determine the direction. For example: If I want to draw a line at a 45 degree angle towards the upper right corner, I would use the angle 45. But if I want to draw a line at a 45 degree angle towards the lower left corner, I would use the angle 225. You may also use Polar Input for Horizontal and Vertical lines using the angles 0, 90, 180 and 270. No negative input is required.

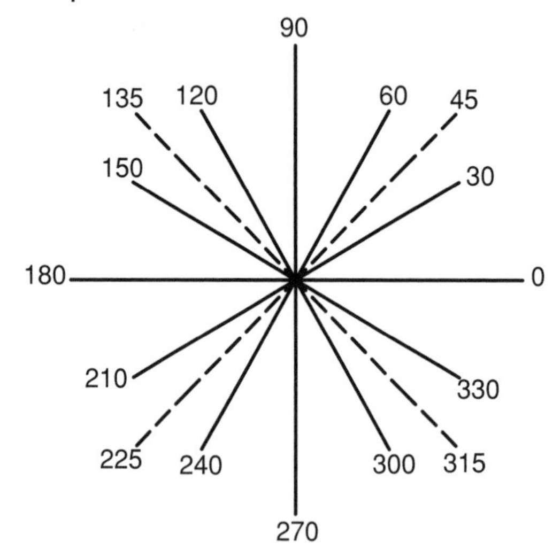

DRAWING WITH *POLAR COORDINATE INPUT*

A Polar coordinate may come ***from the last point entered*** or ***from the Origin,*** depending upon whether you use the @ symbol or not. The first number represents the **Distance** and the second number represents the **Angle**. The two numbers are separated by the **less than (<)** symbol.
The input format is: **distance < angle**

A Polar coordinate of **@6<45** will be 6 units long and at an angle of 45 degrees ***from the last point entered.***

A Polar coordinate of **6<45** will be 6 units and 45 degrees from the ***Origin.***

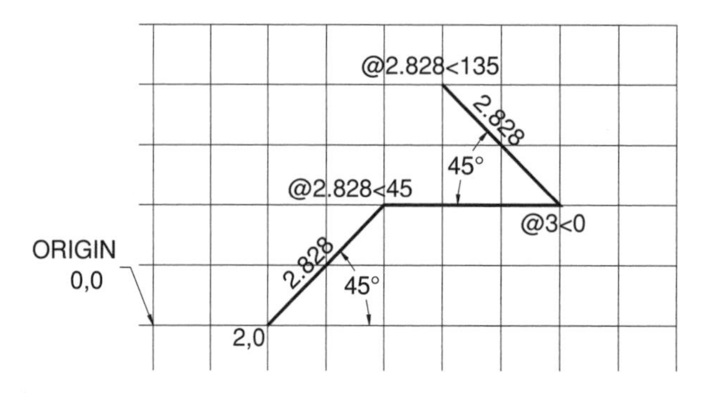

POLAR TRACKING

Polar Tracking can be used instead of **ORTHO**. When *Polar Tracking* is "**ON**", a dotted ***"tracking"*** line and a ***"tool tip"*** box appear and "snaps" to a **preset angle increment** when the cursor approaches one of the preset angles. The word ***"Polar"***, followed by the ***"distance"*** and ***"angle"*** from the last point, appears in the box.
(A step by step example is described on the next page.)

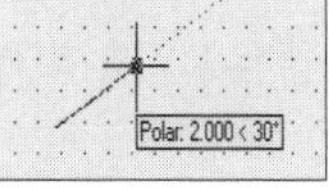

SETTING THE ANGLE INCREMENT

1. Right Click on the POLAR button on the Status Bar and select "**SETTINGS**" or select **Tools / Drafting Settings / Polar Tracking** tab.
   The following dialog box will appear:

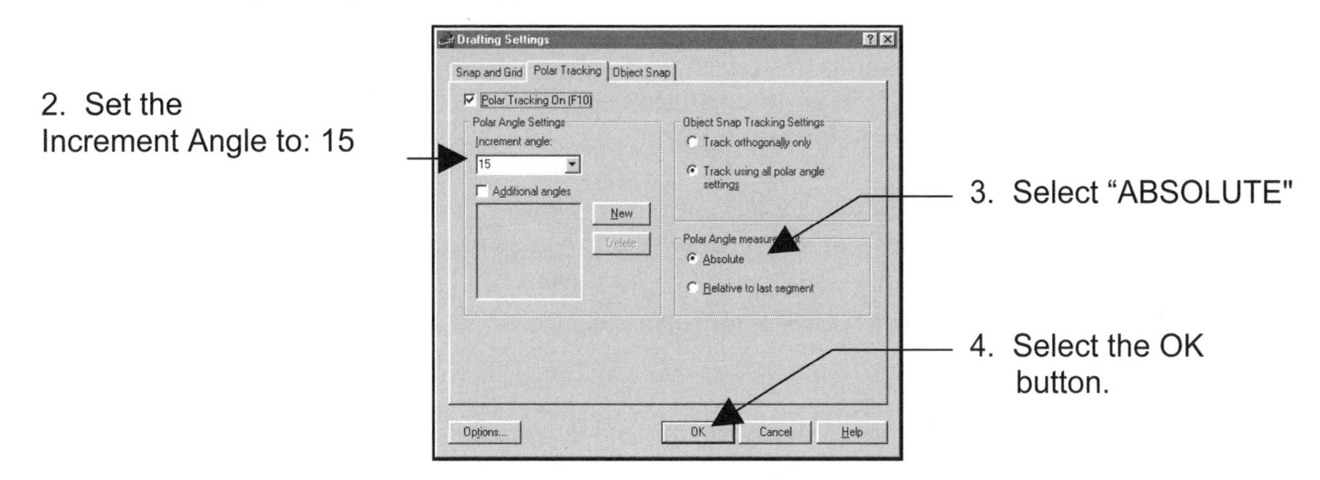

POLAR ANGLE SETTINGS

<u>Increment Angle</u> Choose from a list of Angle increments including 90, 45, 30, 22.5, 18, 15,10 and 5. You will be able to "snap" to multiples of that angle.

<u>Additional Angles</u> Check this box if you would like to use an angle other than one in the Incremental Angle list.

<u>New</u> You may add an angle by selecting the "New" button. You will be able to snap to this new angle in addition to the incremental Angle selected. But you will not be able to snap to it's multiple.

<u>Delete</u> Deletes an Additional Angle. Select the Additional angle to be deleted and then the Delete button.

POLAR ANGLE MEASUREMENT

<u>ABSOLUTE</u> Polar tracking <u>angles</u> are relative to the UCS.

<u>RELATIVE TO LAST SEGMENT</u> Polar tracking <u>angles</u> are relative to the last segment.

USING POLAR TRACKING AND DIRECT DISTANCE ENTRY

1. Use the Polar Tracking Settings from the previous page.
2. Set the Status Bar as follows:

SNAP	GRID	ORTHO	POLAR	OSNAP	OTRACK	LWT	MODEL
OFF	ON	OFF	ON	OFF	OFF	OFF	ON

3. **Select the Line command**:

P1

1. Start the Line anywhere on the in the drawing area.

P2

1. Move the cursor in the direction of P2 until the Tool Tip box displays 30 degrees. (Length is not important yet)
2. Type 2 <enter> (for the length)

P3

1. Move the cursor in the direction of P3 until the Tool Tip box displays 90 degrees. (Length is not important yet)
2. Type 2 <enter> (for the length)

P4

1. Move the cursor in the direction of P4 until the Tool Tip box displays 0 degrees. (Length is not important yet)
2. Type 2 <enter> (for the length)

P5

1. Move the cursor in the direction of P5 until the Tool Tip box displays 150 degrees. (Length is not important yet)
2. Type 2 <enter> (for the length)

P6

1. Move the cursor in the direction of P6 until the Tool Tip box displays 180 degrees. (Length is not important yet)
2. Type 2 <enter> (for the length)
3. Then type C for close.

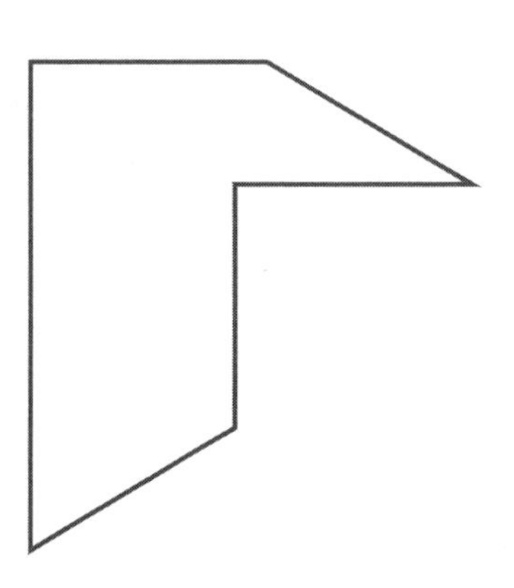

POLAR TRACKING ON or OFF

You may toggle Polar Tracking On or Off using one of the following:

- Left click on the POLAR button on the Status Bar
- Press F10

POLAR SNAP

Polar Snap is used with Polar Tracking to make the cursor snap to specific ***distances*** and ***angles***. If you set Polar Snap distance to 1 and Polar Tracking to angle 30 you can draw lines 1, 2, 3, 4 units… long at an angle of 30, 60, 90 etc. without typing anything on the command line. You just move the cursor and watch the tool tips.
(A step by step example is described on the next page)

SETTING THE ANGLE INCREMENT

1. Right Click on the POLAR button on the Status Bar and select "SETTINGS" or select **Tools / Drafting Settings / Polar Tracking** tab.
   The following dialog box will appear:

2. Select

3. Set the
Increment Angle to: 15

4. Select "ABSOLUTE"

5. Select the OK
button.

SETTING THE POLAR SNAP

1. Right Click on the SNAP button on the Status Bar and select "SETTINGS" or select **Tools / Drafting Settings / Snap and Grid** tab.
   The following dialog box will appear:

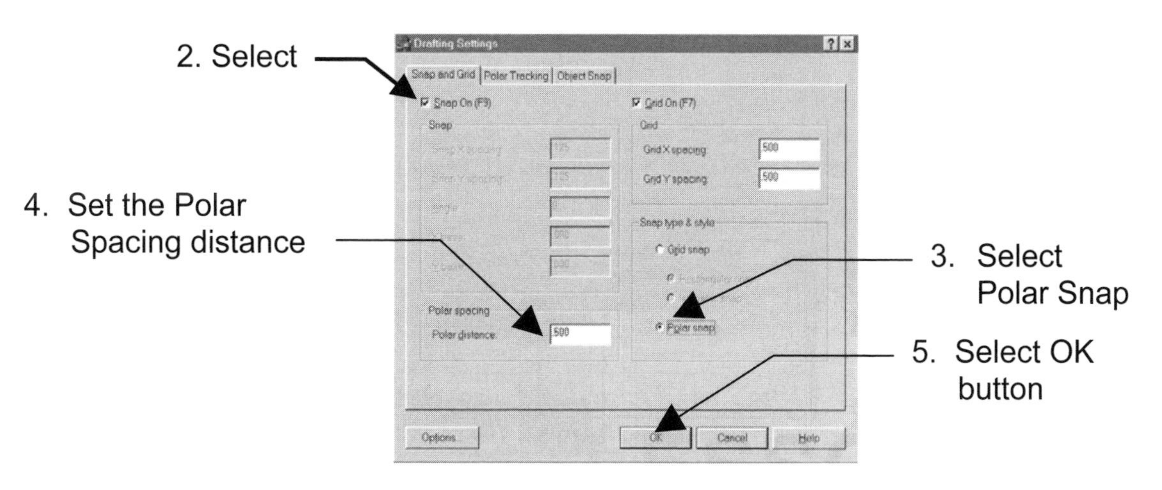

SNAP AND GRID Sets standard snap and grid information
POLAR SPACING Increment Snap distance when Polar Snap is ON.
STYLE This setting will be taught in the Advanced course.
TYPE Sets the Snap to Polar or Grid

USING POLAR TRACKING AND POLAR SNAP

Now let's draw the objects below again, but this time with "Polar Snap" instead of Direct Distance Entry.

1. Use the Polar Tracking and Polar Snap Settings from the previous page.

2. Set the Status Bar as follows:

SNAP	GRID	ORTHO	POLAR	OSNAP	OTRACK	LWT	MODEL
ON	**ON**	**OFF**	**ON**	**OFF**	**OFF**	**OFF**	**ON**

3. Select the Line command:

   P1
   1. Start the Line anywhere in the drawing area.

   P2
   1. Move the cursor in the direction of P2 until the Tool Tip box displays **Polar 2.00 <30°**

   P3
   1. Move the cursor in the direction of P3 until the Tool Tip box displays **Polar 2.00 <90°**

   P4
   1. Move the cursor in the direction of P4 until the Tool Tip box displays **Polar 2.00 <0°**

   P5
   1. Move the cursor in the direction of P5 until the Tool Tip box displays **Polar 2.00 <150°**

   P6
   1. Move the cursor in the direction of P6 until the Tool Tip box displays **Polar 2.00 <180°**
   2. Then type C for close.

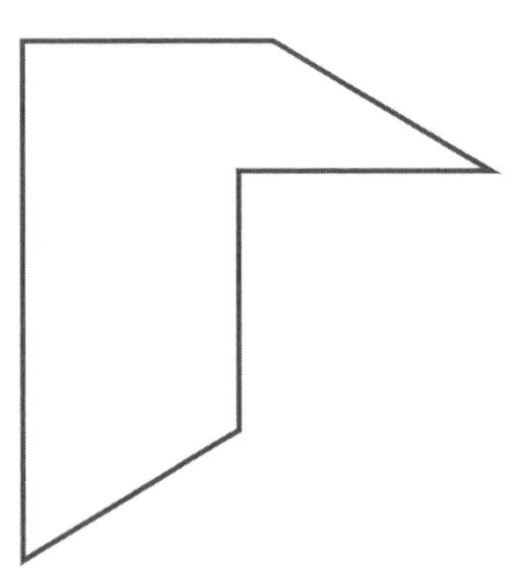

NOTE: You may OVERRIDE the Polar Settings at any time by typing: Polar coordinates (@ Length< Angle) on the Command line.

EXERCISE 11A

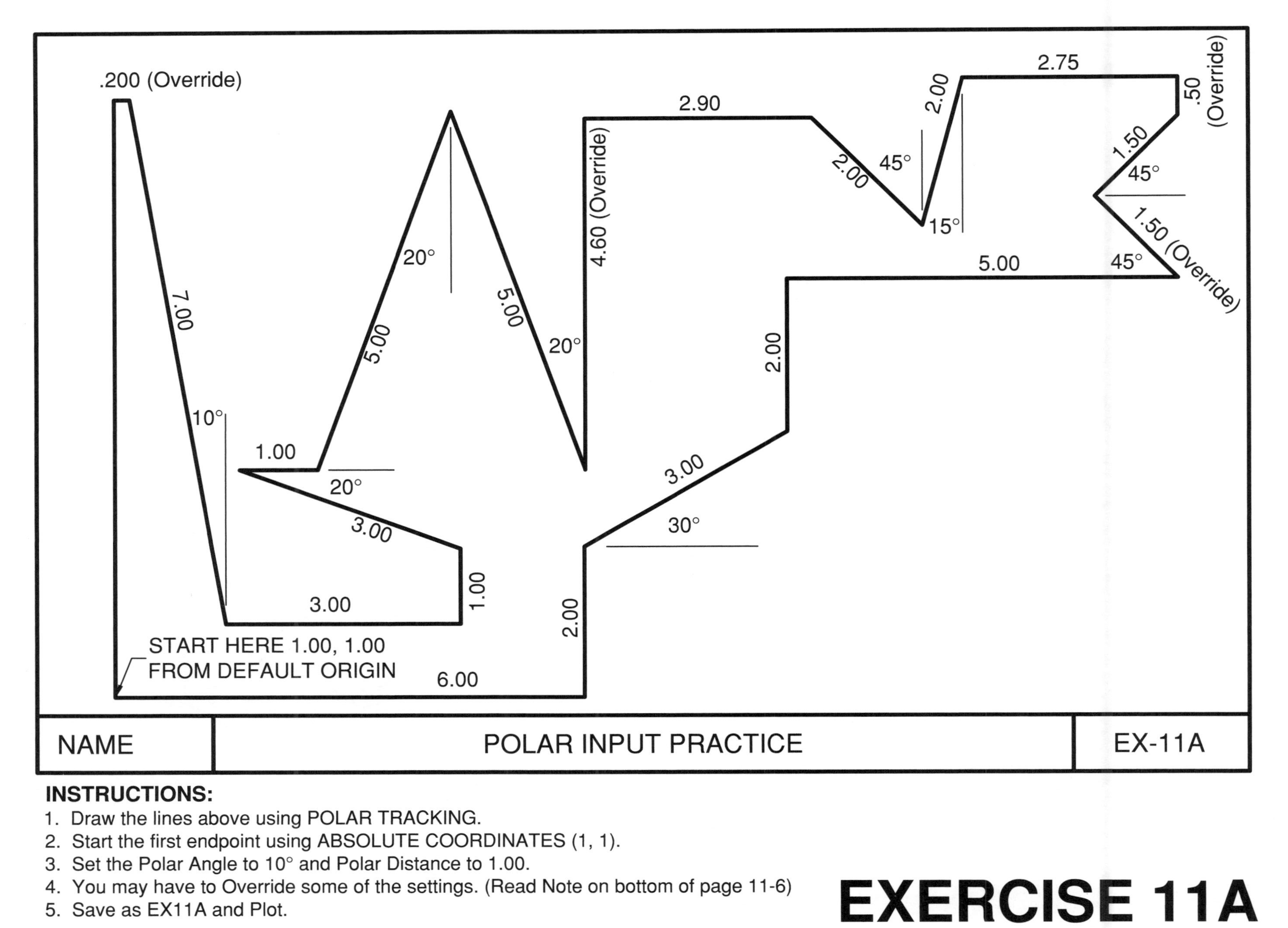

INSTRUCTIONS:

1. Draw the lines above using POLAR TRACKING.
2. Start the first endpoint using ABSOLUTE COORDINATES (1, 1).
3. Set the Polar Angle to 10° and Polar Distance to 1.00.
4. You may have to Override some of the settings. (Read Note on bottom of page 11-6)
5. Save as EX11A and Plot.

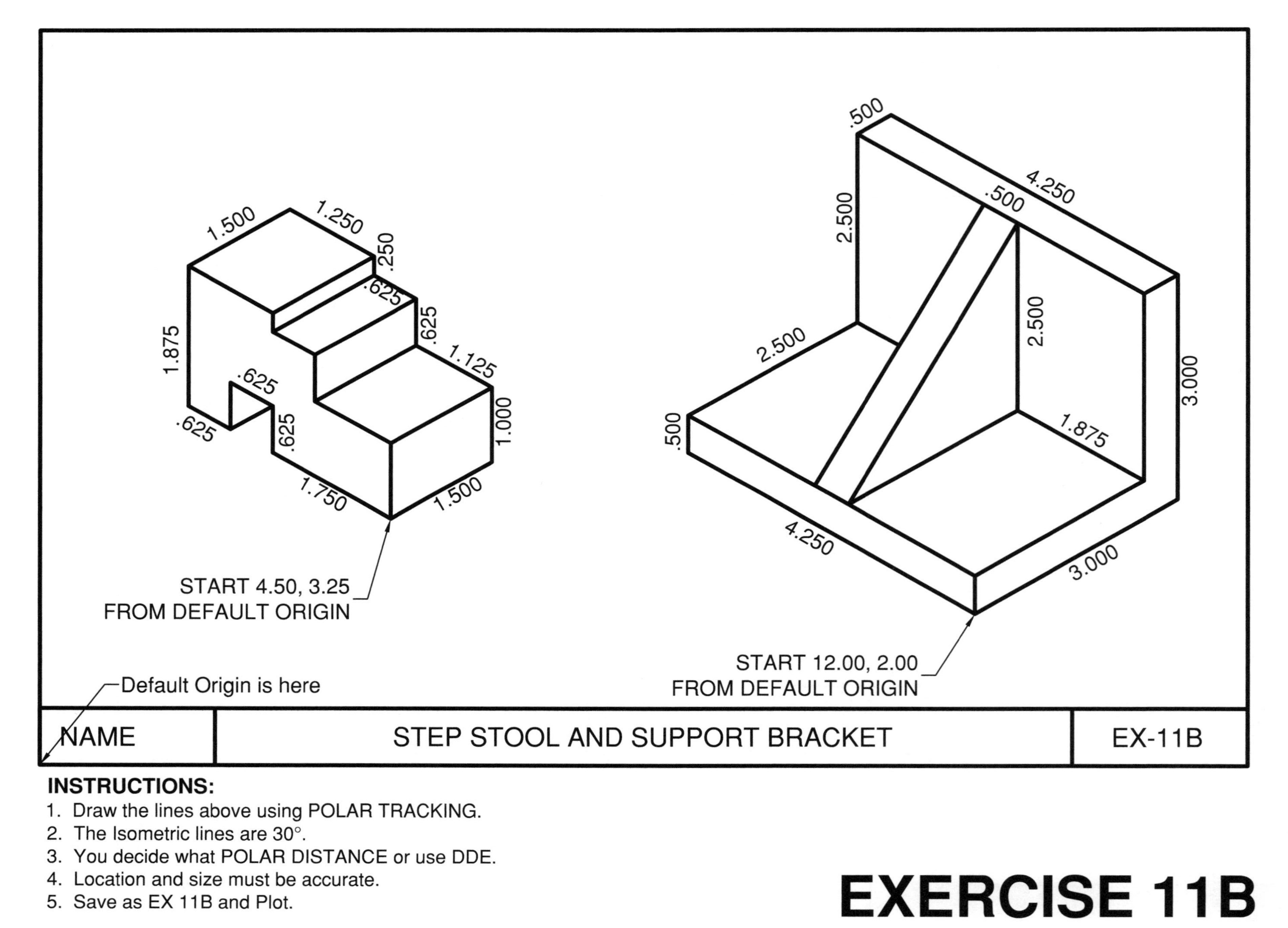

INSTRUCTIONS:

1. Draw the lines above using POLAR TRACKING.
2. The Isometric lines are 30°.
3. You decide what POLAR DISTANCE or use DDE.
4. Location and size must be accurate.
5. Save as EX 11B and Plot.

EXERCISE 11B

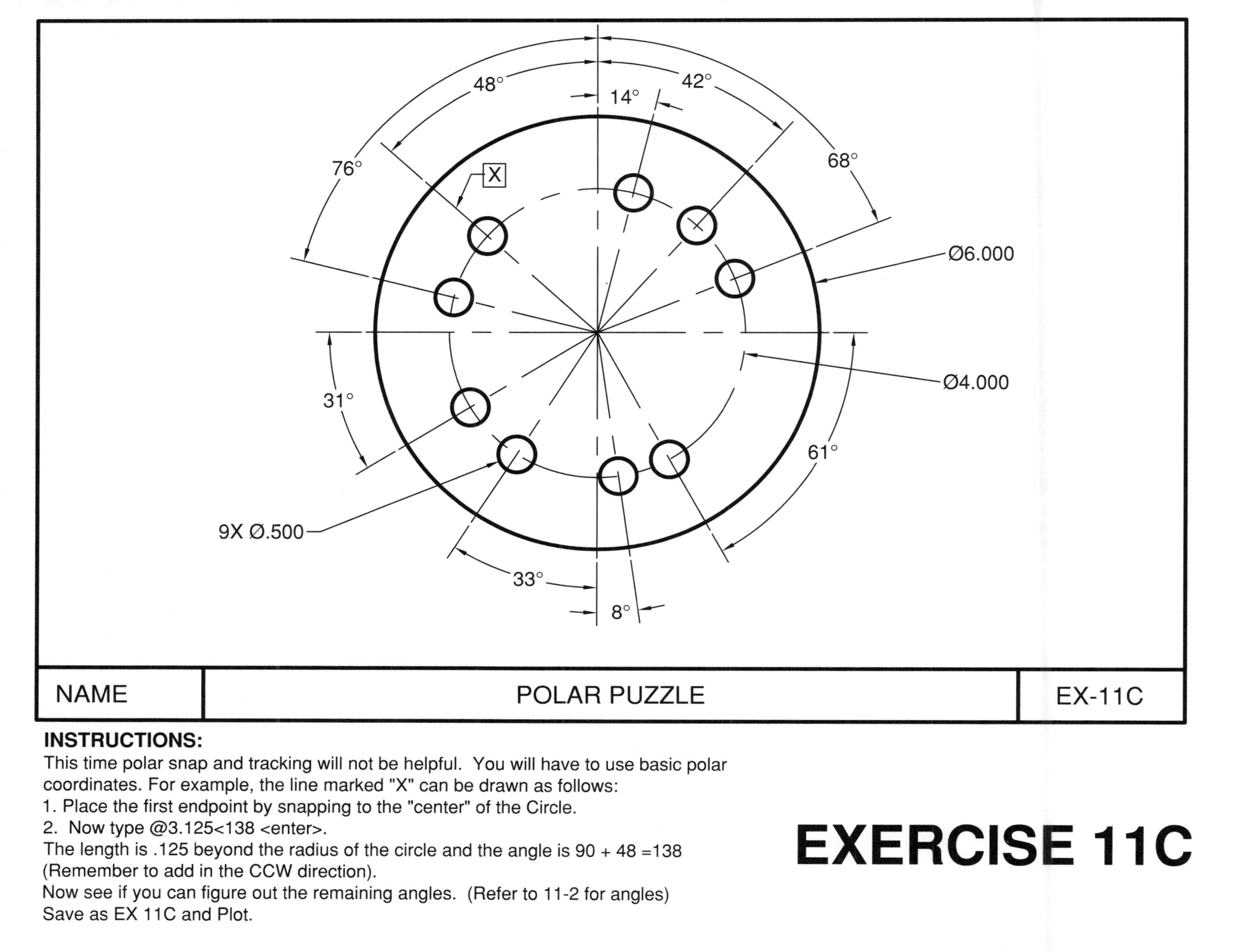

INSTRUCTIONS:

This time polar snap and tracking will not be helpful. You will have to use basic polar coordinates. For example, the line marked "X" can be drawn as follows:

1. Place the first endpoint by snapping to the "center" of the Circle.
2. Now type @3.125<138 <enter>.

The length is .125 beyond the radius of the circle and the angle is 90 + 48 =138 (Remember to add in the CCW direction).

Now see if you can figure out the remaining angles. (Refer to 11-2 for angles)

Save as EX 11C and Plot.

EXERCISE 11C

NOTES:

LEARNING OBJECTIVES

After completing this lesson, you will be able to:

1. Duplicate an object at a specified distance away.
2. Make changes to an object's properties.
3. Create tables with Single Line and Multiline text.

LESSON 12

OFFSET

The **OFFSET** command duplicates an object parallel to the original object at a specified distance away. The new object will retain the same color, layer and linetype of the original.

1. Select the OFFSET command using one of the following:

   TYPING = OFFSET
   PULLDOWN = MODIFY / OFFSET
   TOOLBAR = MODIFY

2. Specify offset distance or [Through] <Through>: ***type the offset distance***
3. Select object to offset or <exit>: ***select the object to offset***
4. Specify point on side to offset: ***Select which side of the original you want the duplicate to appear, by placing your cursor and clicking.***
5. Select object to offset or <exit>: ***Press <enter> to stop.***

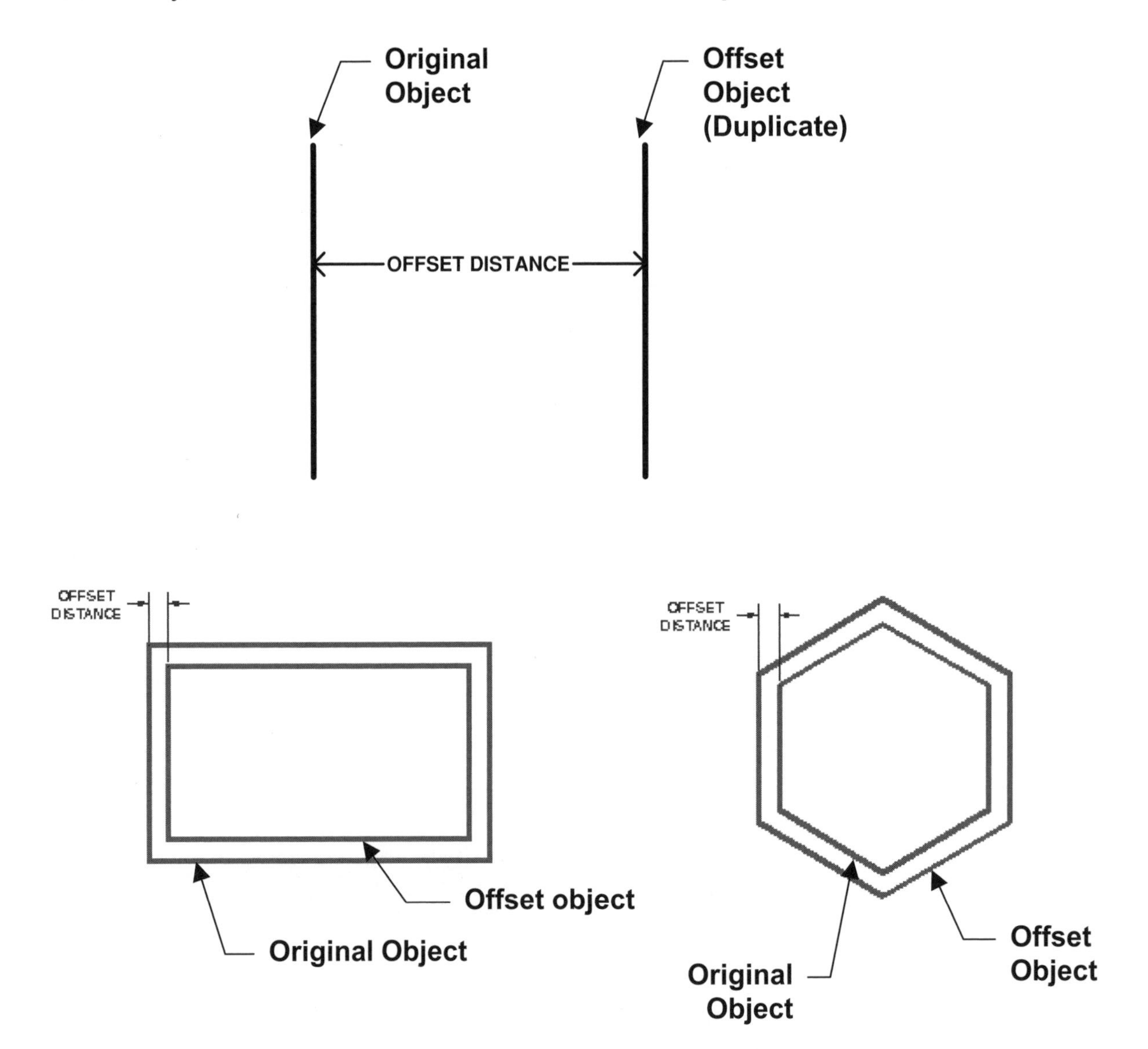

EDITING WITH THE PROPERTIES PALETTE

The **Properties Palette**, shown below, makes it possible to change an object's properties. You simply open the Properties Palette, select an object and you can change any of the properties that are listed.

Open the Properties Palette by double clicking on an object or use one of the following:

TYPING = PROPS or CH
PULLDOWN = MODIFY / PROPERTIES
TOOLBAR = OBJECT PROPERTIES
KEYS = CTRL + 1

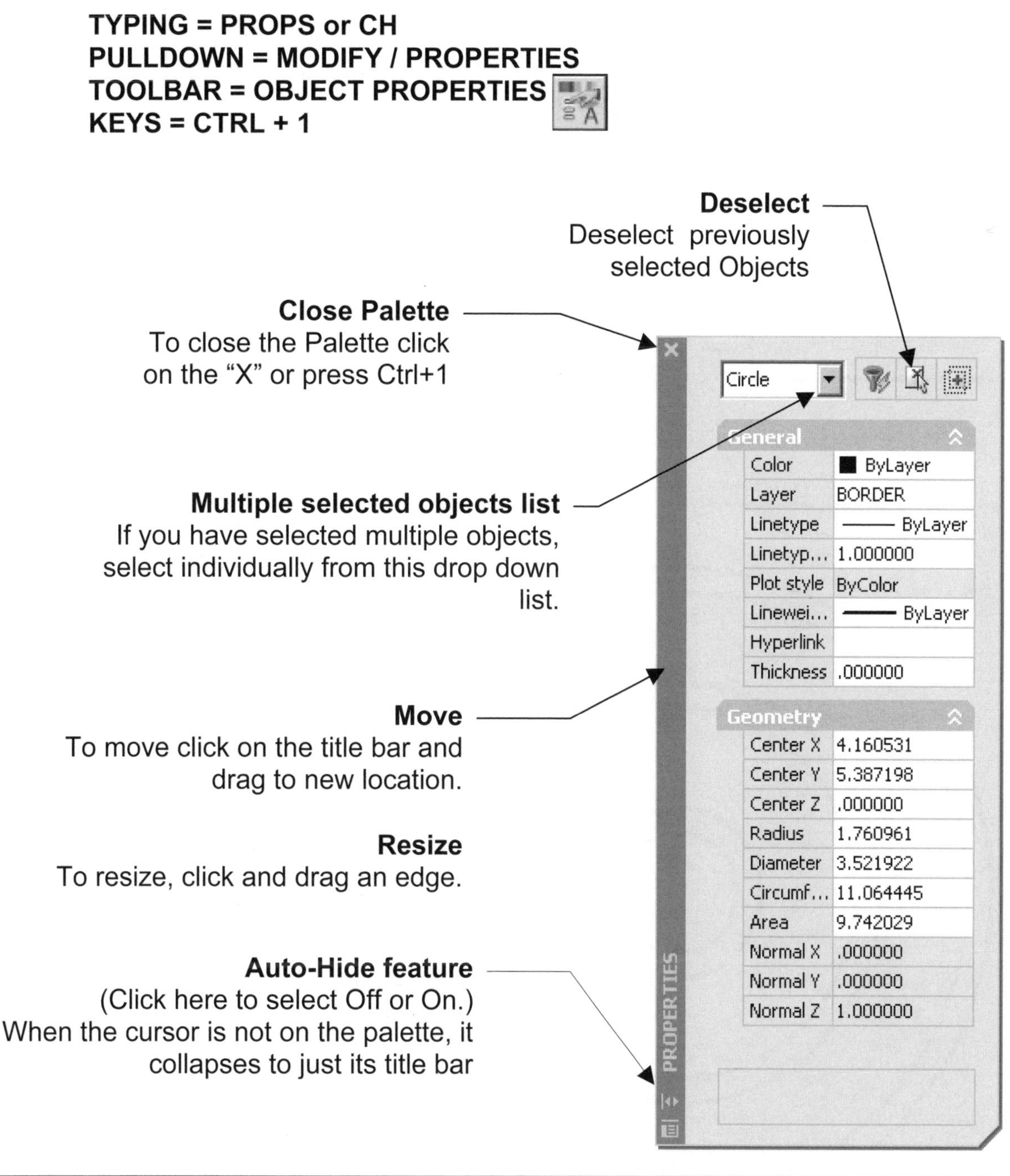

Note: Using the Properties Palette is a great way to change the Lineweight, Linetype and color, of an object, without changing the entire layer. Think about it.......

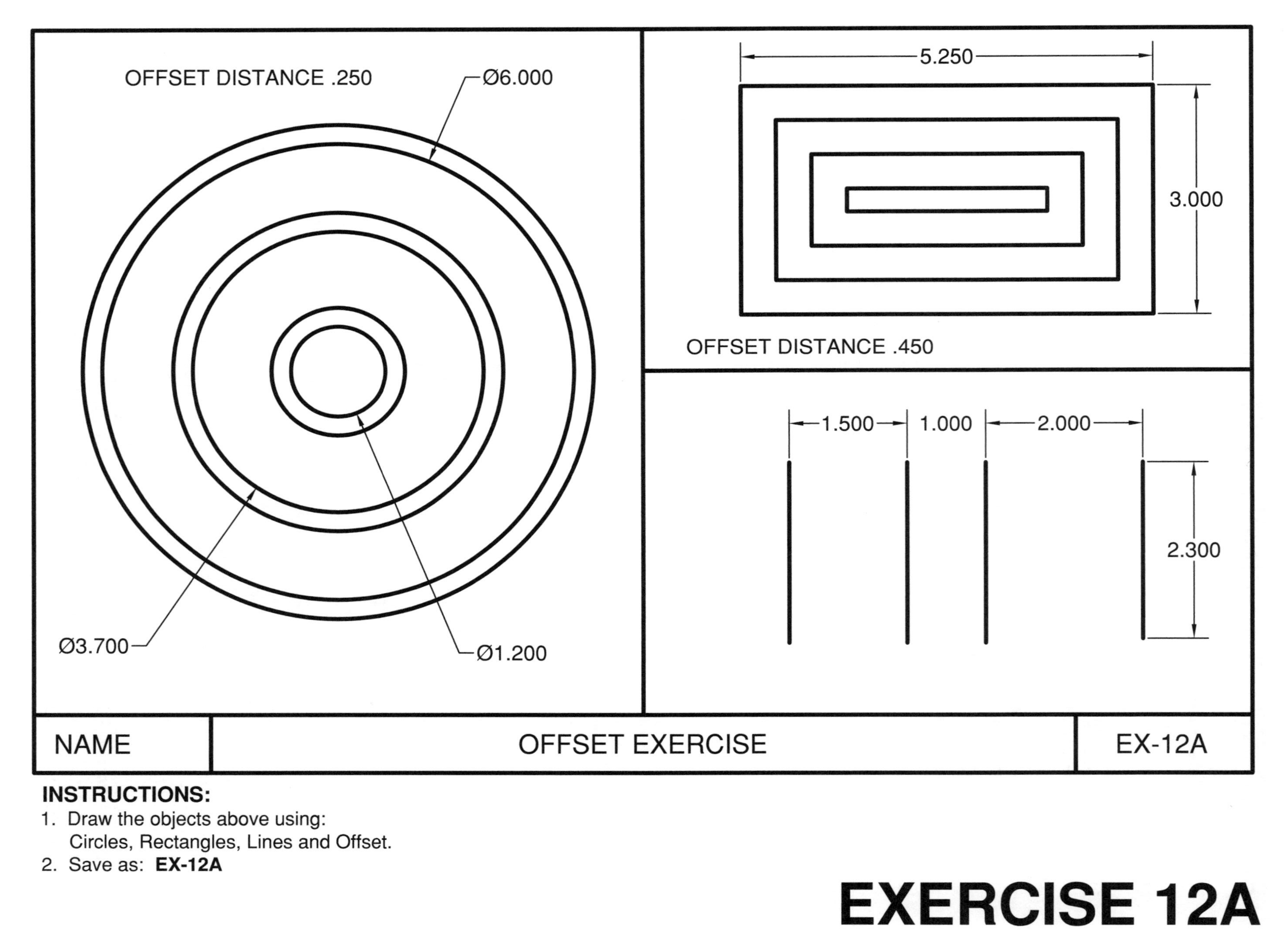

INSTRUCTIONS:

1. Draw the objects above using:
   Circles, Rectangles, Lines and Offset.
2. Save as: **EX-12A**

EXERCISE 12A

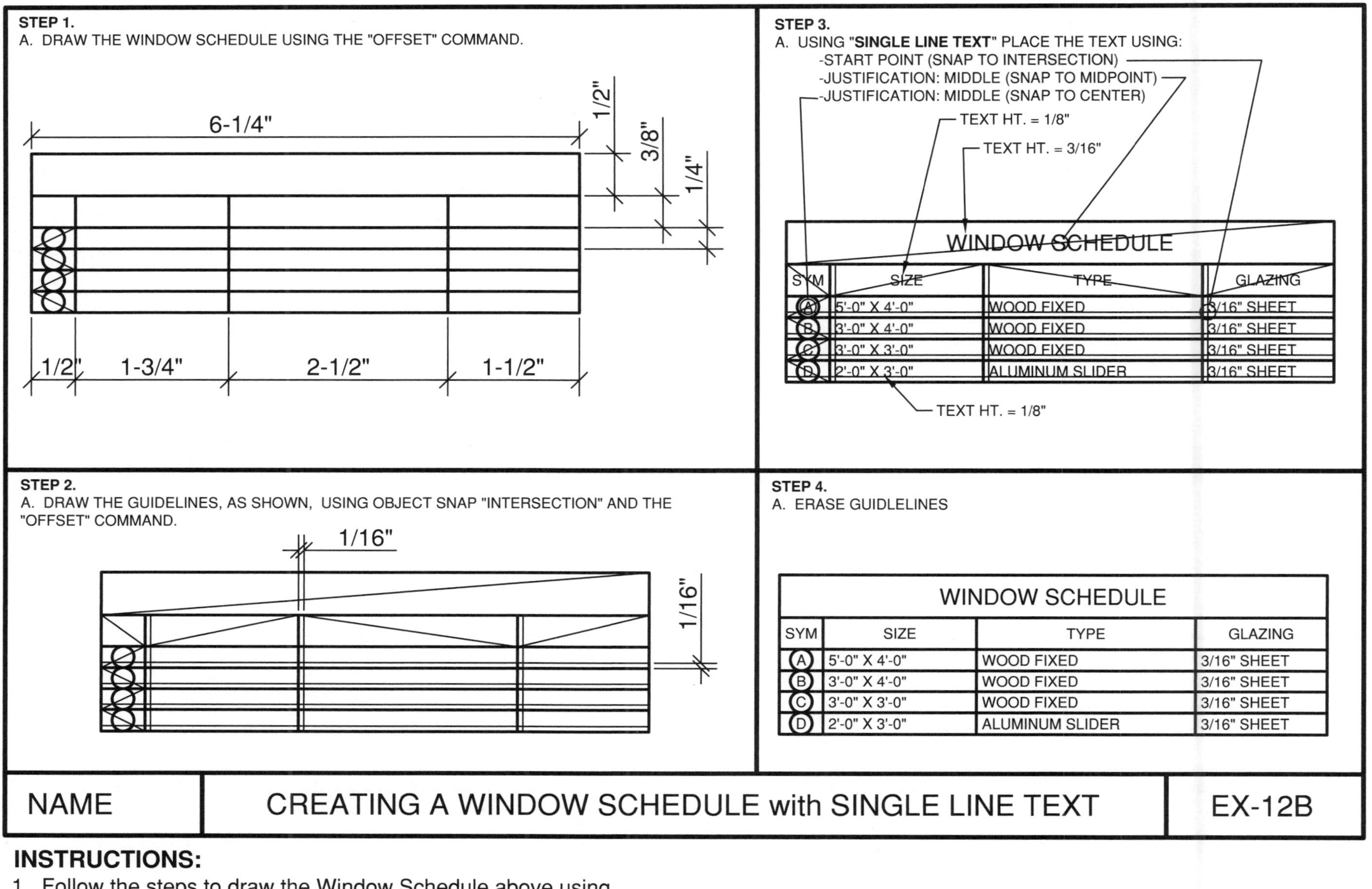

INSTRUCTIONS:

1. Follow the steps to draw the Window Schedule above using.
   (**Use** Single line text. **Do not** use Multiline text)
2. Use Layer: Border for lines. Layer: Text-Hvy for Headings. Layer: Text-lit for descriptions.
3. Save as: **EX-12B**

EXERCISE 12B

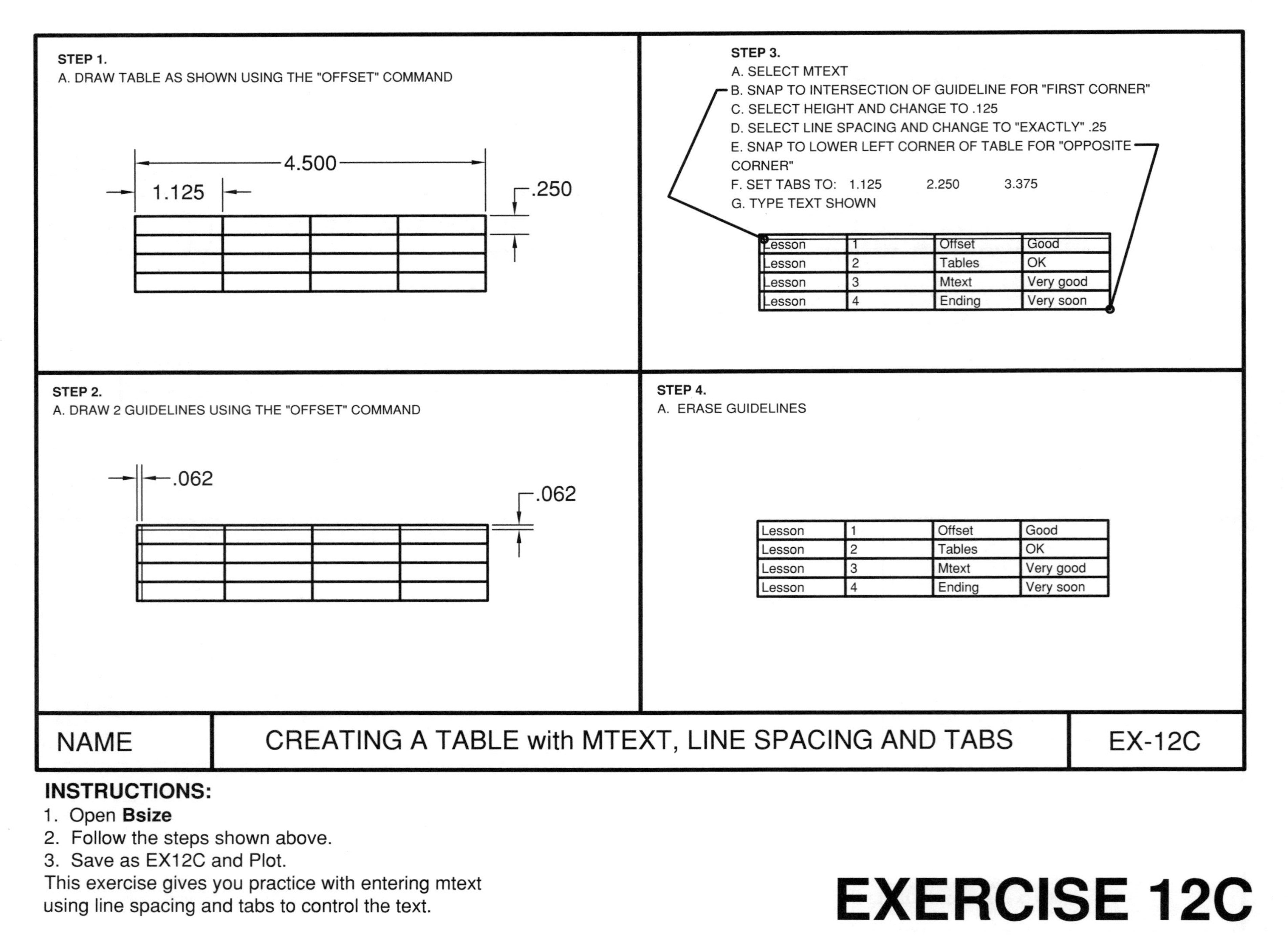

INSTRUCTIONS:

1. Open **Bsize**
2. Follow the steps shown above.
3. Save as EX12C and Plot.

This exercise gives you practice with entering mtext using line spacing and tabs to control the text.

EXERCISE 12C

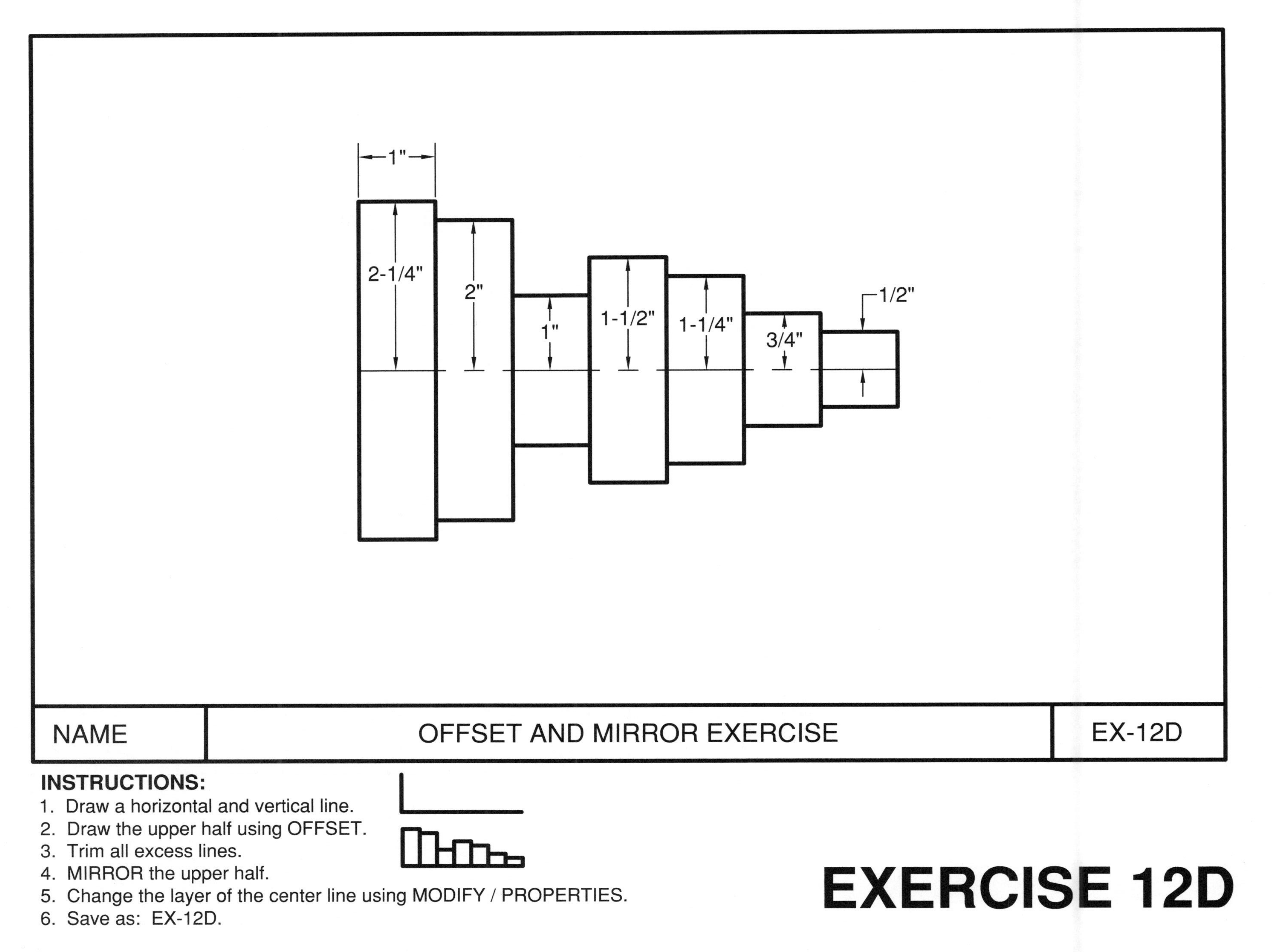
1"
2-1/4"
2"
1"
1-1/2"
1-1/4"
3/4"
1/2"
NAME
OFFSET AND MIRROR EXERCISE
EX-12D
INSTRUCTIONS:
1. Draw a horizontal and vertical line.
2. Draw the upper half using OFFSET.
3. Trim all excess lines.
4. MIRROR the upper half.
5. Change the layer of the center line using MODIFY / PROPERTIES.
6. Save as: EX-12D.
EXERCISE 12D

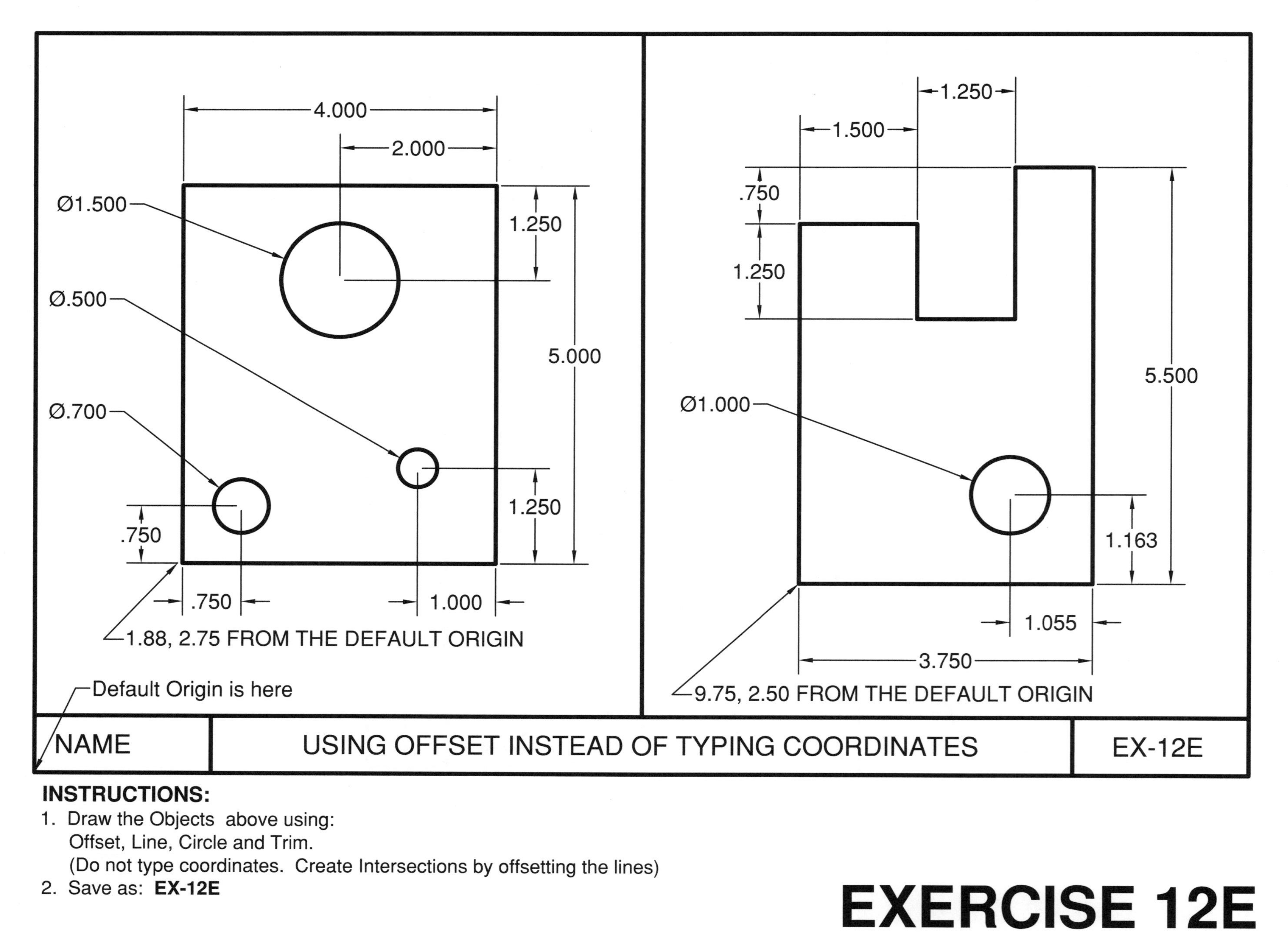

INSTRUCTIONS:

1. Draw the Objects above using:
   Offset, Line, Circle and Trim.
   (Do not type coordinates. Create Intersections by offsetting the lines)
2. Save as: **EX-12E**

EXERCISE 12E

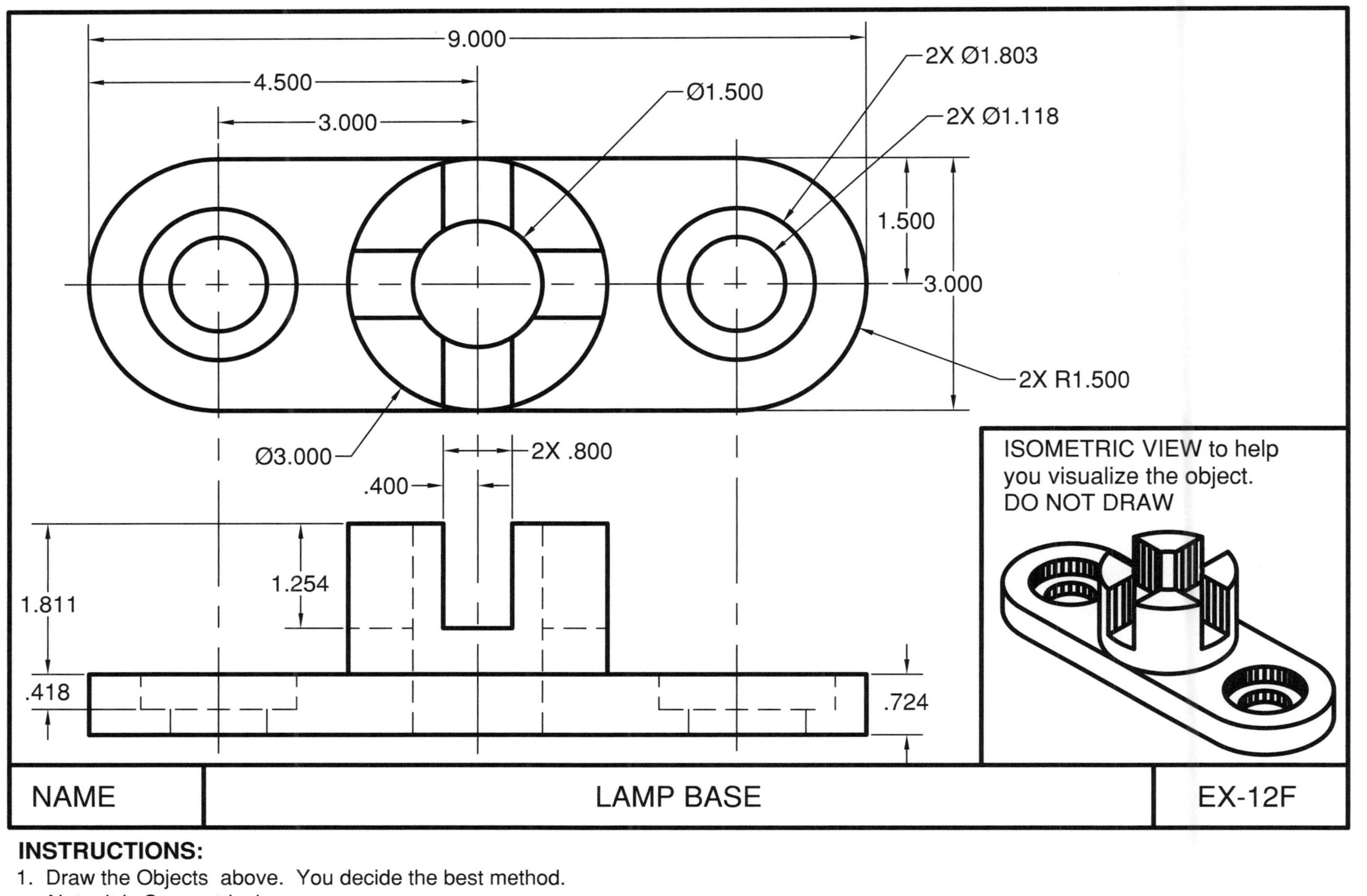

INSTRUCTIONS:

1. Draw the Objects above. You decide the best method.
   Note: It is Symmetrical.
2. Save as: **EX-12F**

EXERCISE 12F

NOTES:

LEARNING OBJECTIVES

After completing this lesson, you will be able to:

1. Create multiple copies in a rectangular or Circular pattern.

LESSON 13

ARRAY

The ARRAY command allows you to make multiple copies in a **RECTANGULAR** or Circular **(POLAR)** pattern. The maximum limit of copies per array is 100,000. This limit can be changed but should accommodate most users.

RECTANGULAR ARRAY

The method allows you to make multiple copies of object(s) in a rectangular pattern. You specify the number of rows (horizontal), columns (vertical) and the offset distance between the rows and columns. The offset distances will be equally spaced.

Offset Distance is sometimes tricky to understand. *Read this carefully.* The offset distance is the distance from a specific location on the original to that same location on the invisible copy. It is not just the space in between the two. Refer to the example below.

To use the rectangular array command you will select the object(s), specify how many rows and columns desired and the offset distance for the rows and the columns.
Step by step instructions on page 13-3.

Example of Rectangular Array:

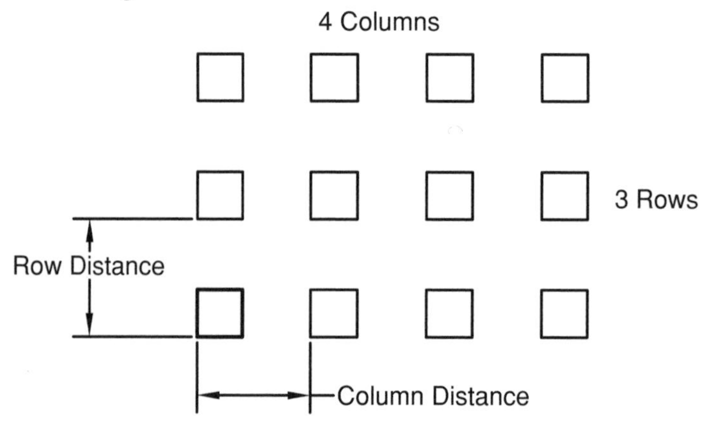

Example of Rectangular Array on an angle.
Notice the copies do not rotate. The Angle is only used to establish the placement.

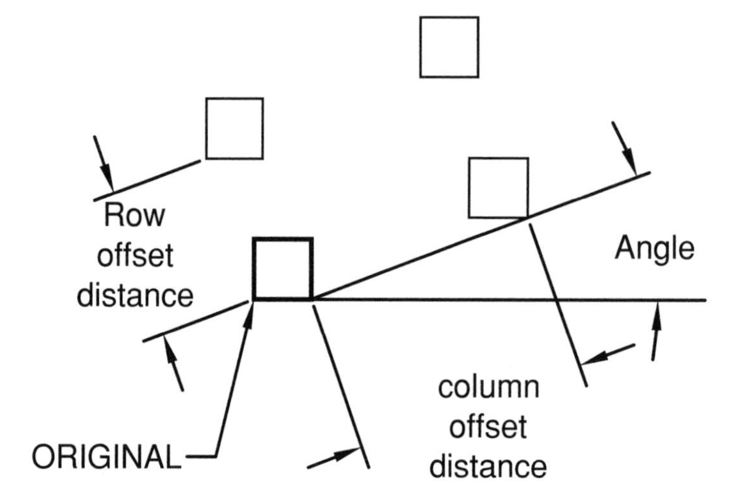

ARRAY (Continued)

RECTANGULAR ARRAY

1. Select the ARRAY command using one of the following:

TYPE = ARRAY
PULLDOWN = MODIFY / ARRAY
TOOLBAR = MODIFY

The dialog shown below will appear.

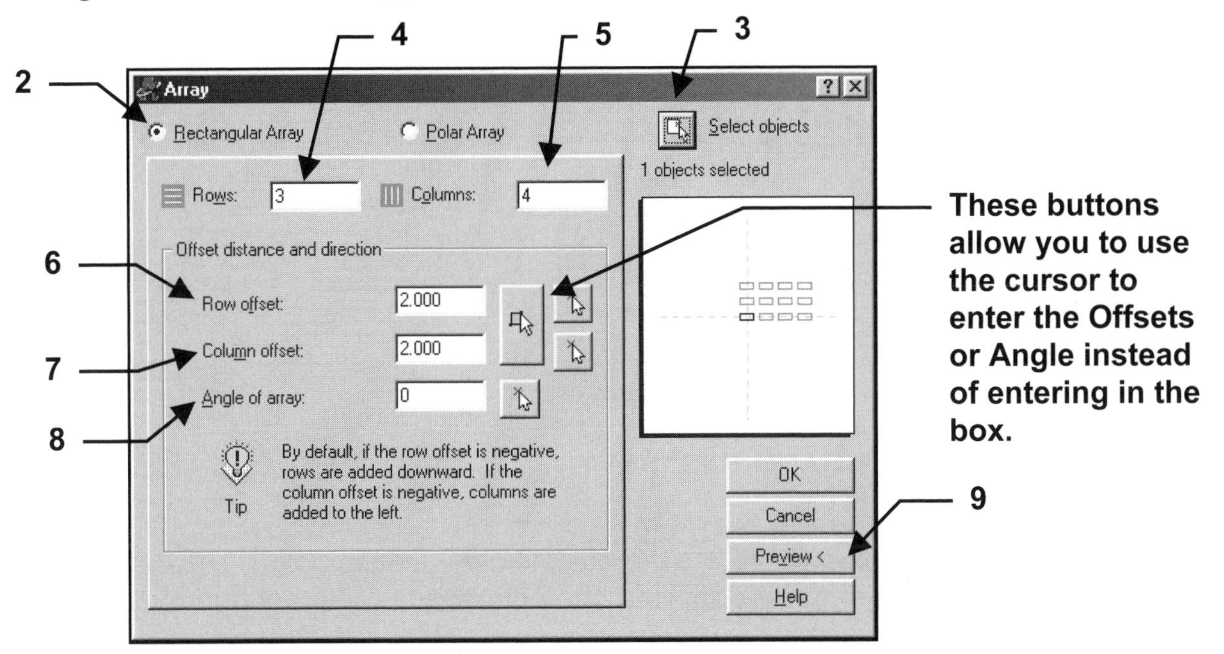

2. Select "**Rectangular Array**"

3. Select the **"Select Objects"** button.
   This will take you back to your drawing. Select the objects to Array then <enter>

4. Enter the number of rows.

5. Enter the number of columns.

6. Enter the row offset. (The distance from a specific location on the original to that same specific location on the future copy) See example on pg. 13-2.

7. Enter the column offset. (The distance from a specific location on the original to that same specific location on the future copy) See example on pg. 13-2.

8. Enter an angle if you would like the array to be on an angle.

9. Select the Preview button.
   If it looks correct, select the Accept button. If it is not correct, select the Modify button, make the necessary corrections and preview again.
   (Note: if the Preview button is gray, you have forgotten to "Select objects"- #3)

ARRAY (continued)

POLAR ARRAY

This method allows you to make multiple copies in a circular pattern. You specify the total number of copies to fill a specific Angle or specify the angle between each copy and angle to fill.

To use the polar array command you select the object(s) to copy, specify the center of the array, specify the number of copies or the angle between the copies, the angle to fill and if you would like the copies to rotate as they are copied.

Example of Polar Array

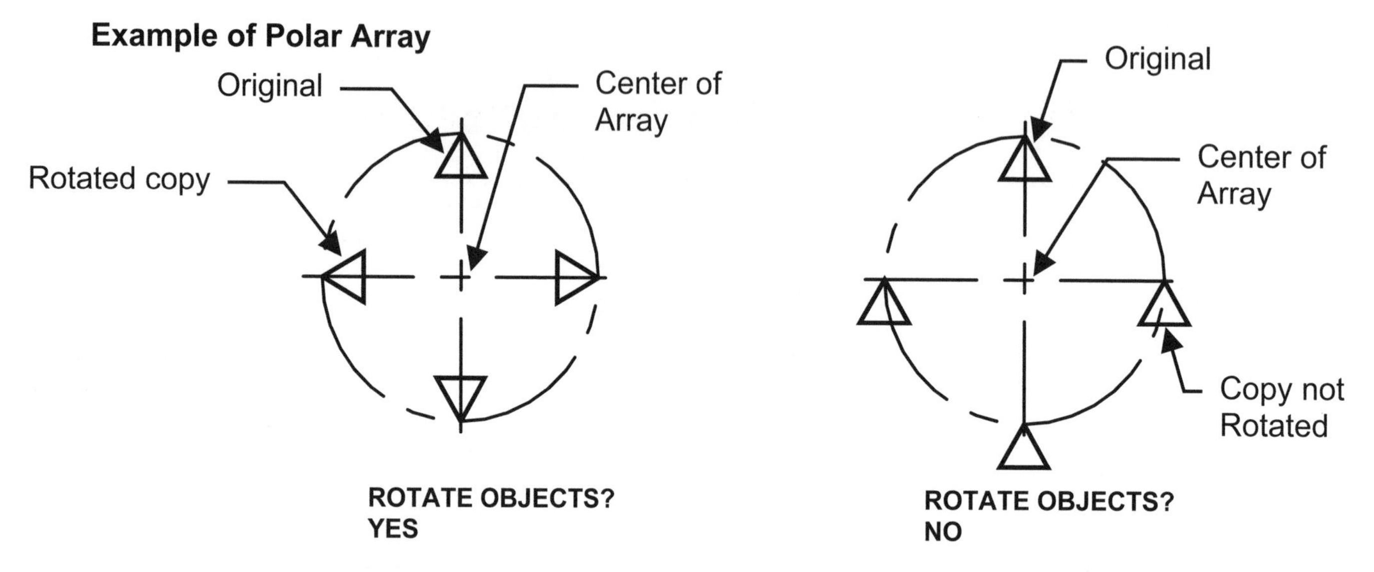

Note: the two examples above use the objects default base point. The example below specifies the base point for the copies. The end result is different from the example above on the right.

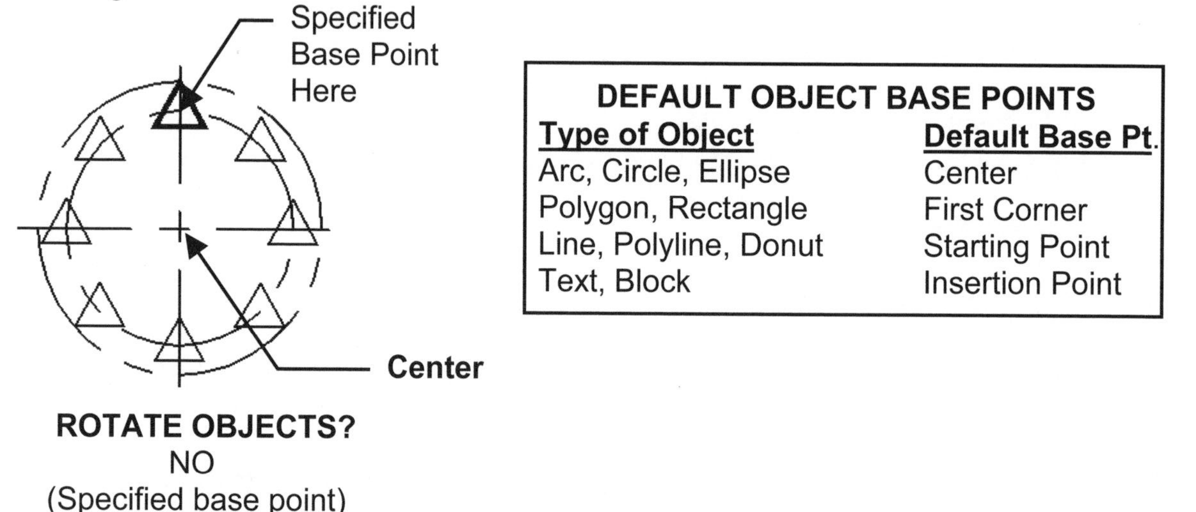

DEFAULT OBJECT BASE POINTS

Type of Object	Default Base Pt.
Arc, Circle, Ellipse	Center
Polygon, Rectangle	First Corner
Line, Polyline, Donut	Starting Point
Text, Block	Insertion Point

The difference between "Center Point" and "Object Base Point" is sometimes confusing. When specifying the Center Point, try to visualize the copies already there. Now, in your mind, try to visualize the center of that array. In other words, it is the Pivot Point from which the copies will be placed around. The Base Point is different. It is located on the original object.

ARRAY (continued)

POLAR ARRAY

1. Select the ARRAY command using one of the following:

   TYPE = ARRAY
   PULLDOWN = MODIFY / ARRAY
   TOOLBAR = MODIFY

The dialog shown below will appear.

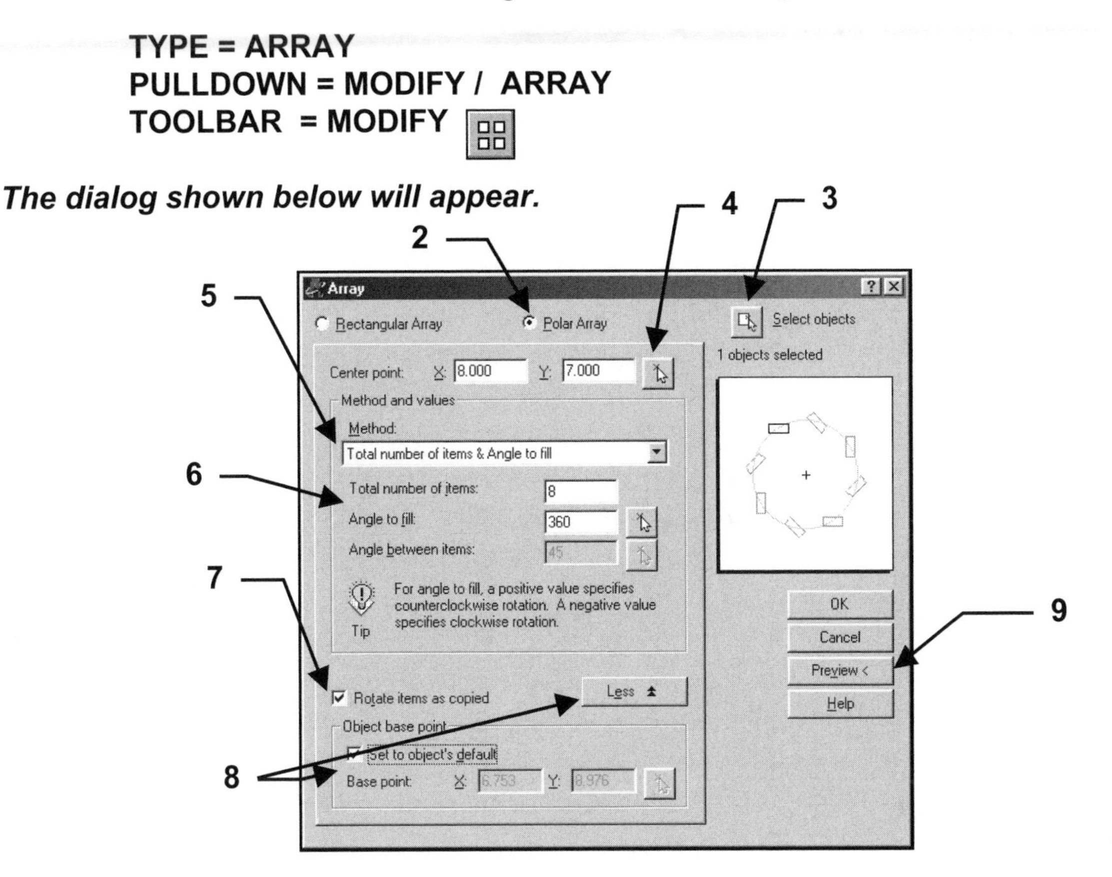

2. Select **" Polar Array"**

3. Select the **"Select Objects"** button.
   This will take you back to your drawing. Select the objects to Array then <enter>

4. Press the "Center Point" button and select the center point with the cursor or enter the X and Y coordinates in the "Center Point" boxes.

5. Select the method.

6. Enter the "Total number of items", "Angle to fill" and or "Angle between items".

7. Select whether you want the items rotated as copied or not. (See pg. 13-4 for explanation)

8. Accept the objects default base point or enter the X and Y coordinates.
   (Select the "More" button to show this area. Select the "Less" button to not show)

9. Select the Preview button.
   If it looks correct, select the Accept button. If it is not correct, select the Modify button, make the necessary corrections and preview again.
   (Note: if the Preview button is gray, you have forgotten to select objects #3)

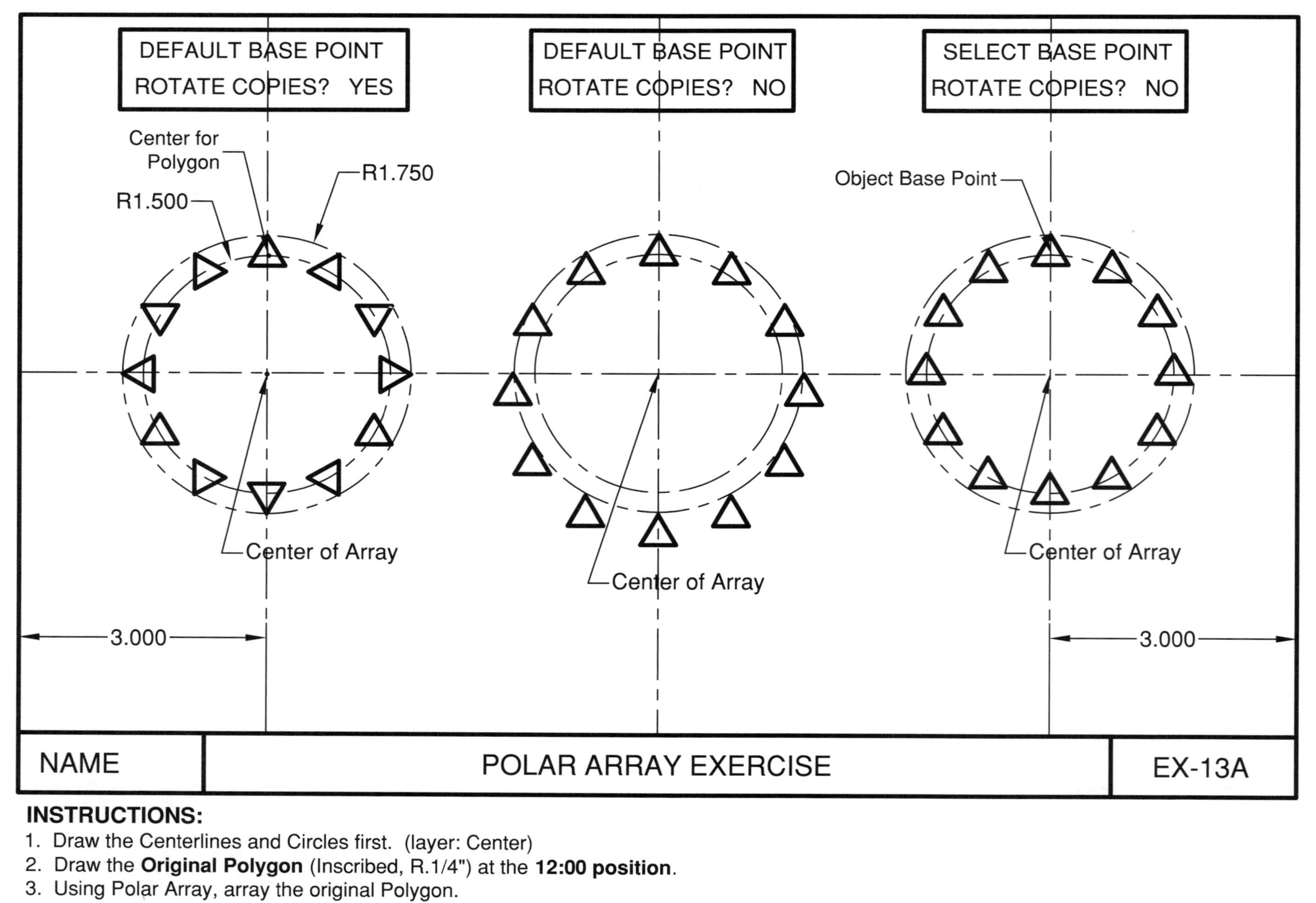

INSTRUCTIONS:

1. Draw the Centerlines and Circles first. (layer: Center)
2. Draw the **Original Polygon** (Inscribed, R.1/4") at the **12:00 position**.
3. Using Polar Array, array the original Polygon.
4. Number of items: **12**
5. Save as: **EX-13A** and Plot

EXERCISE 13A

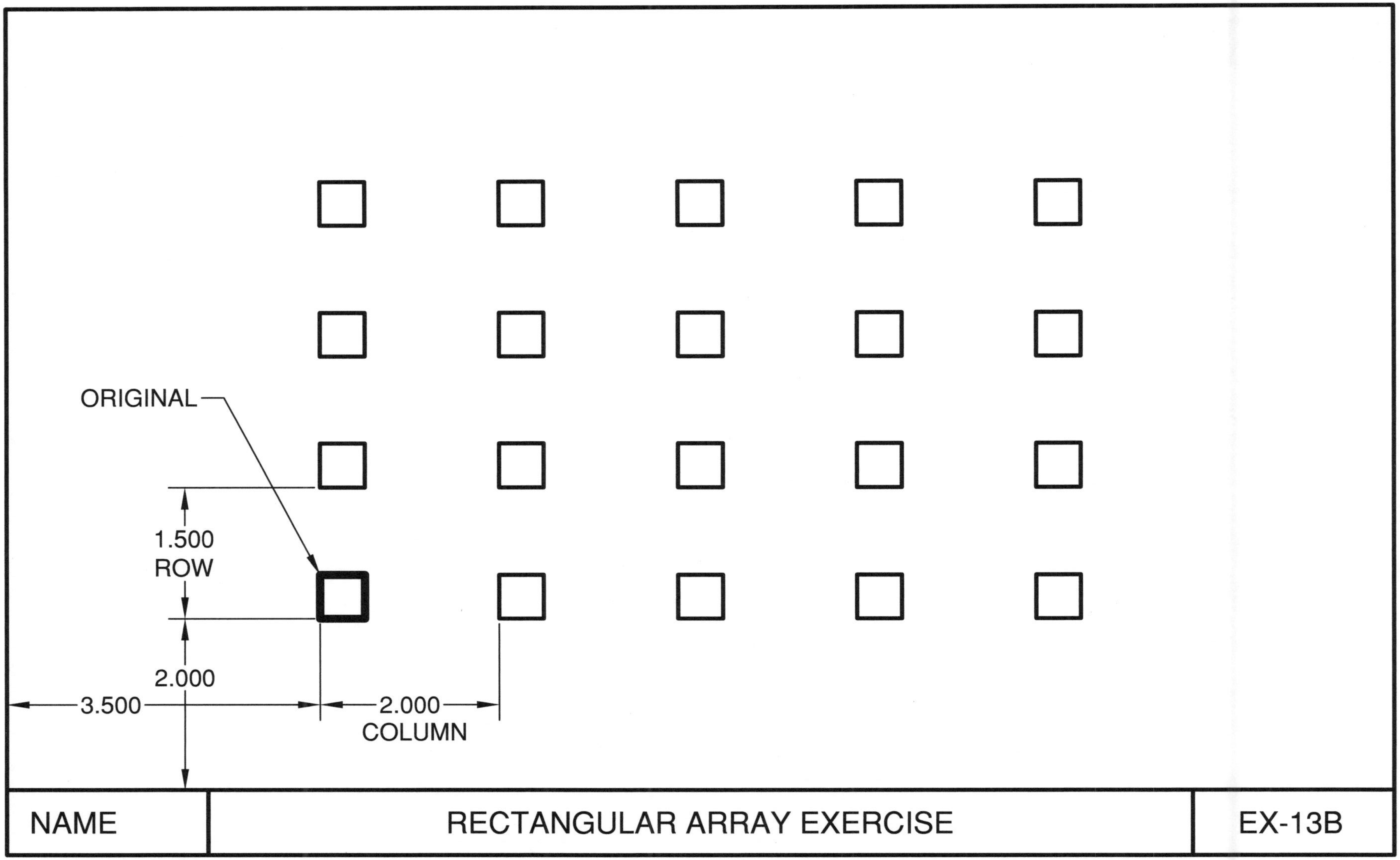

ORIGINAL

1.500
ROW

2.000

3.500

2.000
COLUMN

NAME	RECTANGULAR ARRAY EXERCISE	EX-13B

INSTRUCTIONS:

1. Draw the lower left RECTANGLE first. .500 Square. (layer: Object)
2. Using Rectangular Array, array the original Rectangle as shown.
   No. of Rows = 4 No. of Columns = 5
   Distance between Rows = 1.500 Distance between columns = 2.000
4. Save as: **EX-13B** and Plot.

EXERCISE 13B

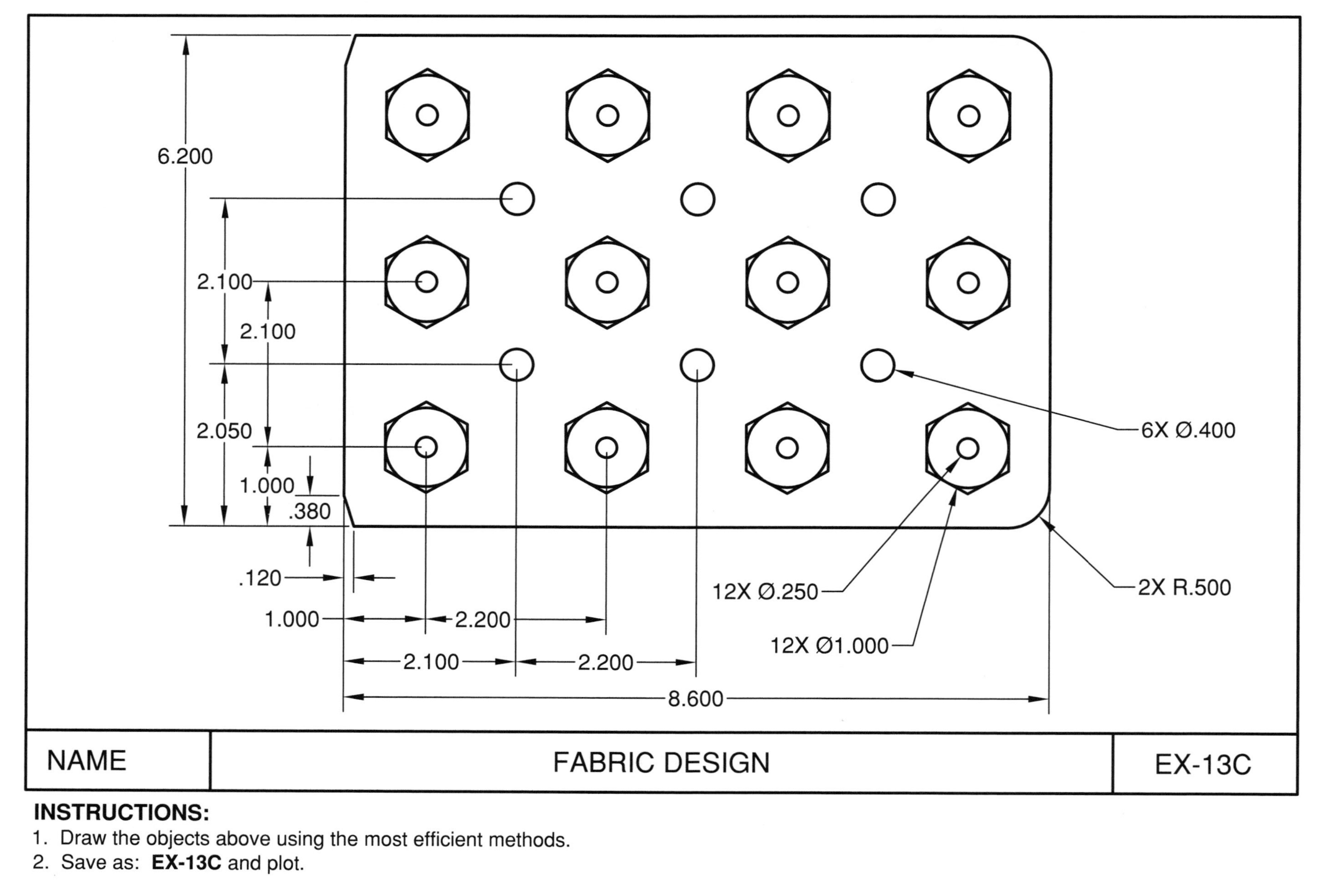

INSTRUCTIONS:

1. Draw the objects above using the most efficient methods.
2. Save as: **EX-13C** and plot.

EXERCISE 13C

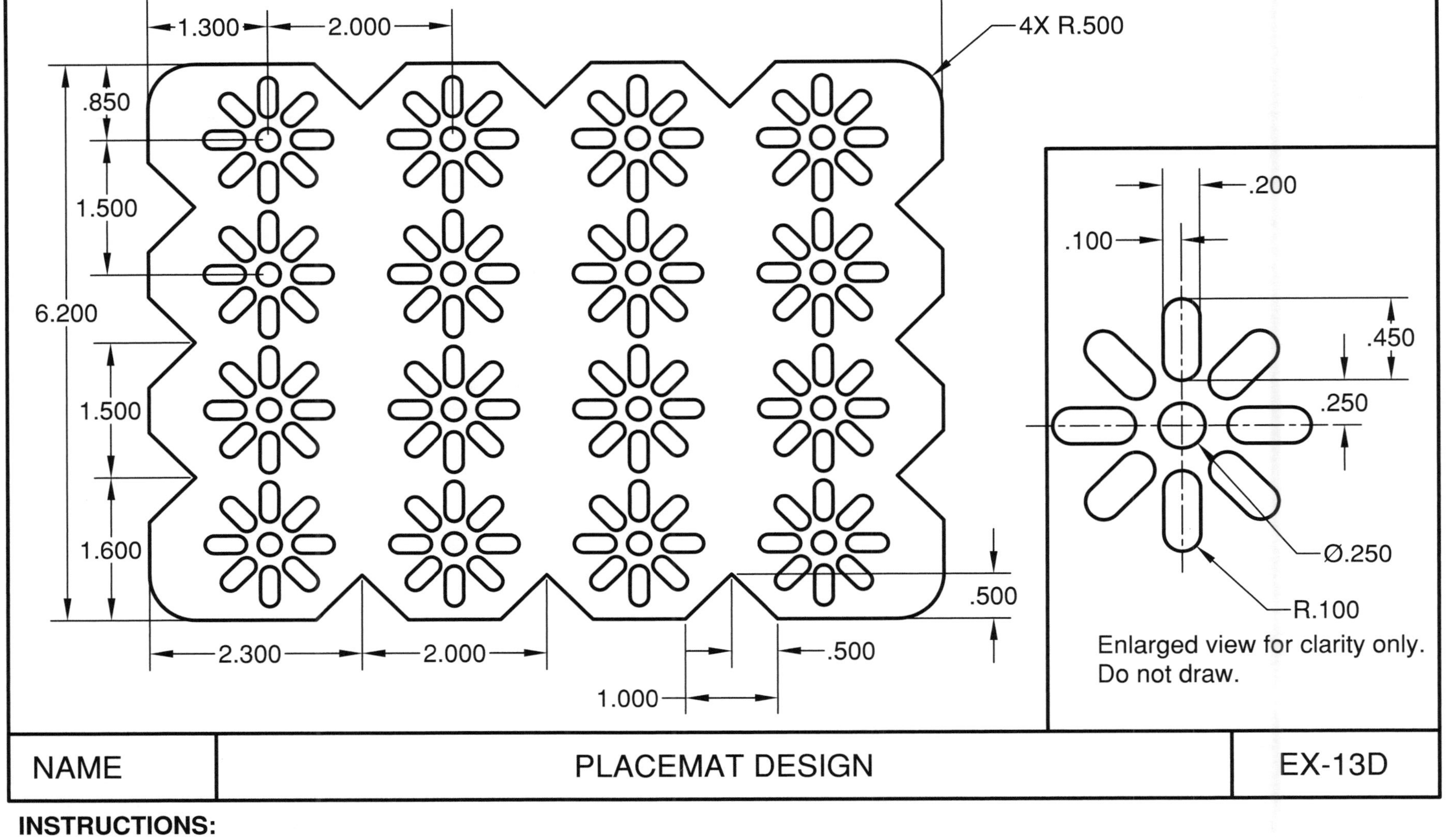

INSTRUCTIONS:

1. Draw the objects above using the most efficient methods.
2. Save as: **EX-13D** and plot.

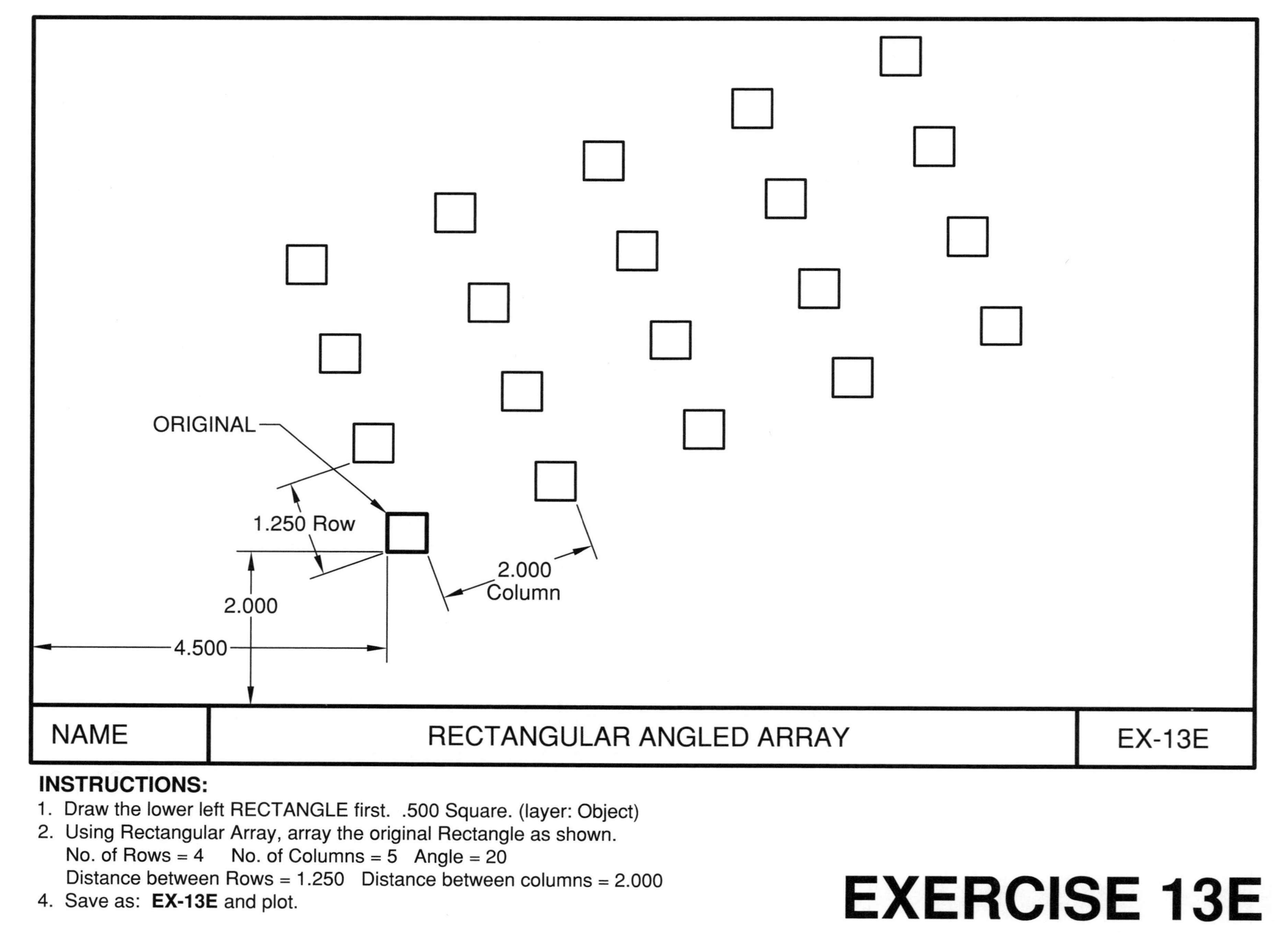

INSTRUCTIONS:

1. Draw the lower left RECTANGLE first. .500 Square. (layer: Object)
2. Using Rectangular Array, array the original Rectangle as shown.
   No. of Rows = 4   No. of Columns = 5   Angle = 20
   Distance between Rows = 1.250   Distance between columns = 2.000
4. Save as: **EX-13E** and plot.

EXERCISE 13E

LEARNING OBJECTIVES

After completing this lesson, you will be able to:

1. Make an existing object larger or smaller proportionately.
2. Stretch or compress an existing object.
3. Rotate an existing object to a specific angle.

LESSON 14

SCALE

The **SCALE** command is used to make objects larger or smaller proportionately. You may scale using a scale factor or a reference length. You must also specify a base point. Think of the base point as a stationary point from which the objects scale. It does not move.

1. Select the SCALE command using one of the following:

   TYPE = SCALE
   PULLDOWN = MODIFY / SCALE
   TOOLBAR = MODIFY

SCALE FACTOR

Command: _scale

2. Select objects: ***select the object(s) to be scaled***
3. Select objects: ***select more object(s) or <enter> to stop***
4. Specify base point: ***select the stationary point on the object***
5. Specify scale factor or [Reference]: ***type the scale factor <enter>***

If the scale factor is greater than 1, the objects will increase in size.
If the scale factor is less than 1, the objects will decrease in size.

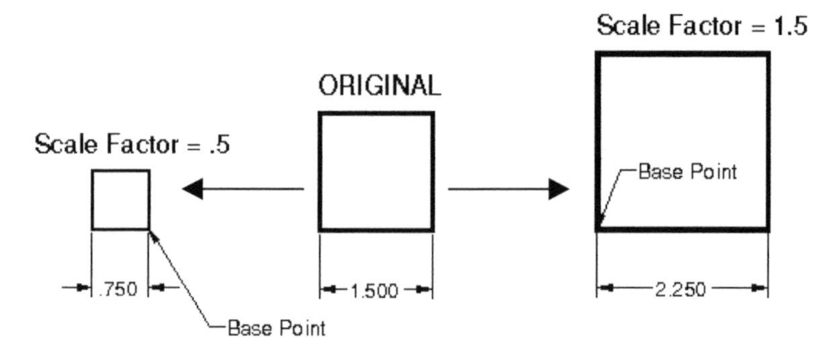

REFERENCE

Command: _scale

2. Select objects: ***select the object(s) to be scaled***
3. Select objects: ***select more object(s) or <enter> to stop***
4. Specify base point: ***select the stationary point on the object***
5. Specify scale factor or [Reference]: ***select Reference***
6. Specify reference length <1>: ***specify a reference length***
7. Specify new length: ***specify the new length***

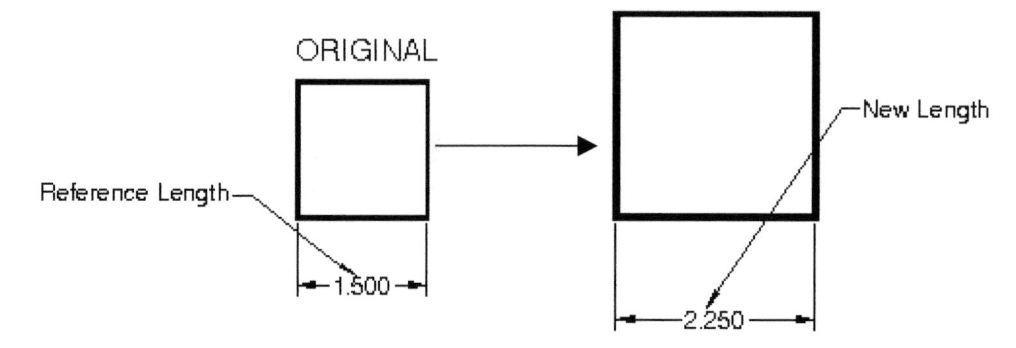

STRETCH

The **STRETCH** command allows you to stretch or compress object(s). Unlike the Scale command, you can alter an objects proportions with the Stretch command. In other words, you may increase the length without changing the width and vice versa.

Stretch is a very valuable tool. Take some time to really understand this command. It will save you hours when making corrections to drawings.

When selecting the object(s) you must use a **CROSSING** window.
Objects that are crossed, will ***stretch.***
Objects that are totally enclosed, will ***move***.

1. Select the STRETCH command using one of the following:

Command: _stretch
2. Select objects to stretch by crossing-window or crossing-polygon...
3. Select objects: ***select the first corner of the crossing window***
4. Specify opposite corner: ***specify the opposite corner of the crossing window***
5. Specify base point or displacement: ***select a base point (where it stretches from)***
6. Specify second point of displacement: ***type coordinates or place location with cursor***

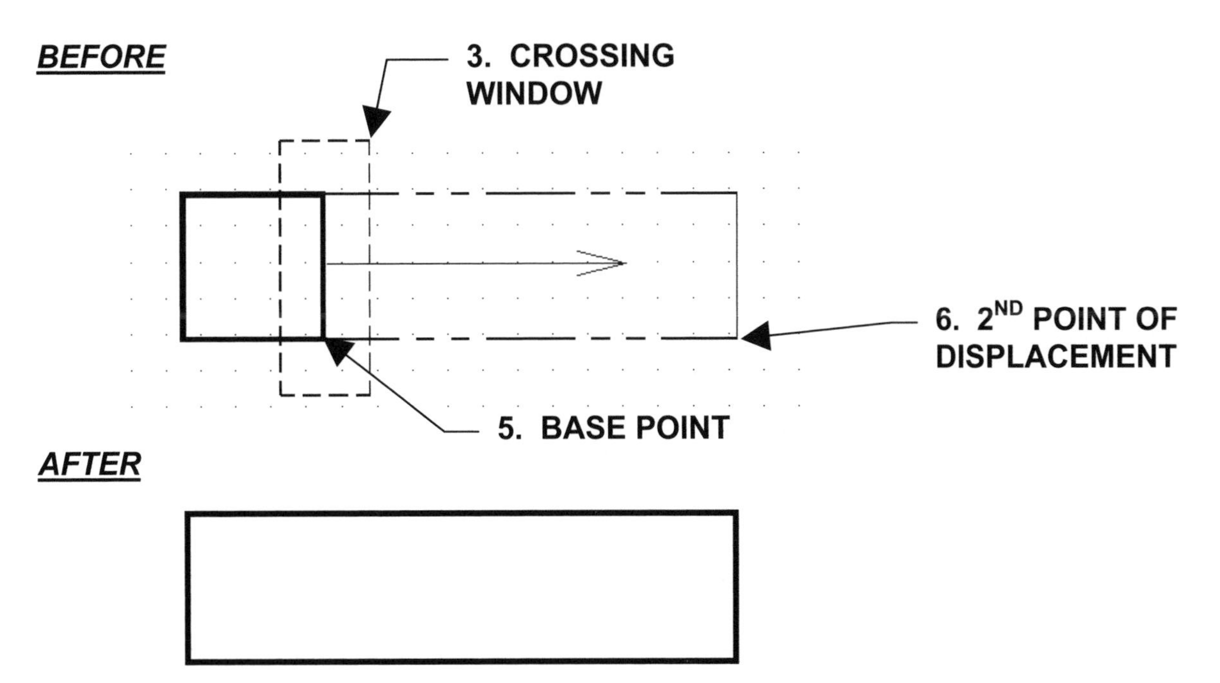

ROTATE

The **ROTATE** command is used to rotate objects around a Base Point. (pivot point)
After selecting the objects and the base point, you will enter the rotation angle or select a reference angle followed by the new angle.
A **Positive** rotation angle revolves the objects **Counter- Clockwise**.
A **Negative** rotation angle revolves the objects **Clockwise**.

Select the ROTATE command using one of the following:

TYPE =RO
PULLDOWN = MODIFY / ROTATE
TOOLBAR = MODIFY

ROTATION ANGLE OPTION

Command: _rotate
1. Current positive angle in UCS: ANGDIR=counterclockwise ANGBASE=0
2. Select objects: ***select the object to rotate***
3. Select objects: ***select more object(s) or <enter> to stop***
4. Specify base point: ***select the base point (pivot point)***
5. Specify rotation angle or [Reference]: ***type the angle of rotation***

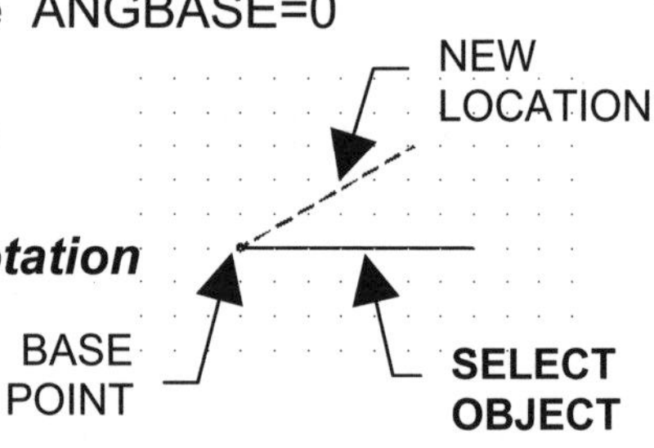

REFERENCE OPTION

Command: _rotate
1. Current positive angle in UCS: ANGDIR=counterclockwise ANGBASE=0
2. Select objects: ***select the object to rotate***
3. Select objects: ***select more object(s) or <enter> to stop***
4. Specify base point: ***select the base point (pivot point)***
5. Specify rotation angle or [Reference]: ***select Reference***
6. Specify the reference angle <0>: ***Snap to the reference object (1) and (2)***
7. Specify the new angle: ***drag the object and snap to the new angle***

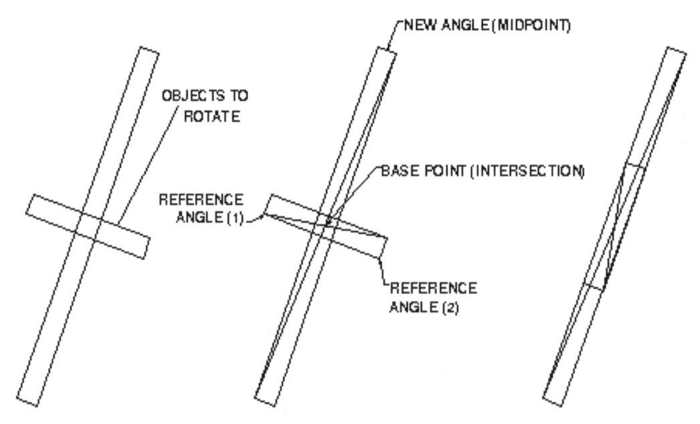

ANGDIR and **ANGBASE** are system variables. The values shown indicate the current status of these system variables.
ANGDIR controls the **CCW** or **CW** rotation.
ANGBASE controls the use of a base angle.
These variables can be changed, but it is best to leave them in the default mode for now.

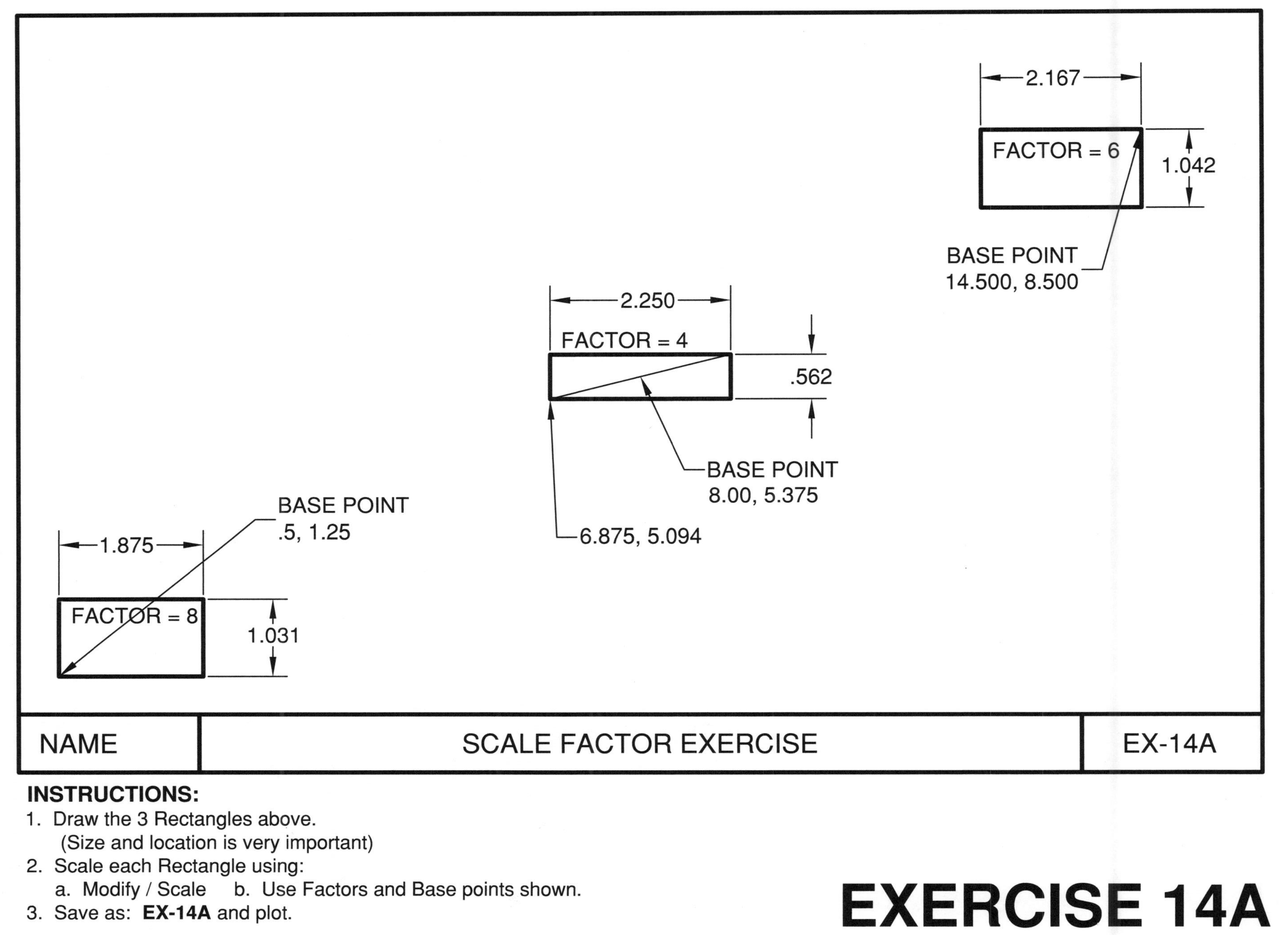

INSTRUCTIONS:

1. Draw the 3 Rectangles above.
   (Size and location is very important)
2. Scale each Rectangle using:
   a. Modify / Scale b. Use Factors and Base points shown.
3. Save as: **EX-14A** and plot.

EXERCISE 14A

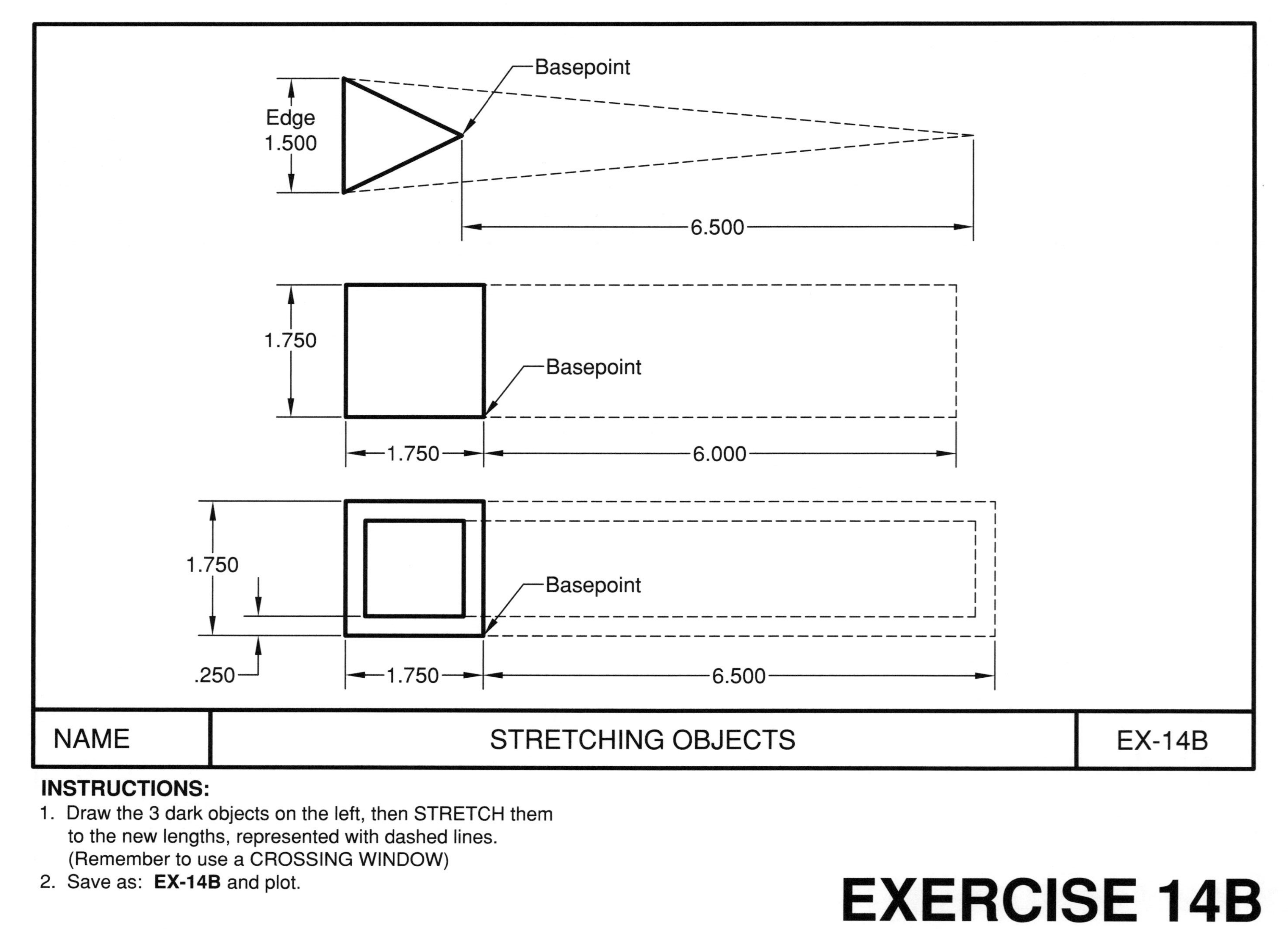

INSTRUCTIONS:

1. Draw the 3 dark objects on the left, then STRETCH them to the new lengths, represented with dashed lines. (Remember to use a CROSSING WINDOW)
2. Save as: **EX-14B** and plot.

EXERCISE 14B

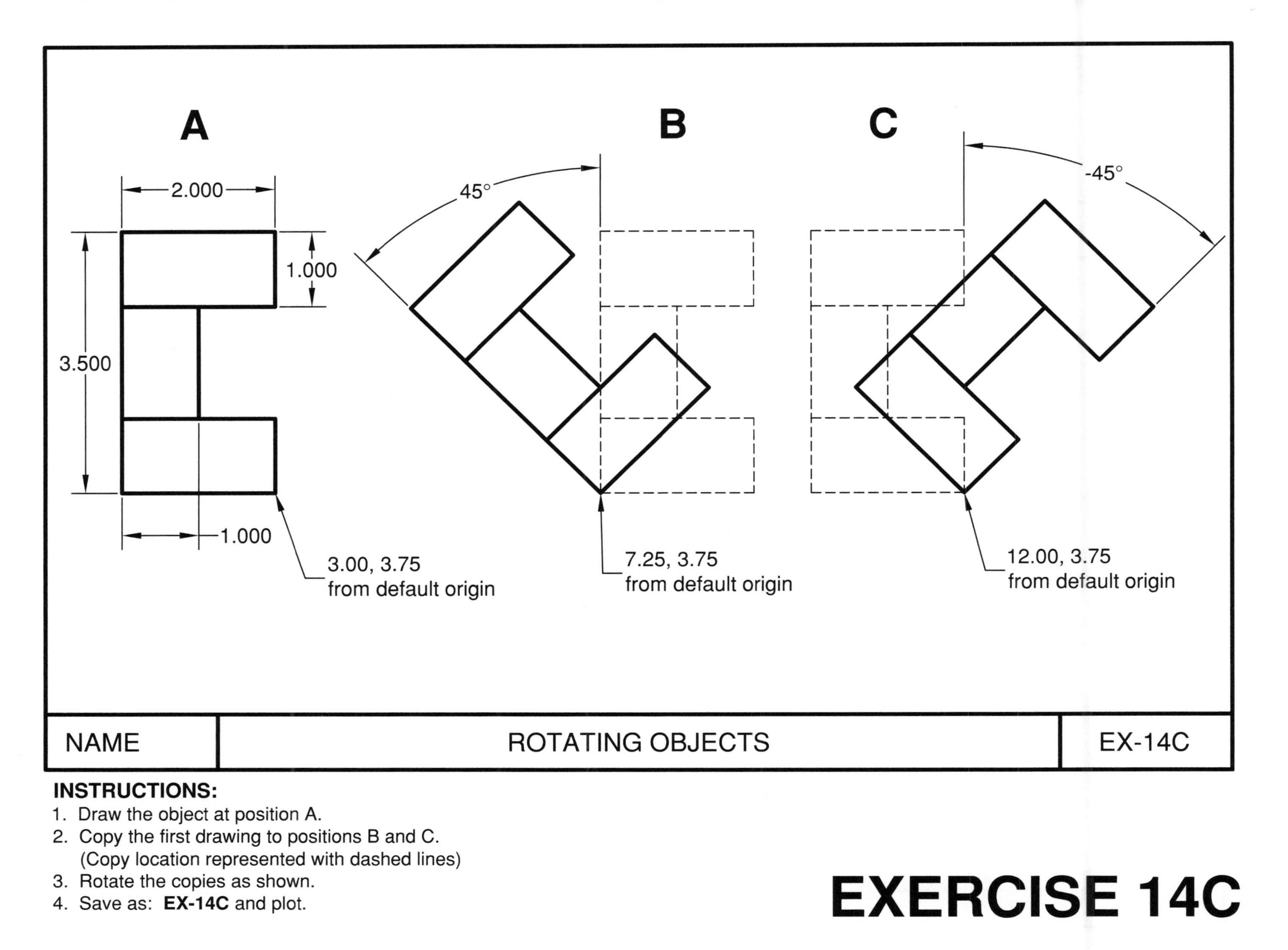

INSTRUCTIONS:

1. Draw the object at position A.
2. Copy the first drawing to positions B and C.
   (Copy location represented with dashed lines)
3. Rotate the copies as shown.
4. Save as: **EX-14C** and plot.

EXERCISE 14C

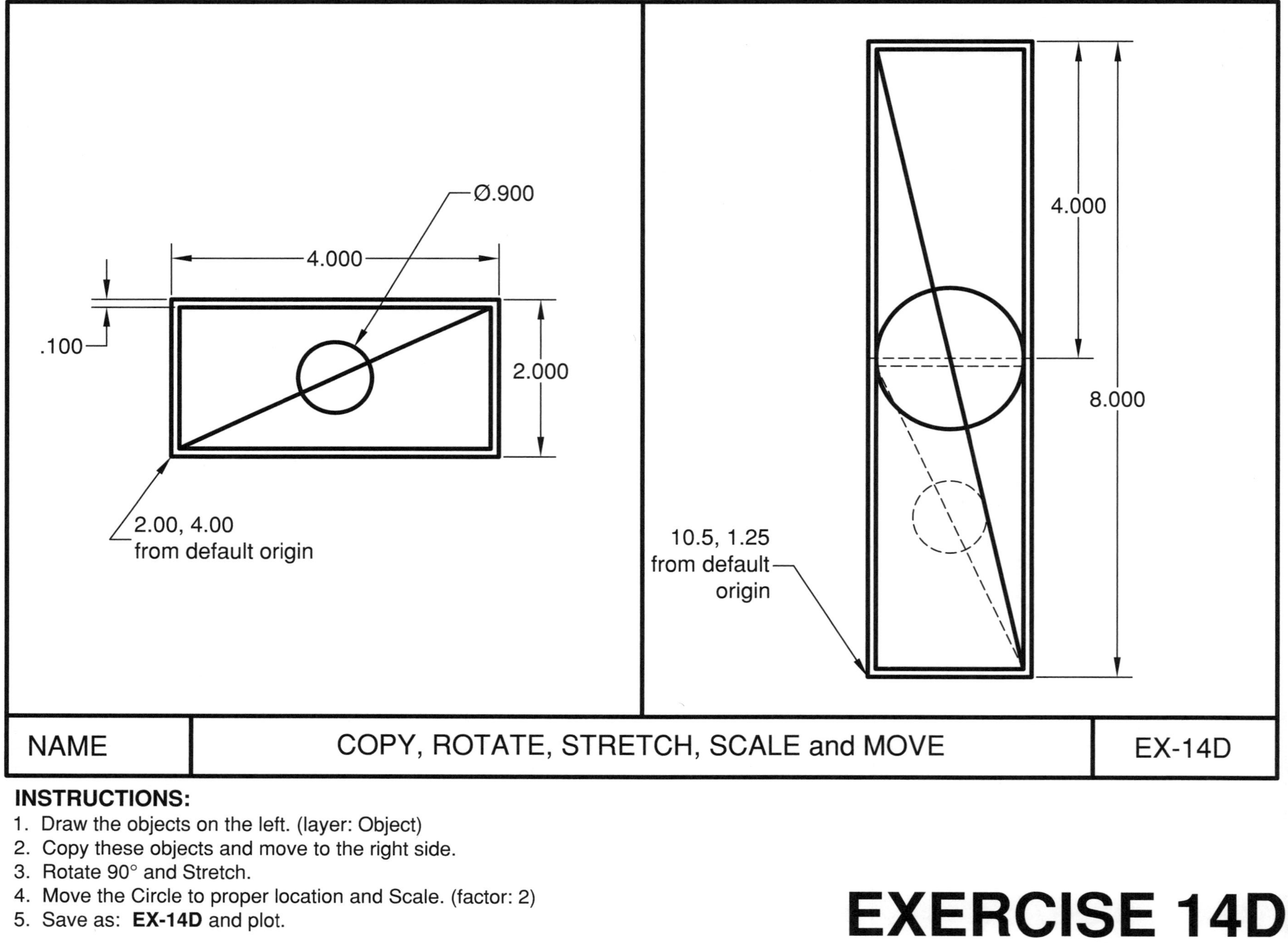

EXERCISE 14D

INSTRUCTIONS:

1. Draw the objects on the left. (layer: Object)
2. Copy these objects and move to the right side.
3. Rotate 90° and Stretch.
4. Move the Circle to proper location and Scale. (factor: 2)
5. Save as: **EX-14D** and plot.

LEARNING OBJECTIVES

After completing this lesson, you will be able to:

1. Cross hatch a section view.
2. Add Gradient filled areas.
3. Solidly fill an area.
4. Make changes to cross hatch already in the drawing.

LESSON 15

HATCH

The **BHATCH** command is used to create hatch lines for section views or filling areas with specific patterns.

To draw **hatch** you must start with a closed boundary. A closed boundary is an area completely enclosed by objects. A rectangle would be a closed boundary. You simply pick inside the closed boundary. BHATCH locates the area and automatically creates a temporary polyline around the outline of the hatch area. After the hatch lines are drawn in the area, the temporary polyline is automatically deleted. (Polylines are discussed in Lesson 23)

A Hatch set is one object. If you explode it, it will return to many objects.

1. Select the BHATCH command using one of the following:

TYPE = BH
PULLDOWN = DRAW / HATCH
TOOLBAR = DRAW

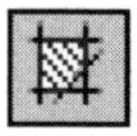

The following dialog box appears:

Gradient tab not available in version "LT"

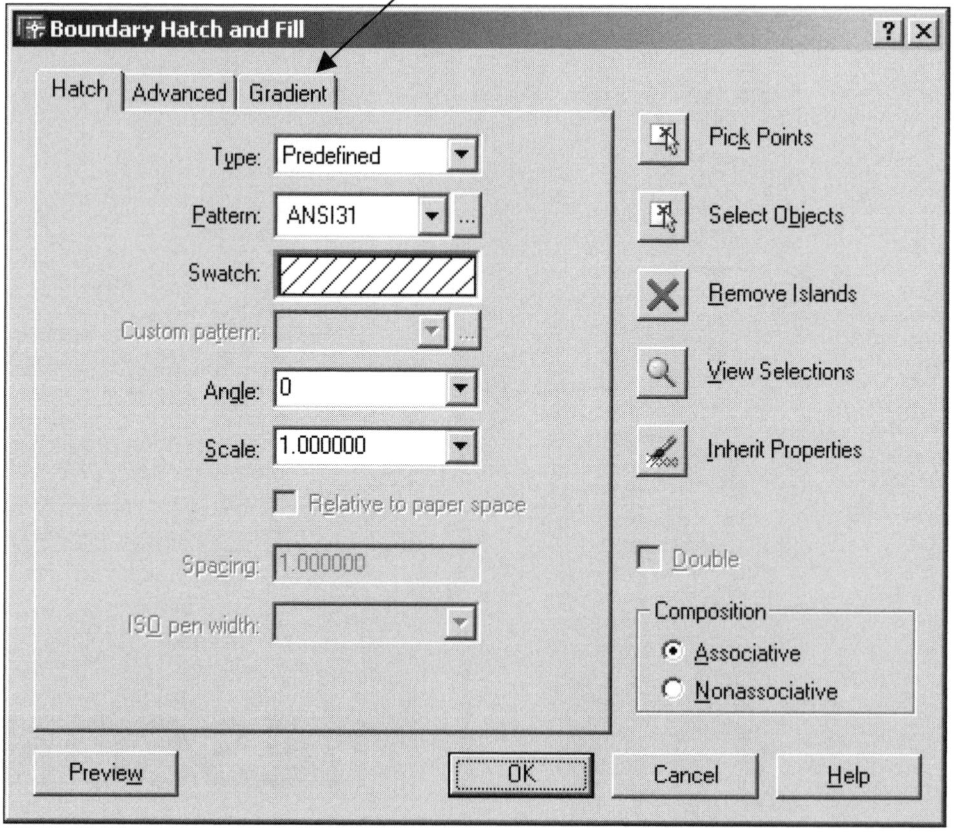

HATCH (continued)

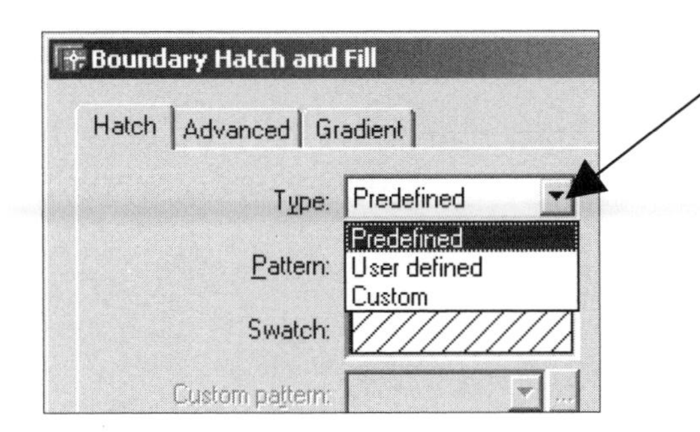

2. Select the hatch “TYPE”

Select one of the following:
PREDEFINED, USER DEFINED or CUSTOM
(Descriptions of each are listed below.)

a. ***PREDEFINED***
AutoCAD has many predefined hatch patterns. These patterns are stored in the acad.pat and acadiso.pat files. (You may also purchase patterns from other software companies.)

> **Note:**Using Hatch patterns will greatly increase the size of the drawing file. So use them conservatively.

To select a pattern by name, click on the “Pattern” down arrow. A drop-down list of available patterns will appear.

To select a pattern by appearance, click on the (…) button. This will display the “Hatch Pattern Palette” dialog box. (Examples on page 15-6)

Boundary Hatch and Fill
Hatch | Advanced | Gradient
Type: Predefined
Pattern: ANSI31
Swatch:
Custom pattern:
Angle: 0
Scale: 1.000000
Relative to paper spa
Spacing: 1.000000
ISO pen width:
Preview

Predefined Pattern Properties

Pattern
This box displays the name of the Pattern you selected.

Swatch
The selected pattern is displayed here.

Angle
This determines the rotation angle of the pattern. A pre-designed pattern has a default angle of 0. If you change this Angle it will rotate the pattern relative to it’s original design.

Scale
The value in this box is the scale factor. A good starting point would be to enter the scale factor of the drawing. Such as: If the drawing will be plotted at ¼” = 1, “4” is the scale factor. (Drawing Scale factors will be discussed in Lesson 27)

HATCH (continued

b. *USER DEFINED*
This selection allows you to simply draw continuous lines. (No special pattern) You specify the Angle and the Spacing between the lines. (This selection does not increase the size of the drawing file like Predefined)

User-defined Pattern Properties

Swatch
A sample of the angle and spacing settings is displayed here.

Angle
Specify the actual angle of the hatch lines.
(0 to 180)

Spacing
Specify the actual distance between each hatch line.

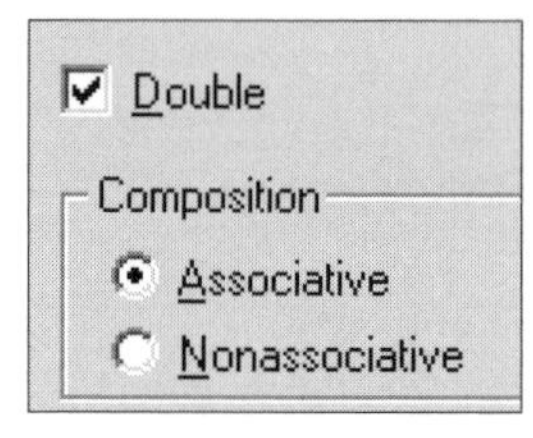

Double
If this box is checked, the hatch set will be drawn first at the angle specified, then a second hatch set will be drawn rotated 90 degrees relative to the first hatch set, creating a criss-cross affect.
(This option is only available when using "User Defined")

c. *CUSTOM*
See the AutoCAD Customization Guide for information on creating and saving custom hatch patterns.

3. Select the Composition Associative or Non Associative.

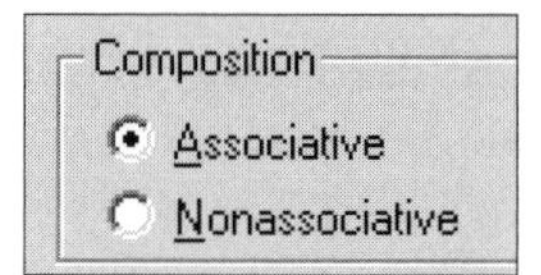

a. Associative: The hatch set is one entity and if the boundary size is changed the hatch will automatically change to the new boundary shape.

b. Non Associative: The hatch set is exploded into multiple entities and if the boundary shape is changed the hatch set will not change.

HATCH (continued)

4. Select the Area you want to Hatch using "Pick Points or Select Objects.

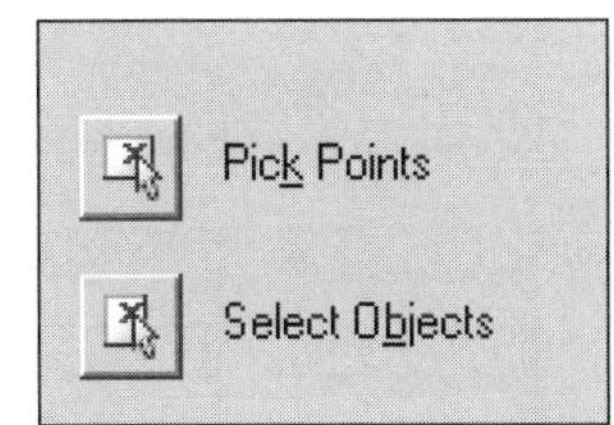

a. PICK POINTS

Select the **PICK POINTS** box then select a point inside the area you want to hatch. A boundary will automatically be determined.

b. SELECT OBJECTS

Select the boundary by selecting the object(s). The objects must form a closed shape with no gaps or overlaps.

5. Preview the Hatch.

a. After you have selected the boundary, press the right mouse button and select "**Preview**" from the short cut menu. This option allows you to preview the hatch set before it is actually applied to the drawing.

b. If the preview is not what you expected, press the **ESC** key, make the changes and preview again.

c. When you are satisfied, **right click** or press **<enter>** to accept.

Note: It is always a good idea to take the extra time to preview the hatch. It will actually save you time in the long run.

AutoCAD Hatch Patterns

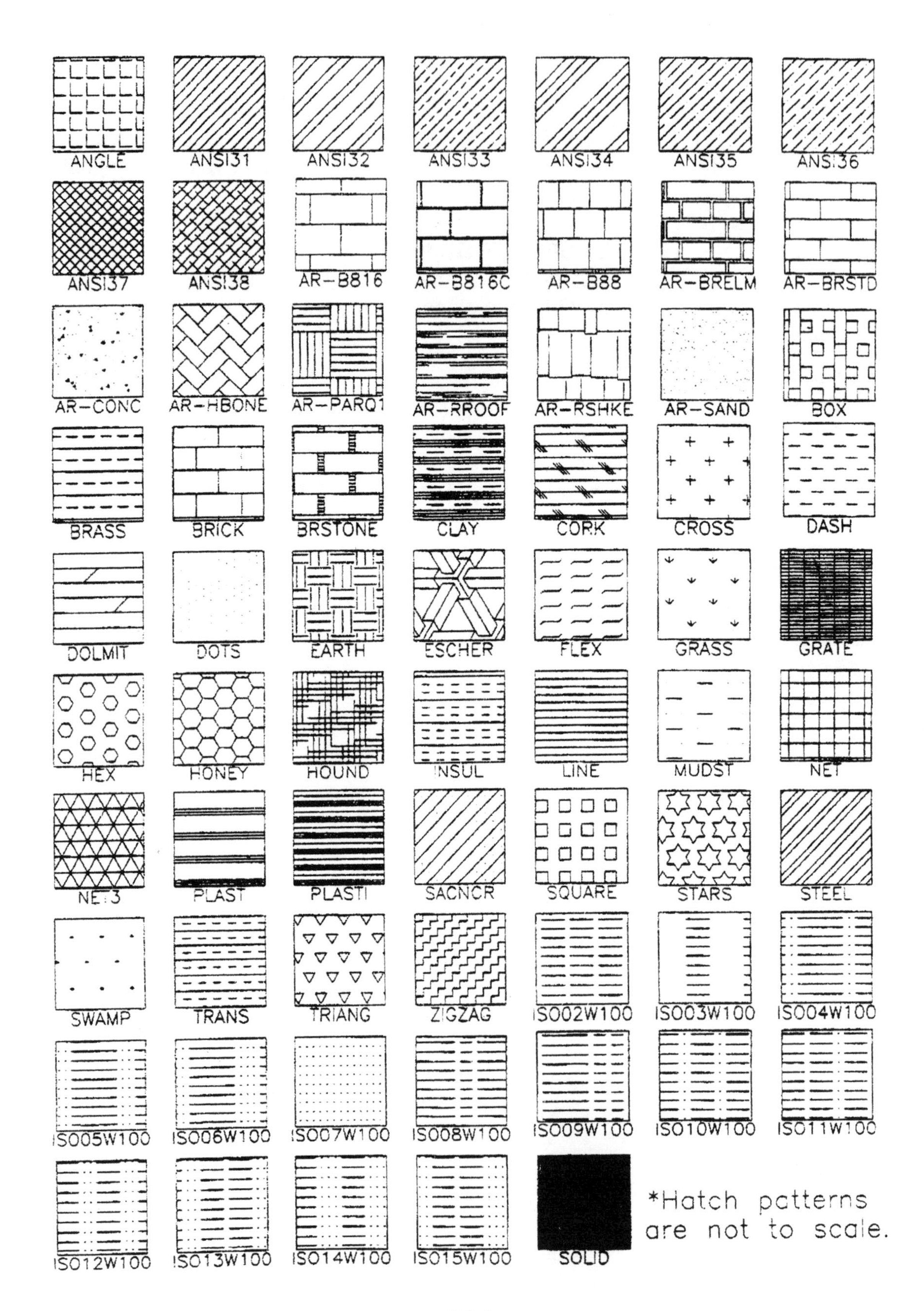

GRADIENT FILLS (not available in the LT version)

Gradients are fills that gradually change from dark to light or from one color to another. Gradient fills can be used to enhance presentation drawings, giving the appearance of light reflecting on an object, or creating interesting backgrounds for illustrations.

Gradients are definitely fun to experiment with but you will have to practice a lot to achieve complete control.

1. Select the BHATCH command using one of the following:

TYPE = BH
PULLDOWN = DRAW / HATCH
TOOLBAR = DRAW

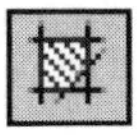

The following dialog box appears:

2. Select the Gradient tab.
Note: Gradient tab not available in version "LT"

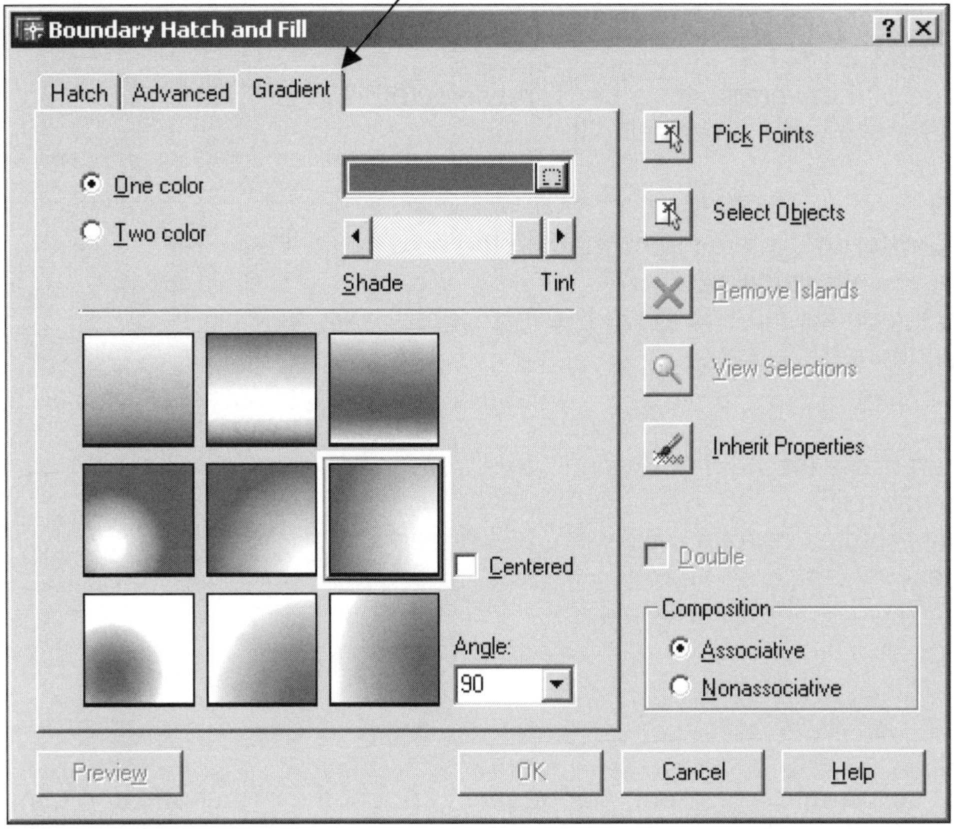

Boundary Hatch and Fill
Hatch
Advanced
Gradient
One color
Two color
Shade
Tint
Centered
Angle:
90
Pick Points
Select Objects
Remove Islands
View Selections
Inherit Properties
Double
Composition
Associative
Nonassociative
Preview
OK
Cancel
Help

2. Select the Gradient tab.
3. Select the area for the gradient fill. (Pick Points or Select Objects)
4. Choose the Gradient settings. (Refer to page 15-8)
5. Preview
6. Accept or make changes.

GRADIENT FILLS continued...

ONE COLOR

Click the ... button to the right of the color swatch to open the Select Color dialog box.
Choose the color you want.

Use the "**Shade and Tint**" slider to choose the gradient range from lighter to darker.

TWO COLOR

Click each ... button to choose a color.
When you choose two colors the transition is both from light to dark and from the first color to the second.

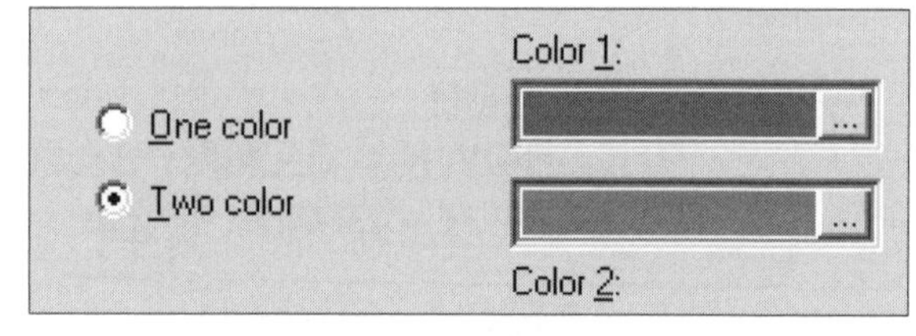

9 GRADIENT STYLES

Select one of the 9 gradient styles. (The selected style will have a white box around it.)

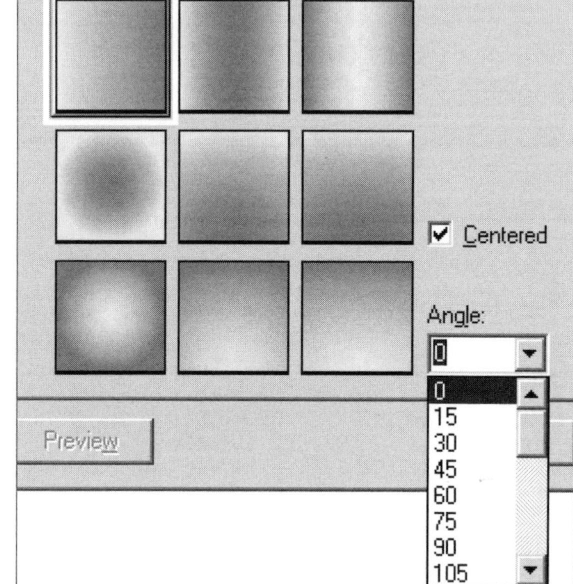
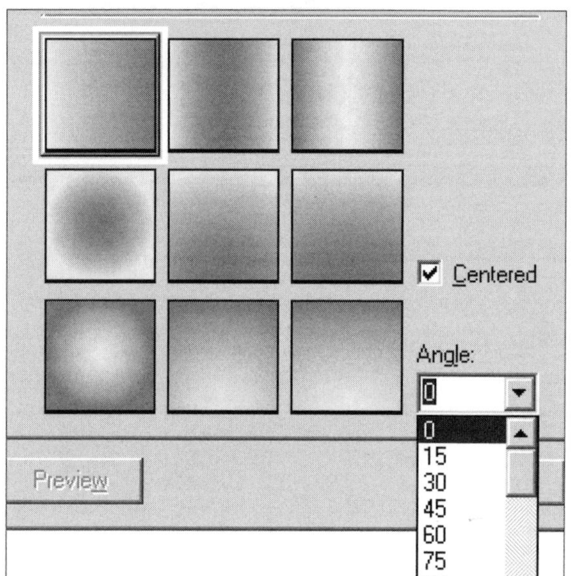

CENTER

Select "**Centered**", with a check mark in the box, to create a symmetrical fill.
Remove the check mark to move the "highlight" up and to the left.

ANGLE

Specify an angle for the "highlighted" area from the drop down list.
(The angle rotates counter clockwise, 0 is upper left corner)

NOTES:

1. When you create a gradient, it appears in front of the object and sometimes obscures the objects outline. To move the object's outline to the front, select:
Tools / Display Order / Bring to Front and select the object's outline.

2. It is good drawing management to always place gradient fill on it's own layer.

3. You may also make gradient fill appear or disappear with the **FILL** command.
Type "**FILL**" **<enter>** on the command line. Then type "**ON**" or "**OFF**".
Select **VIEW / REGEN** to see the effect.

EDITING HATCH

HATCHEDIT allows you to edit an existing hatch pattern in the drawing.
You simply select the hatch pattern that you want to change and the hatch dialog box will appear. Make the changes, preview and accept.

1. Double click on the Hatch that you wish to edit or select the Hatch Edit command using one of the following and then select the Hatch to edit:

TYPE = HE
PULLDOWN = MODIFY / OBJECT / HATCH
TOOLBAR = MODIFY II

2. Make the changes to the settings and preview. Your changes should be displayed.

3. Right click to accept or ESC to make additional changes.

Note:
If the hatch is "Associative" and you change the size of the object (hatch boundary), the hatch will change with the area.

EXERCISE 15A

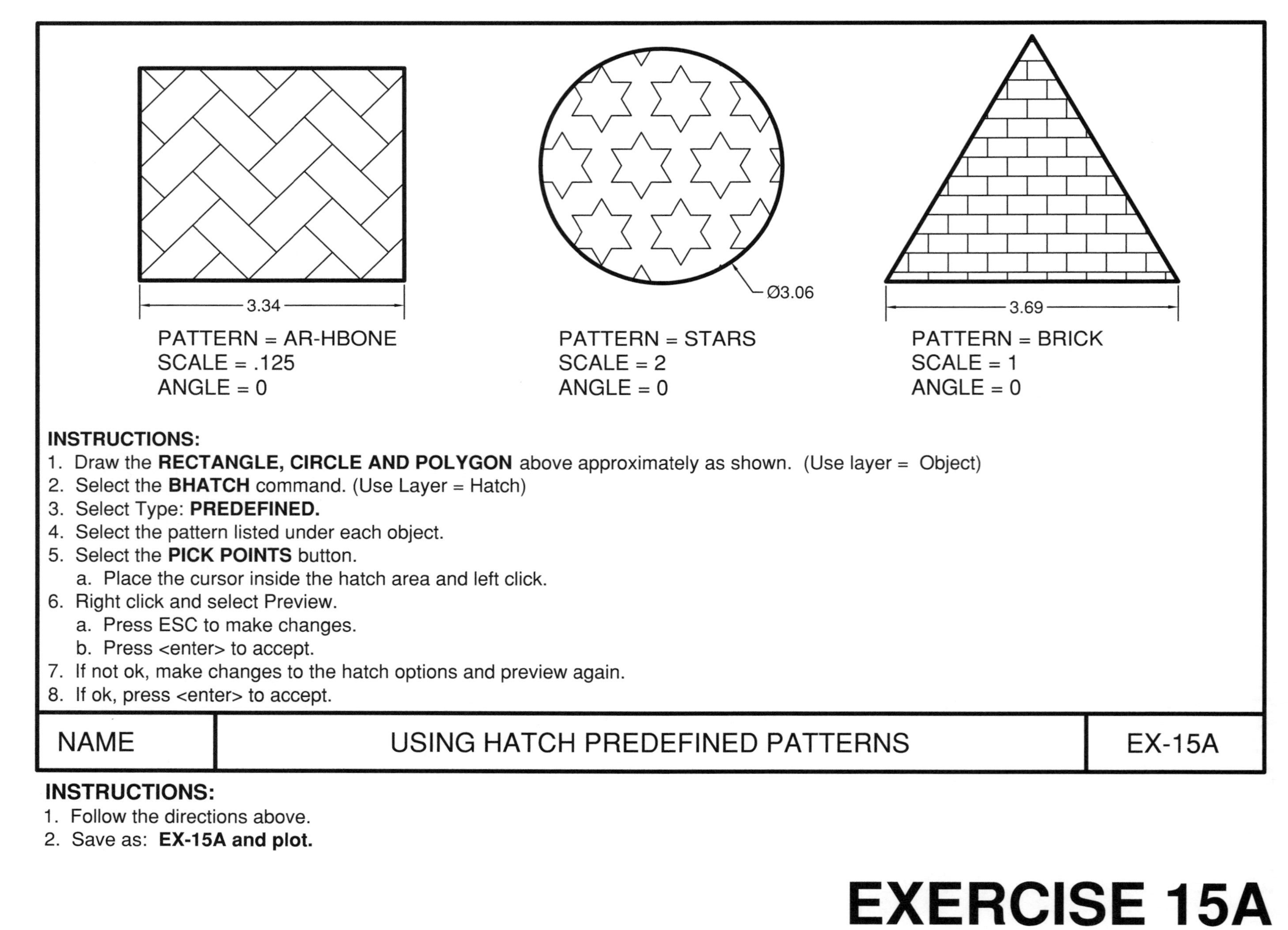

INSTRUCTIONS:

1. Draw the **RECTANGLE, CIRCLE AND POLYGON** above approximately as shown. (Use layer = Object)
2. Select the **BHATCH** command. (Use Layer = Hatch)
3. Select Type: **PREDEFINED.**
4. Select the pattern listed under each object.
5. Select the **PICK POINTS** button.
   a. Place the cursor inside the hatch area and left click.
6. Right click and select Preview.
   a. Press ESC to make changes.
   b. Press <enter> to accept.
7. If not ok, make changes to the hatch options and preview again.
8. If ok, press <enter> to accept.

INSTRUCTIONS:

1. Follow the directions above.
2. Save as: **EX-15A and plot.**

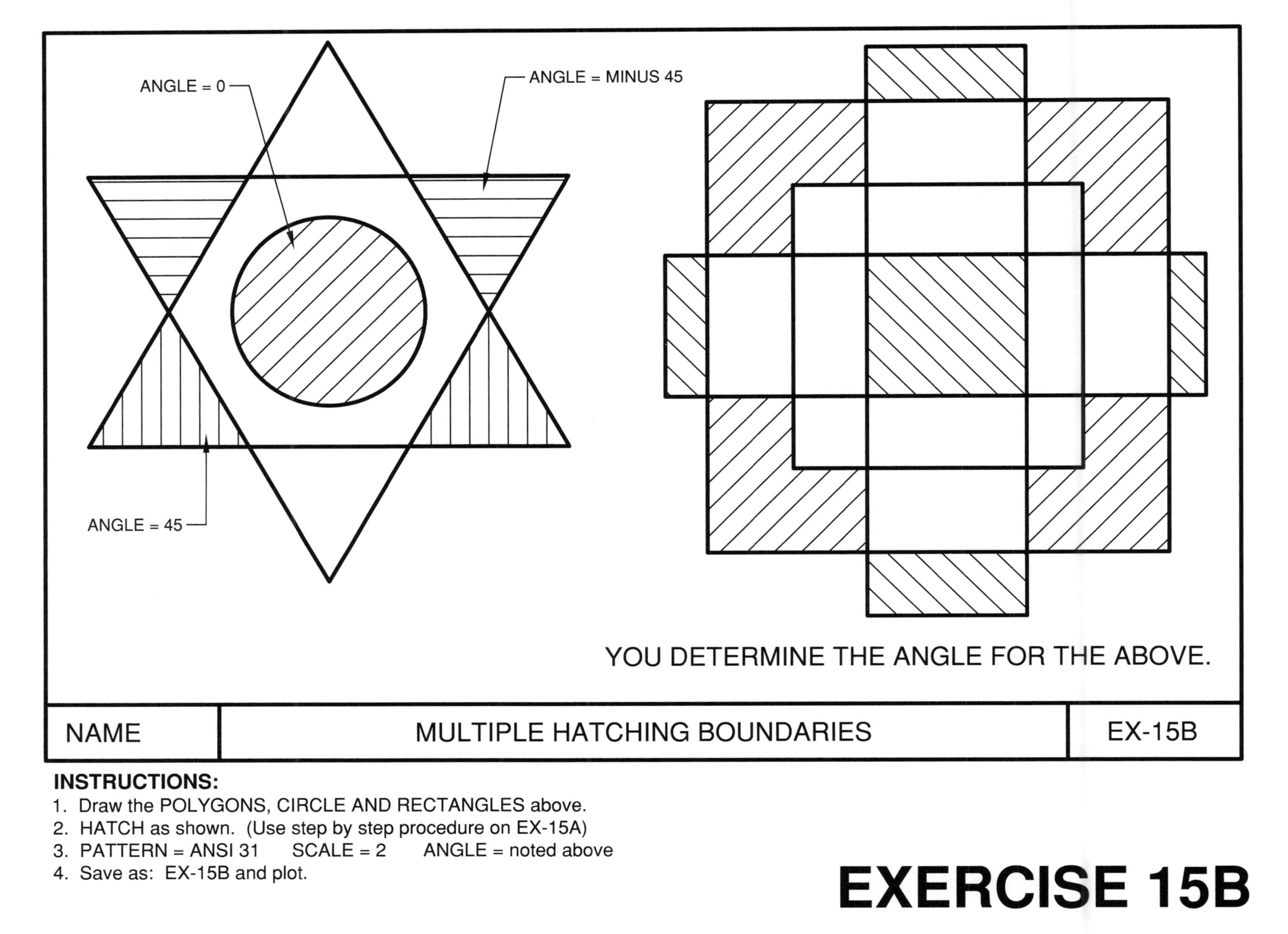

INSTRUCTIONS:

1. Draw the POLYGONS, CIRCLE AND RECTANGLES above.
2. HATCH as shown. (Use step by step procedure on EX-15A)
3. PATTERN = ANSI 31 SCALE = 2 ANGLE = noted above
4. Save as: EX-15B and plot.

EXERCISE 15B

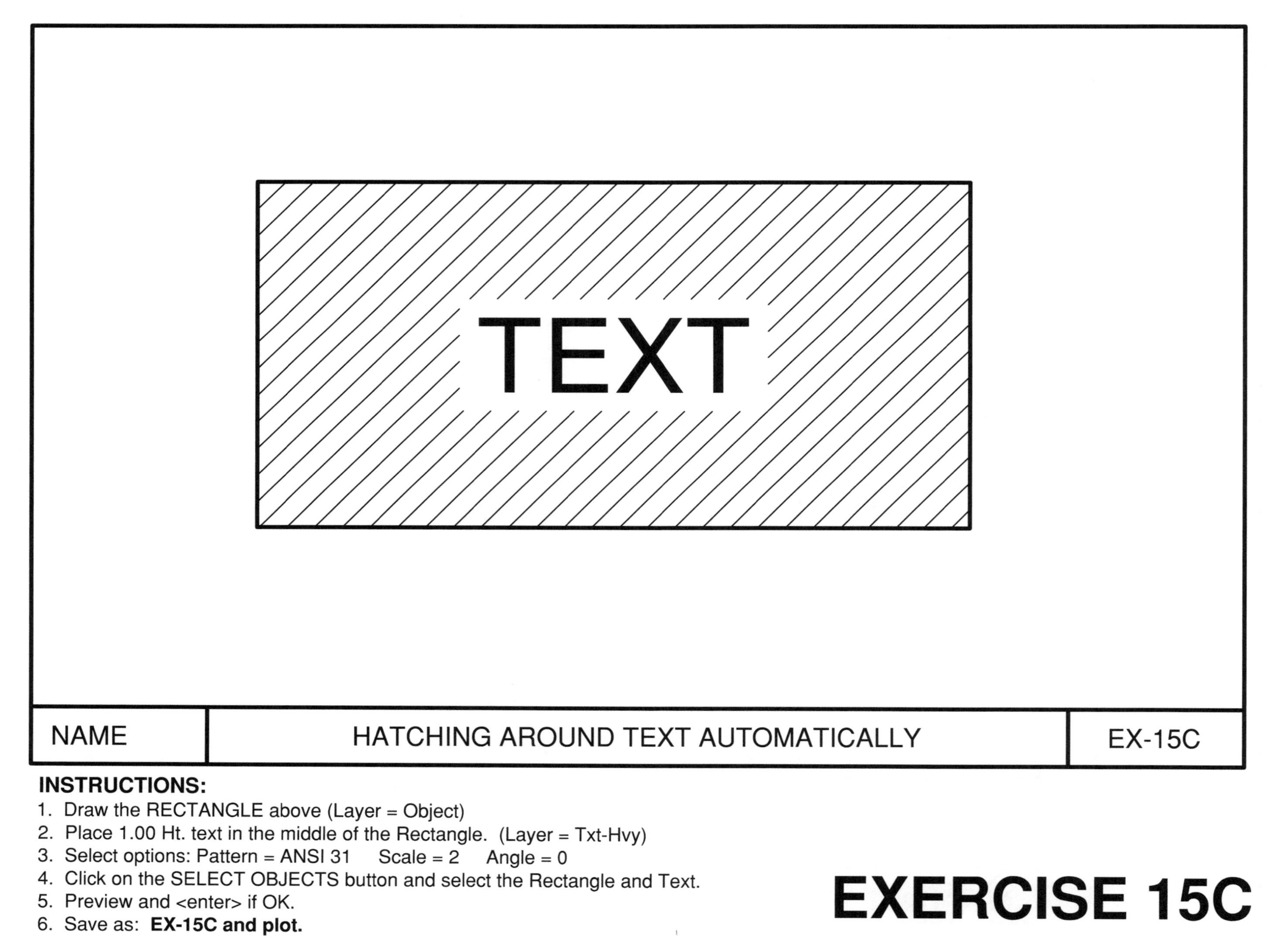

INSTRUCTIONS:

1. Draw the RECTANGLE above (Layer = Object)
2. Place 1.00 Ht. text in the middle of the Rectangle. (Layer = Txt-Hvy)
3. Select options: Pattern = ANSI 31   Scale = 2   Angle = 0
4. Click on the SELECT OBJECTS button and select the Rectangle and Text.
5. Preview and <enter> if OK.
6. Save as: **EX-15C and plot.**

EXERCISE 15C

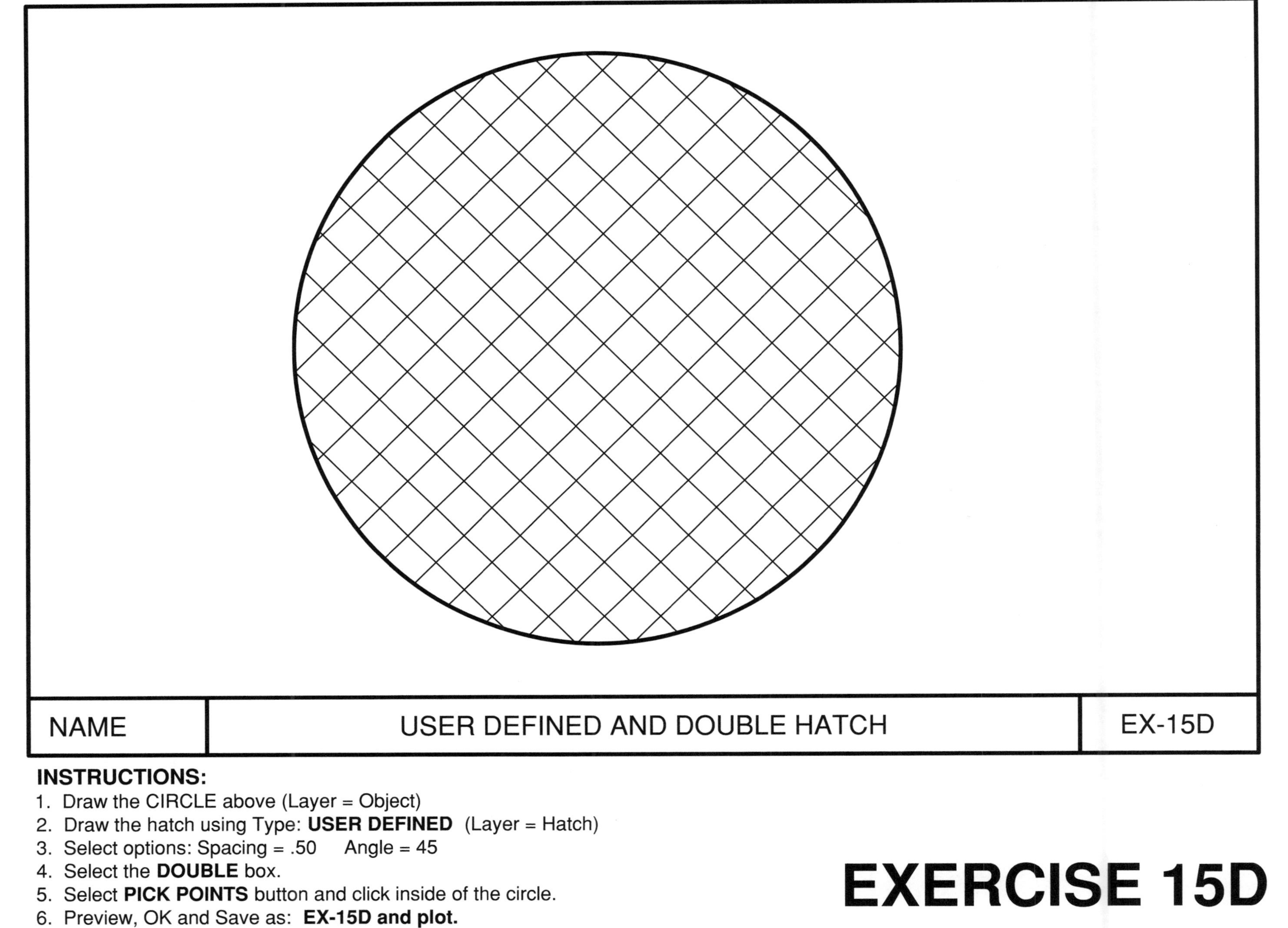

INSTRUCTIONS:

1. Draw the CIRCLE above (Layer = Object)
2. Draw the hatch using Type: **USER DEFINED** (Layer = Hatch)
3. Select options: Spacing = .50 Angle = 45
4. Select the **DOUBLE** box.
5. Select **PICK POINTS** button and click inside of the circle.
6. Preview, OK and Save as: **EX-15D and plot.**

EXERCISE 15D

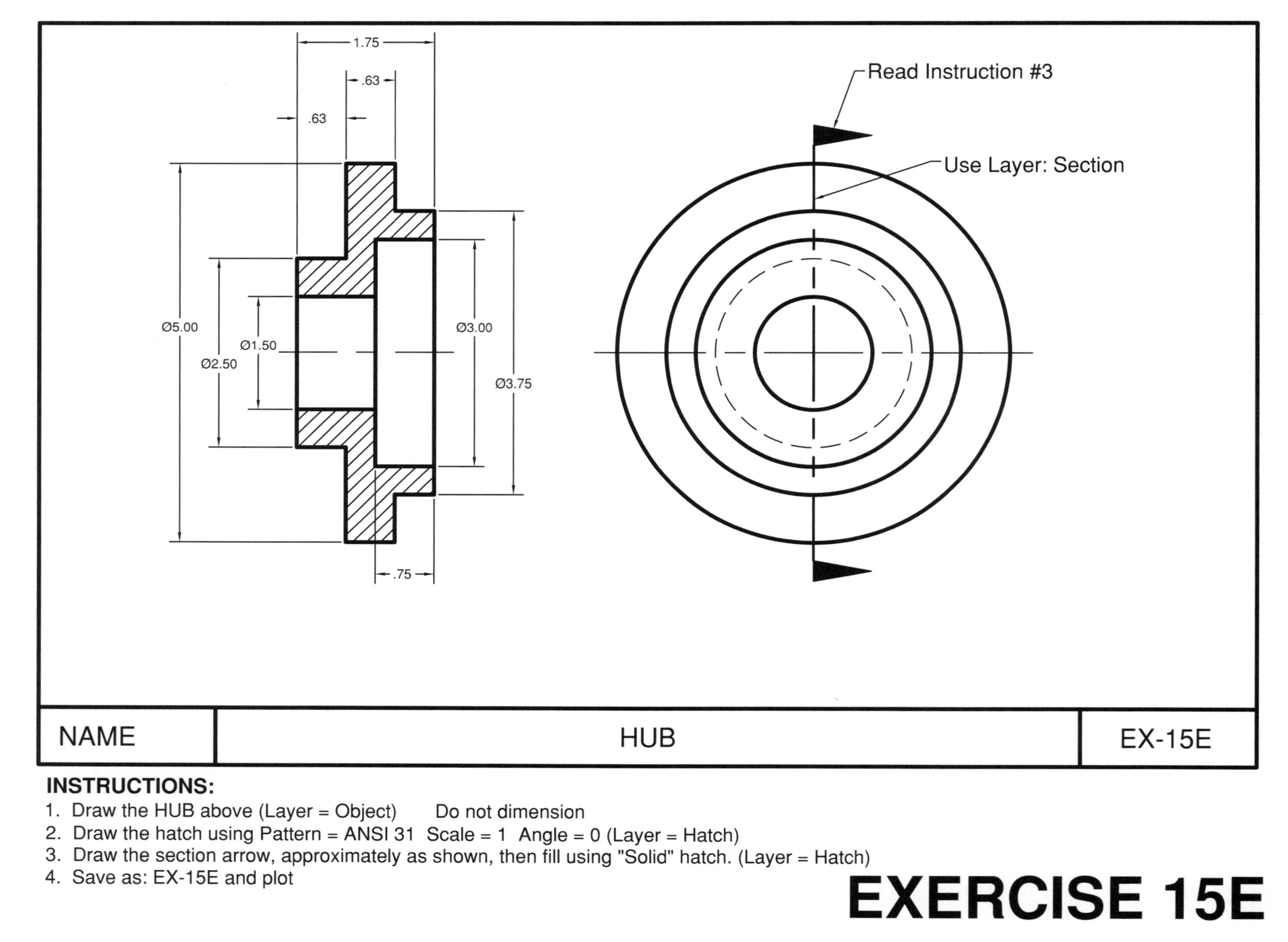

INSTRUCTIONS:

1. Draw the HUB above (Layer = Object) Do not dimension
2. Draw the hatch using Pattern = ANSI 31 Scale = 1 Angle = 0 (Layer = Hatch)
3. Draw the section arrow, approximately as shown, then fill using "Solid" hatch. (Layer = Hatch)
4. Save as: EX-15E and plot

EXERCISE 15E

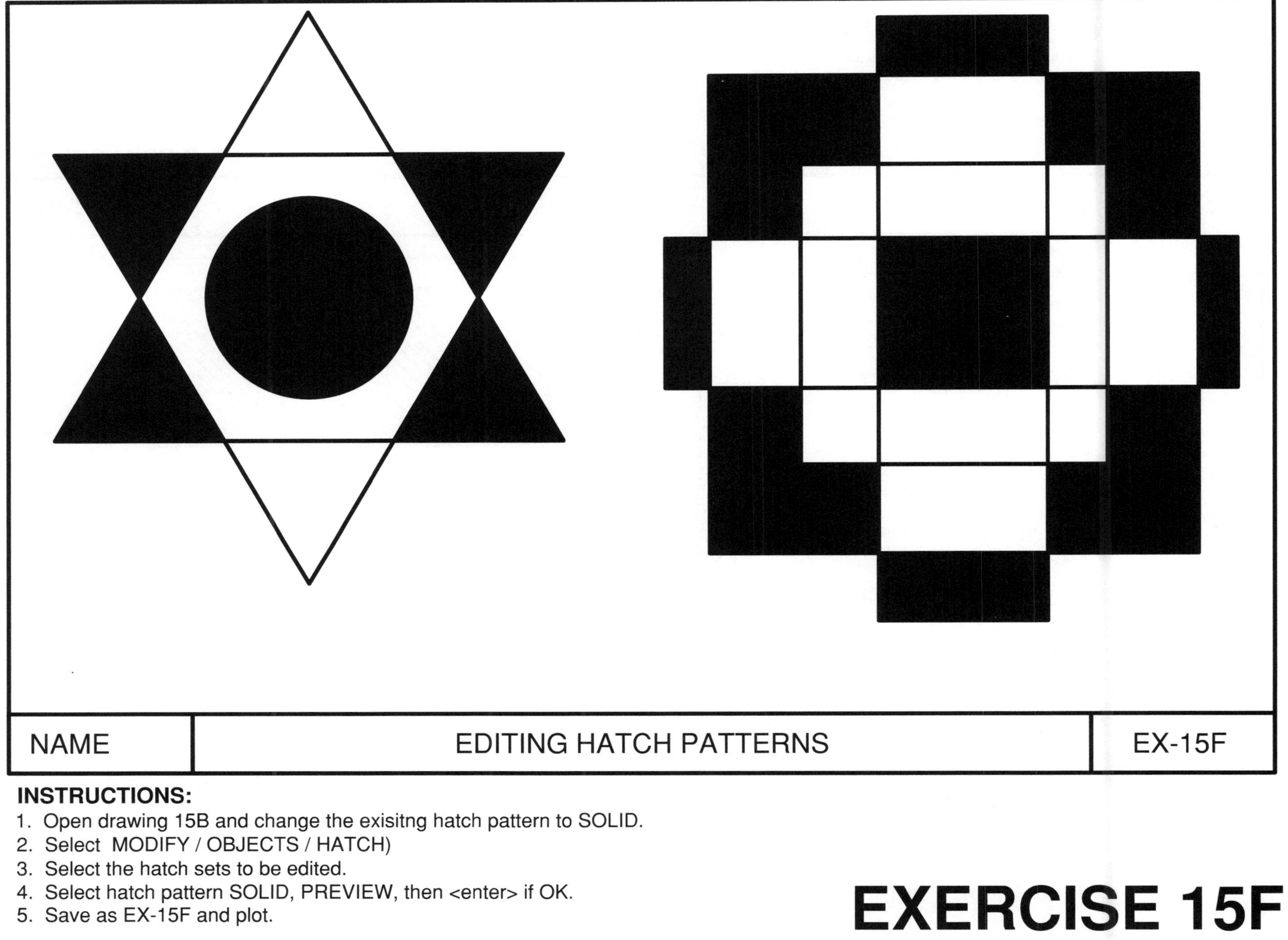

INSTRUCTIONS:

1. Open drawing 15B and change the exisitng hatch pattern to SOLID.
2. Select MODIFY / OBJECTS / HATCH)
3. Select the hatch sets to be edited.
4. Select hatch pattern SOLID, PREVIEW, then <enter> if OK.
5. Save as EX-15F and plot.

EXERCISE 15F

Step 1.

1. Draw the Rectangle below. (Layer: Object)

2. Hatch with Ansi 31, scale: 1

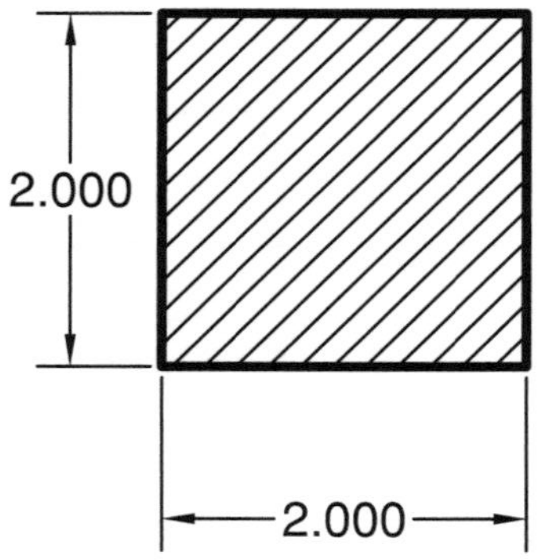

2.000

2.000

Step 2.

1. "Stretch" the Rectangle to a length of 5 inches.

Notice the Hatch automatically changed with the Rectangle.

How did that happen?? Because the Hatch is "Associative".

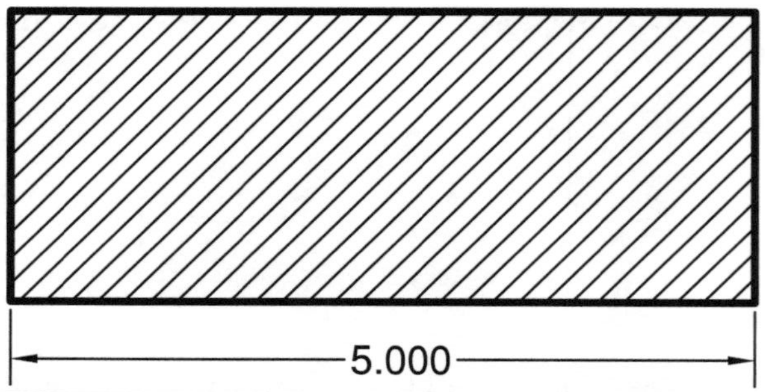

5.000

Step 1.

1. Draw the Rectangle below. (Layer: Object)

2. Hatch again but this time select "Nonassociative"

(Refer to page 15-4)

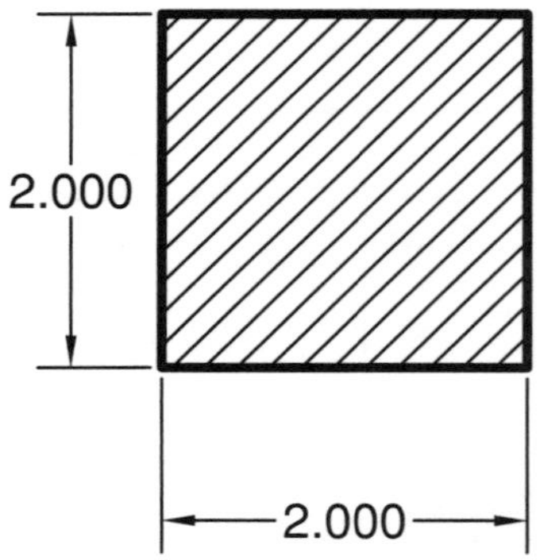

2.000

2.000

Step 2.

1. "Stretch" this Rectangle to a length of 5 inches.

What happened to the hatch? and Why?

NAME	ASSOCIATIVE HATCH	EX-15G

INSTRUCTIONS:

1. Follow the steps above.
2. Save as EX 15G do not Plot.

EXERCISE 15G

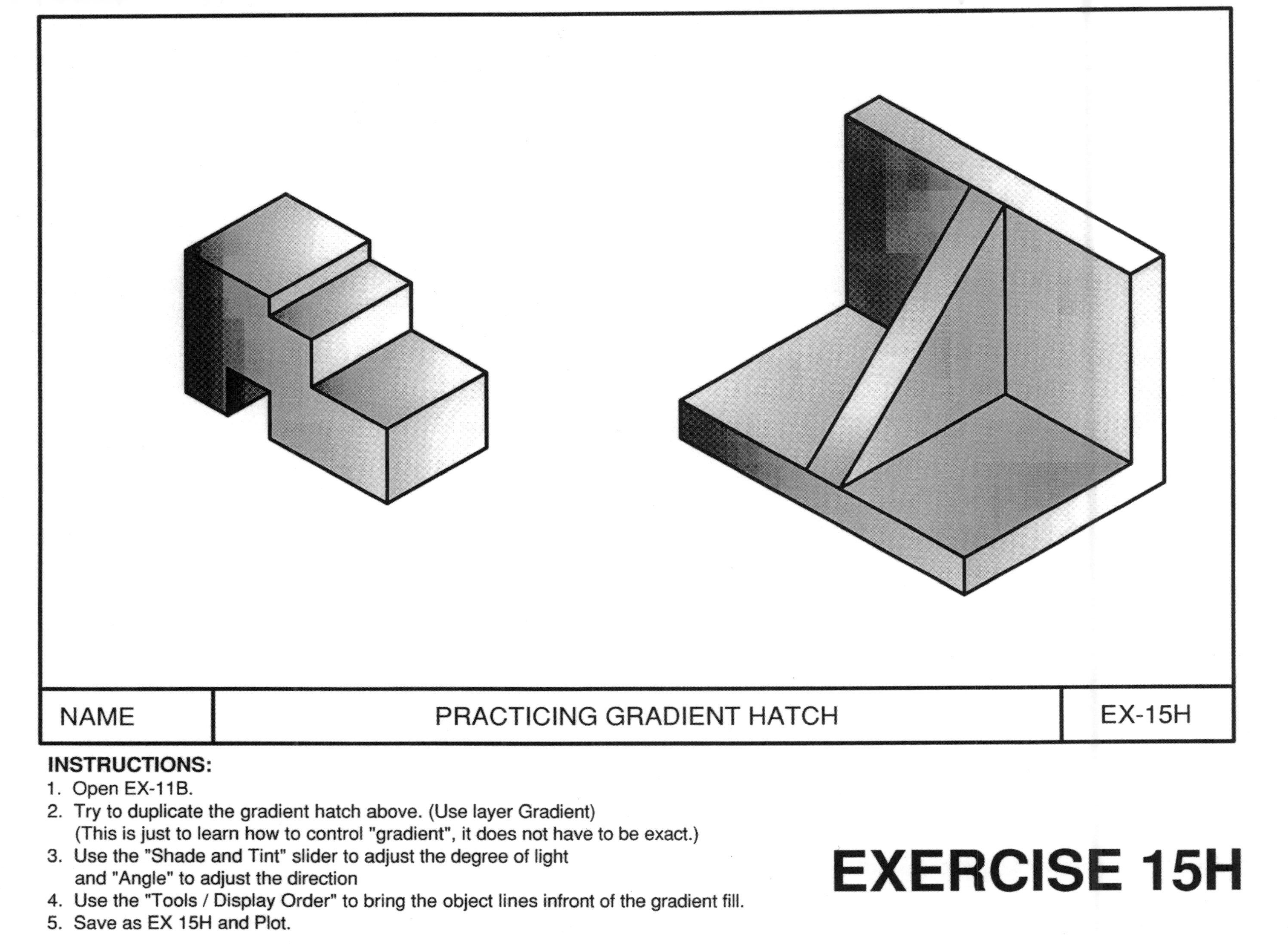

INSTRUCTIONS:

1. Open EX-11B.
2. Try to duplicate the gradient hatch above. (Use layer Gradient)
   (This is just to learn how to control "gradient", it does not have to be exact.)
3. Use the "Shade and Tint" slider to adjust the degree of light
   and "Angle" to adjust the direction
4. Use the "Tools / Display Order" to bring the object lines infront of the gradient fill.
5. Save as EX 15H and Plot.

EXERCISE 15H

NOTES:

LEARNING OBJECTIVES

After completing this lesson, you will be able to:

1. Understand the importance of True Associative dimensioning.
2. Use Grips.
3. Add Linear, Baseline and Continued dimensions to your drawing.
4. Control the appearance of dimensions.
5. Create a New Dimension Style.
6. Compare two Dimension Styles.

LESSON 16

DIMENSIONING

Dimensioning is basically easy, but as always, there are many options to learn. As a result, I have divided the dimensioning process into 5 lessons. (Lessons 16 through 20) So relax and just take it one lesson at a time.

In this Lesson you will learn how to create a dimension style and how to create horizontal and vertical dimensions. But first you need to understand about AutoCAD 2002's new true associative and trans-spatial dimensioning feature.

True Associative

True Associative Dimensioning means that the dimensions are actually attached to the objects that they dimension. If you move the object, the dimension will move with it. If you scale or stretch the object, the dimension text value will change also. (Note: It is not parametric. This means, you can not change the dimension text value and expect the object to change.)

True Associative Dimensioning can be set to ON, OFF or Exploded.

I strongly suggest that you keep true associative dimensioning on. It is truly a very powerful feature and will make editing the objects and dimensions much easier.

On = Dimensions are truly associative. The dimensions are associated to the objects and will change if the object is changed. (Setting 2)

Off = Dimensions are non-associative. The dimensions are not associated to the objects and will not change if the object changes. (Setting 1)

Exploded = Dimensions are not associated to the objects and are totally separate objects. (arrows, lines and text) (Setting 0)

How to turn Associative dimensioning On or OFF.

There are 2 methods:

Method 1.
On the command line type: ***dimassoc <enter>,***
then enter the number ***2 for On, 1 for Off*** or ***0 for Exploded*** **<enter>**

Method 2.
Select **Tools / Options / User Preferences tab.**

Associative Dimensioning
Make new dimensions associative

Checked box = On

Unchecked box = Off

Note: You can not set Exploded here.

The Default setting is "2" or "ON" but check to make sure

When you first enter AutoCAD, the True Associative dimensioning feature is "ON". This setting is saved with each individual drawing. It is not a system setting for all drawings. This means, when you open a drawing, it is important to check the **dimassoc** command to verify that True Associative dimensioning is ON. Especially if you open a drawing created with an AutoCAD release previous to 2002.

How to Re-associate a dimension

If a dimension was created with a previous version of AutoCAD or True Associative dimensioning was turned off, you may use the **dimreassociate** command to change the non-associative dimensions into associative dimensions.
(You may use the **Tools / Inquiry / List** command to determine whether a dimension is associative or non-associative.)

1. Select **Dimension / Reassociate Dimensions**. (No icon available)

   Command: _dimreassociate
   Select dimensions to reassociate ...

2. Select objects: ***select the dimension to be reassociated***

3. Select objects: ***select more dimensions or <enter> to stop***

4. Specify first extension line origin or [Select object] <next>: ***an "X" will mark the first extension line; use object snap to select the exact location of the extension line point (If the "X" has a box around it, just press <enter>***

5. Specify second extension line origin <next>: ***the "X" will move to another location. Use object snap or <enter>***

6. ***Continue until all extension line points are selected.***

Regenerating Associative dimensions

Sometimes after panning and zooming, the associative dimensions seem to be floating or not following the object. The **DIMREGEN** command will move the associative dimensions back into their correct location.

*You must type **dimregen <enter>** on the command line. Sorry, AutoCAD did not provide an icon or a pull-down menu.*

GRIPS

Grips are little boxes that appear if you select an object when no command is in use. Grips must be enabled by typing "grips" <enter> then 1 <enter> on the command line or selecting the "enable grips" box in the **OPTIONS / SELECTION** dialog box.

Grips can be used to quickly edit objects. You can move, copy, stretch, mirror, rotate, and scale objects using grips.

The following is a brief overview on how to use three of the most frequently used options. Grips have many more options and if you like the example below, you should research them further in the AutoCAD help menu.

1. Select the object (no command can be in use while using grips)
2. Select one of the **blue** grips. It will turn to "**red**". This indicates that it is "**hot**". The "**Hot**" grip is the **basepoint**.
3. The editing modes will be displayed on the command line. You may cycle through these modes by pressing the SPACEBAR or ENTER key or use the shortcut menu.
4. After editing you must press the ESC key to deactivate the grips on that object.

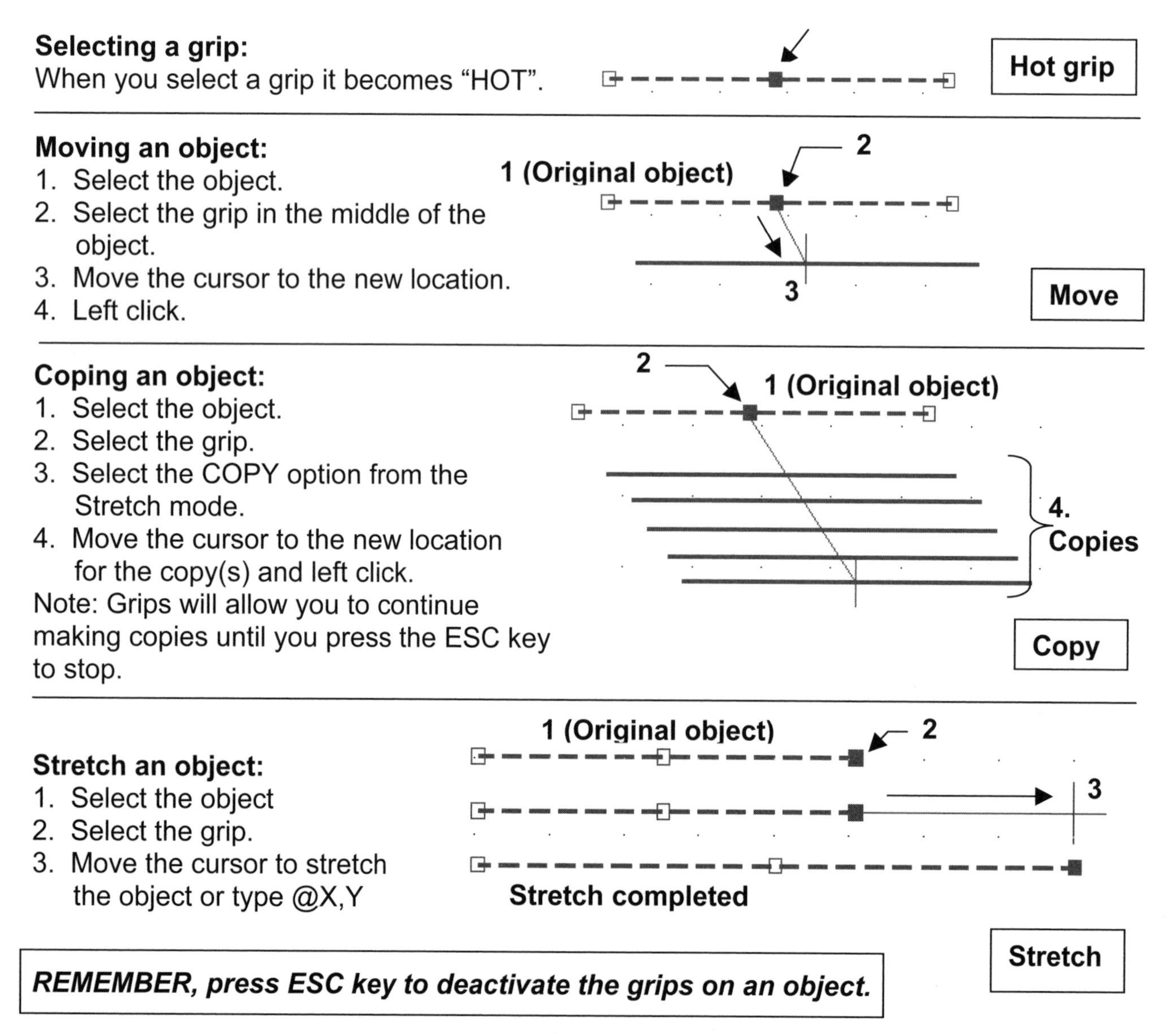

Selecting a grip:
When you select a grip it becomes "HOT".

Moving an object:
1. Select the object.
2. Select the grip in the middle of the object.
3. Move the cursor to the new location.
4. Left click.

Coping an object:
1. Select the object.
2. Select the grip.
3. Select the COPY option from the Stretch mode.
4. Move the cursor to the new location for the copy(s) and left click.

Note: Grips will allow you to continue making copies until you press the ESC key to stop.

Stretch an object:
1. Select the object
2. Select the grip.
3. Move the cursor to stretch the object or type @X,Y

REMEMBER, press ESC key to deactivate the grips on an object.

LINEAR DIMENSIONING 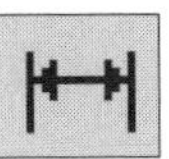

First open the dimension toolbar. Using the toolbar icons to select the dimension commands is the most efficient method. (Refer to page 2-6 to open toolbar)

Linear dimensioning allows you to create horizontal and vertical dimensions.

1. Select the **LINEAR** command using the icon or Dimension / Linear.
   Command:_Linear
2. Specify first extension line origin or <select object>: ***snap to first extension line origin (P1)***
3. Specify second extension line origin: ***snap to second extension line origin (P2)***
4. Specify dimension line location or [Mtext/Text/Angle/Horizontal/Vertical/Rotated]: ***select where you want the dimension line placed. (P3)***
   Dimension text = 2.000 (***the dimension text value will be displayed on the last line)***

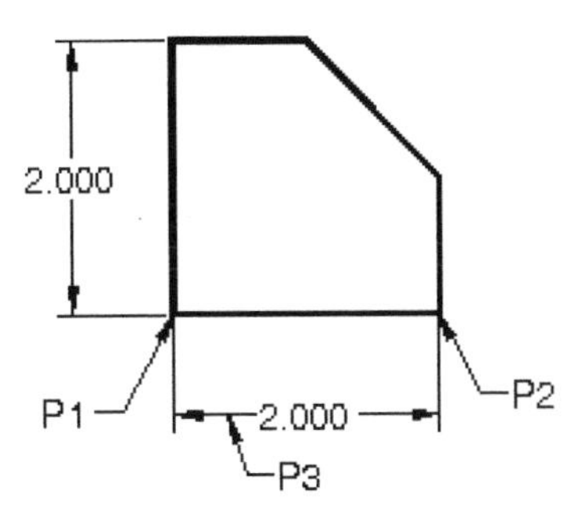

BASELINE DIMENSIONING

Baseline dimensioning allows you to establish a **baseline** for successive dimensions. The spacing between dimensions is automatic and should be set in dimension styles. A Baseline dimension must be used with an existing dimension. If you use Baseline dimensioning immediately after a Linear dimension, you do not have to specify the baseline origin.

1. Create a linear dimension first. (1.400 P1 and P2)
2. Select the **BASELINE** command using the icon or Dimension / Baseline.
   Command: _dimbaseline
3. Specify a second extension line origin or [Undo/Select] <Select>: ***snap to the second extension line origin (P3)***
   Dimension text = 2.588
4. Specify a second extension line origin or [Undo/Select] <Select>: ***snap to P4***
   Dimension text = 3.633
5. Specify a second extension line origin or [Undo/Select} <Select>: ***select <enter> twice to stop***

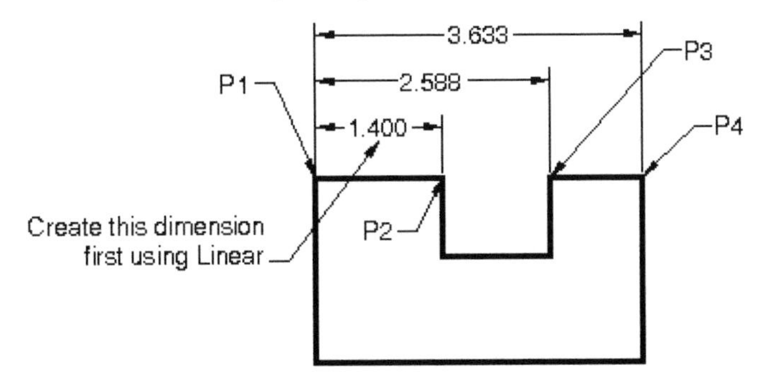

CONTINUE DIMENSIONING

Continue creates a series of dimensions in-line from an existing dimension. If you use the continue dimensioning immediately after a Linear dimension, you do not have to specify the continue extension origin.

1. Create a linear dimension first. (1.400 P1 and P2)
2. Select the Continue command using the icon or Dimension / Continue.
   Command: _dimcontinue
3. Specify a second extension line origin or [Undo/Select] <Select>: ***snap to the second extension line origin (P3)***
   Dimension text = 1.421
4. Specify a second extension line origin or [Undo/Select] <Select>: ***snap to the second extension line origin (P4)***
   Dimension text = 1.364
5. Specify a second extension line origin or [Undo/Select] <Select>: ***press <enter> twice to stop***

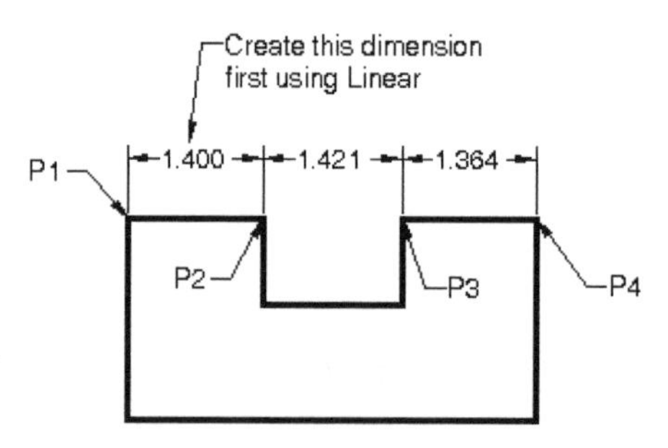

DIMENSION STYLES

Using the "Dimension Style Manager", you can change the appearance of the dimension features such as length of arrowheads, size of the dimension text, etc. There are over 70 different settings.
You can also Create New, Modify, Override and Compare Dimension Styles. All of these are simple by using the Dimension Style Manager described below.

Select the "Dimension Style Manager" using one of the following:

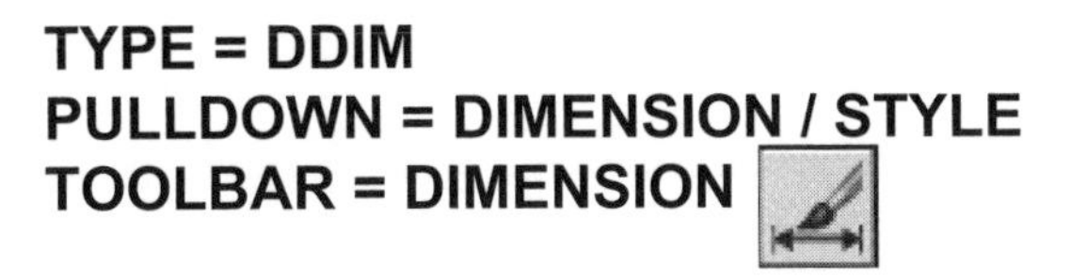

The following dialog box will appear.

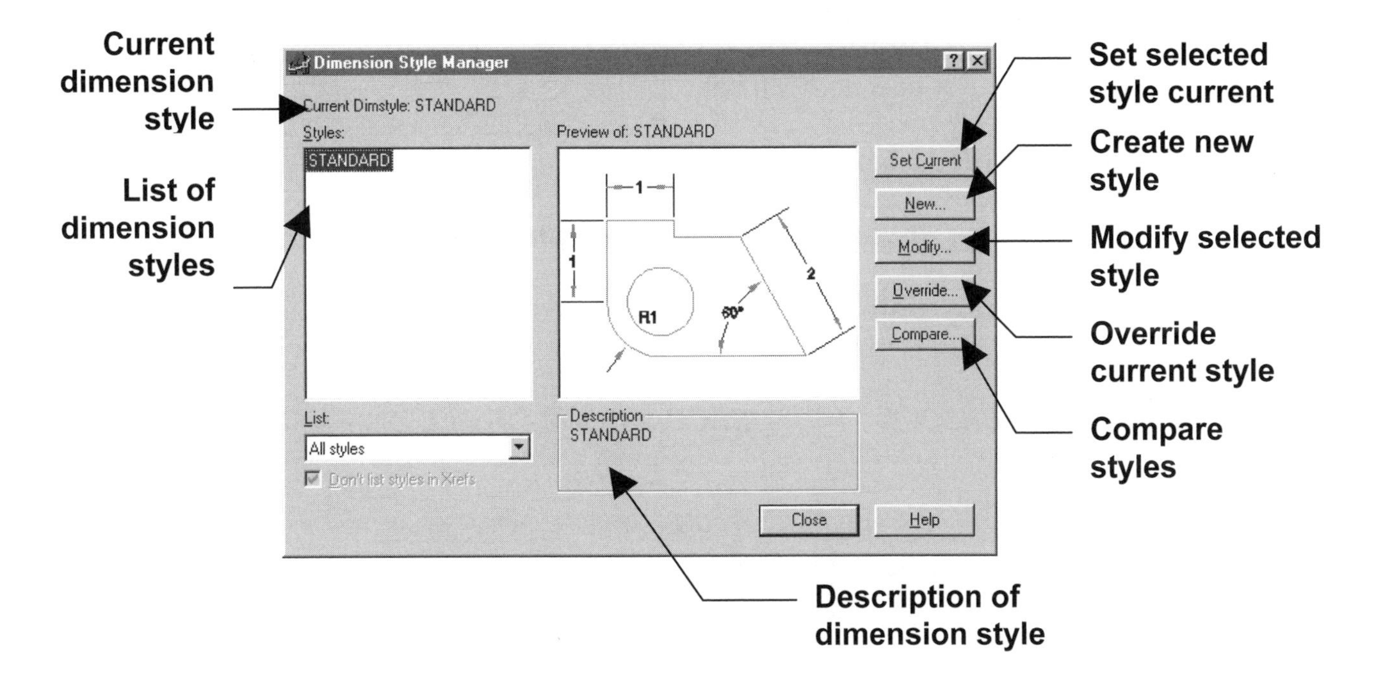

Set Current Select a style from the list of styles and select the **set current** button. (Only Standard is shown unless you have previously created other styles)

New Select this button to create a new style. When you select this button, the **Create New Dimension Style** dialog box is displayed. (See page 16-8)

Modify Selecting this button opens the **Modify dimension Style** dialog box which allows you to make changes to the current style. (See page 17-4)

Override An override is a temporary change to the current style. Selecting this button opens the Override Current Style dialog box. (See page17-5)

Compare Compares two styles. (See page 16-11)

CREATING A NEW DIMENSION STYLE

When creating a new style you must start with an existing style, such as Standard. Next, assign it a new name, make the desired changes and when you select the OK button, the new style will have been successfully created.

LET'S CREATE A NEW STYLE
A dimension style is a group of settings that has been saved under a name you assign.

1. Open your drawing **BSIZE** and set **DIMASSOC** to **2**
2. Select the **DIMENSION STYLE** command (Ref. page 16-7)
3. Select the **NEW** button. (Ref. page 16-7)
4. Enter **CLASS STYLE** in the "New Style Name" box.
5. **Start With**: we will start with the settings in the STANDARD style and then make some changes.
6. **Use For**: is for creating "family" dimension styles and will be discussed later. For now, leave it set to "All dimensions".
7. Select the **CONTINUE** box.

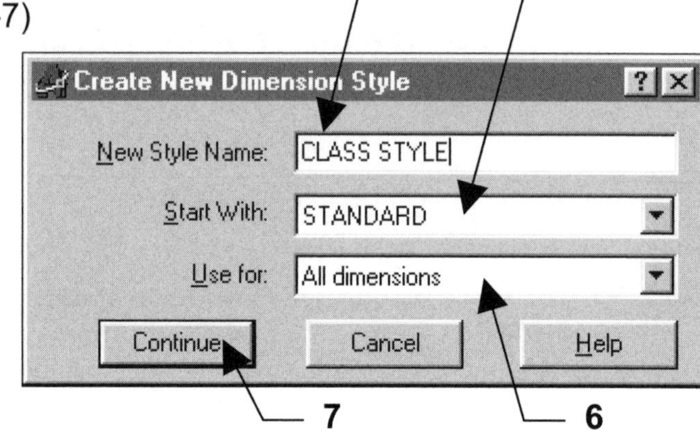

The "New Dimension Style: Class Style" dialog box will appear.

8. Select the **"Primary Units"** tab and make the changes shown below.

NOTE:
REFER TO "APPENDIX B" FOR DESCRIPTIONS OF ALL DIMENSION SETTINGS

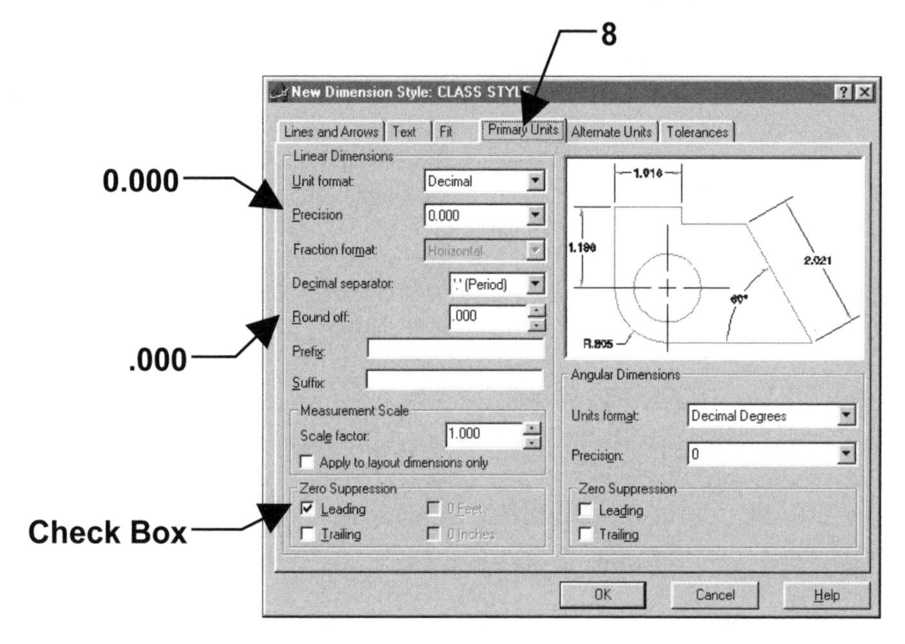

DO NOT SELECT THE OK BUTTON YET

9. Select the ***"Lines and Arrows"*** tab and make the changes shown below.

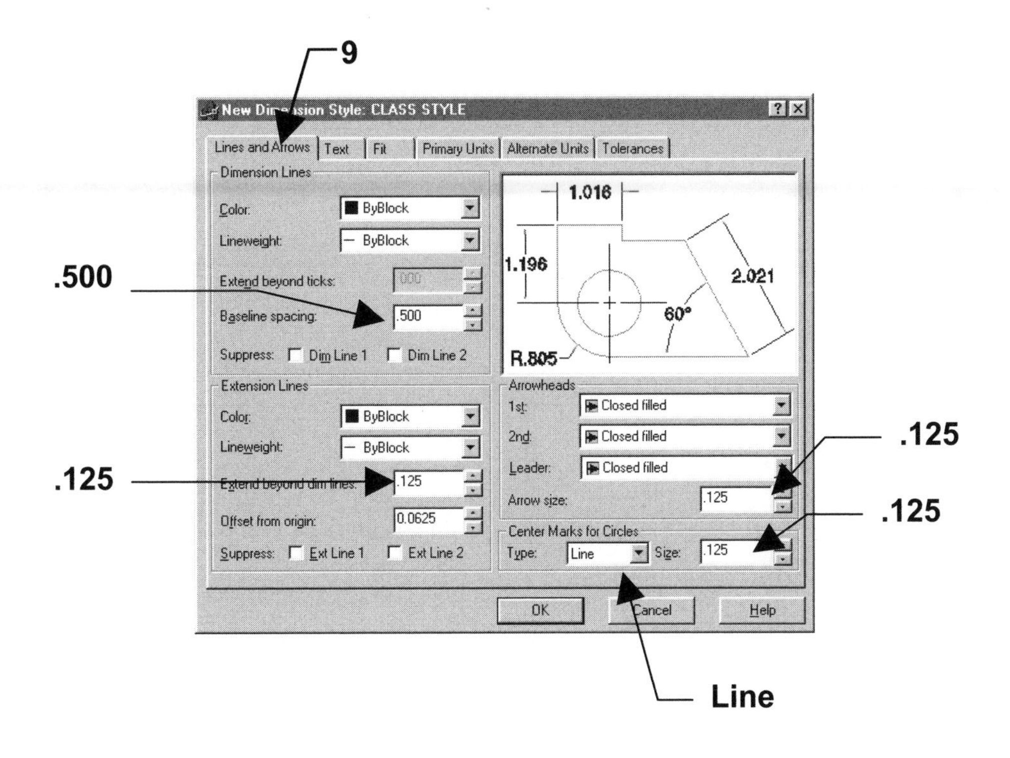

<u>DO NOT SELECT THE OK BUTTON YET.</u>

10. Select the **"Text"** tab and make the changes shown below.

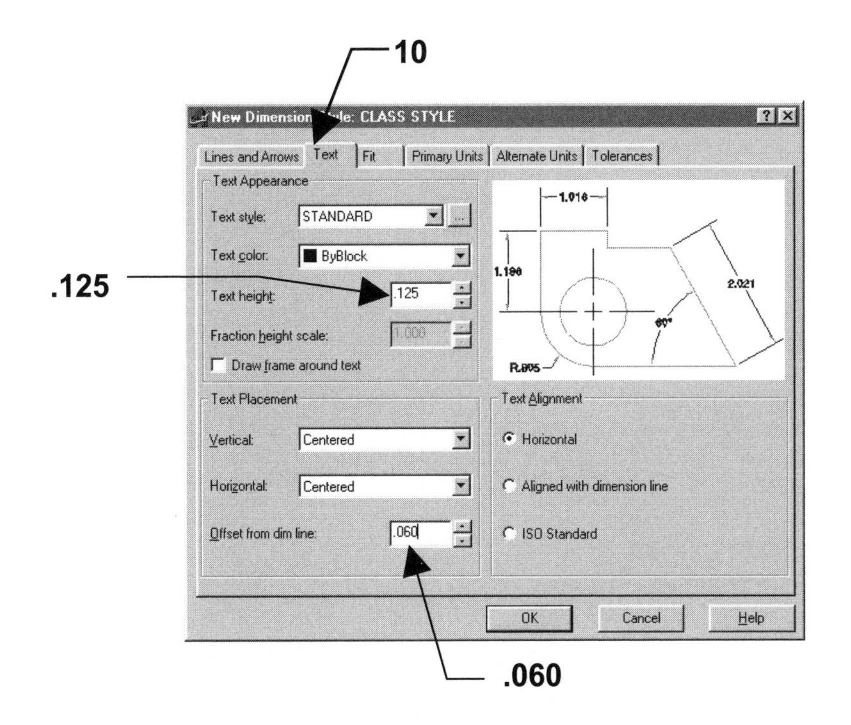

<u>DO NOT SELECT THE OK BUTTON</u>

11. Select the **"Fit"** tab. (Look at it but No changes are necessary.)

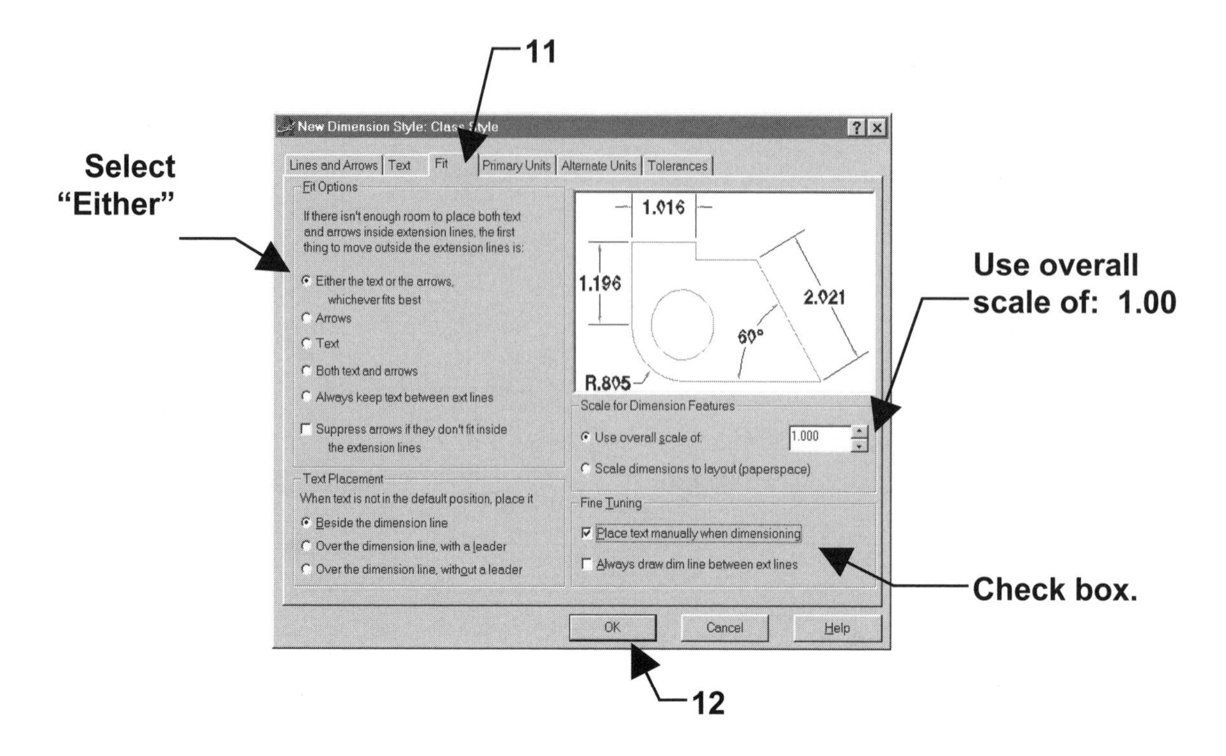

12. **NOW** you may select the **OK** button.

13. *Your new style **"Class Style"** should be listed.* Select the **"Set Current"** button to make your new style "Class Style" the style that we will use.

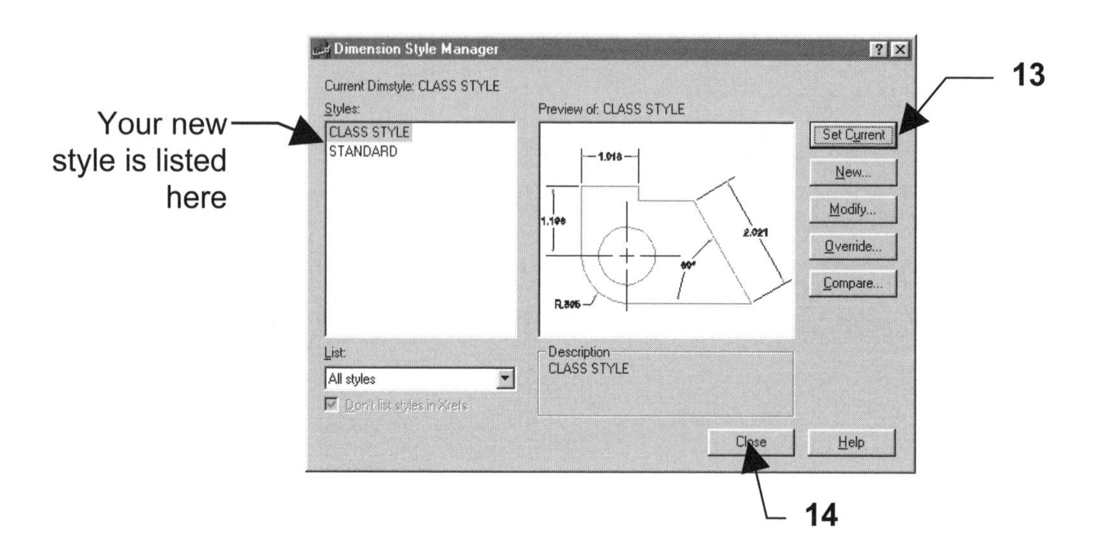

14. Select the **Close** button to **exit**.
15. Save your drawing as **BSIZE.**

Note: You have successfully created a new "Dimension Style" called "Class Style". This style will be saved in your BSIZE drawing after you save the drawing. It is important that you understand that this dimension style resides only in the BSIZE drawing. If you open another drawing, this dimension style will not be there.

COMPARE TWO DIMENSION STYLES

Sometimes it is useful to compare the settings of two styles. Compare will compare the two styles and list the differences.

LET'S COMPARE "CLASS STYLE" AND "STANDARD"

1. Select the Dimension Style command.
2. Select the Compare button.

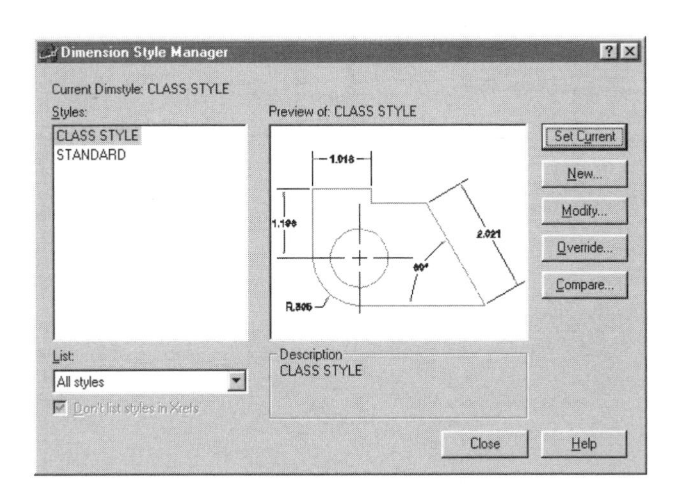

The "Compare Dimension Style" dialog box will appear.

3. Select "Class Style" in the Compare box.
4. Select "Standard" in the With box.
5. AutoCAD found differences and listed them.

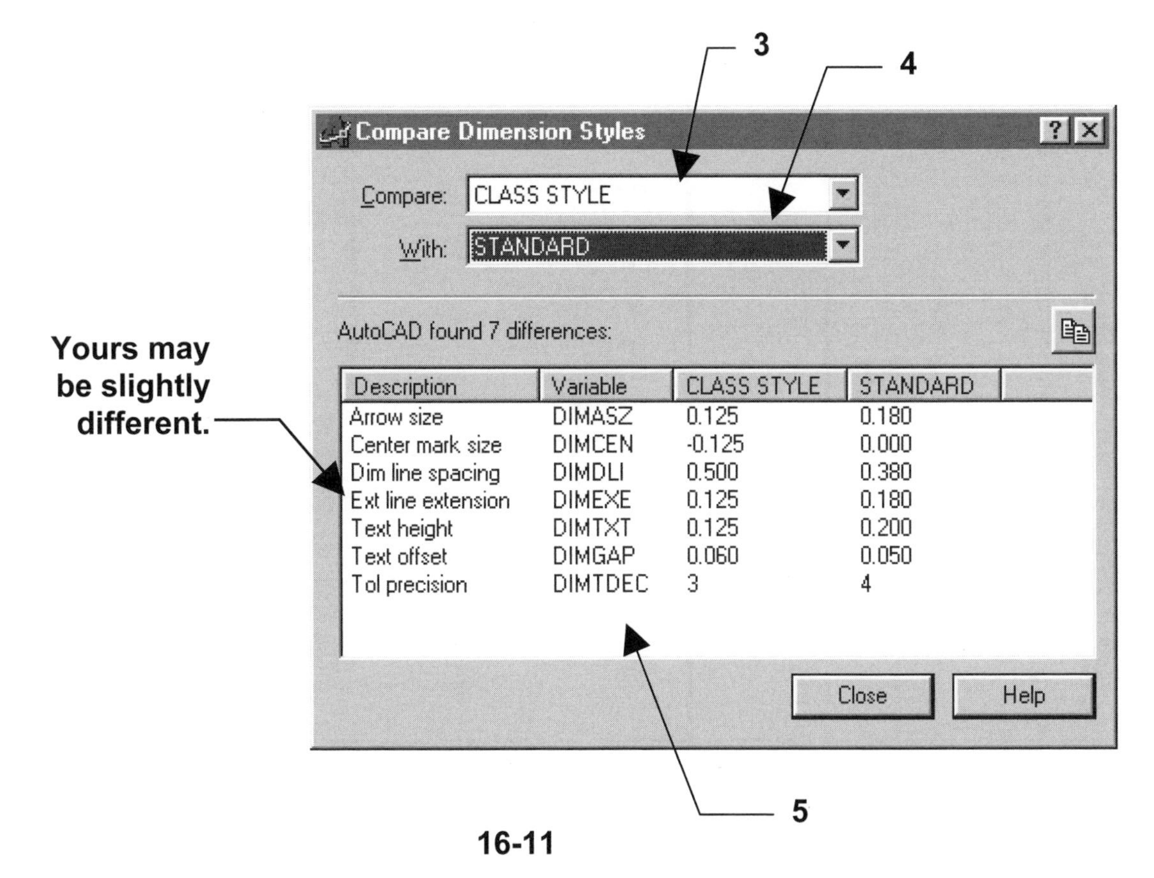

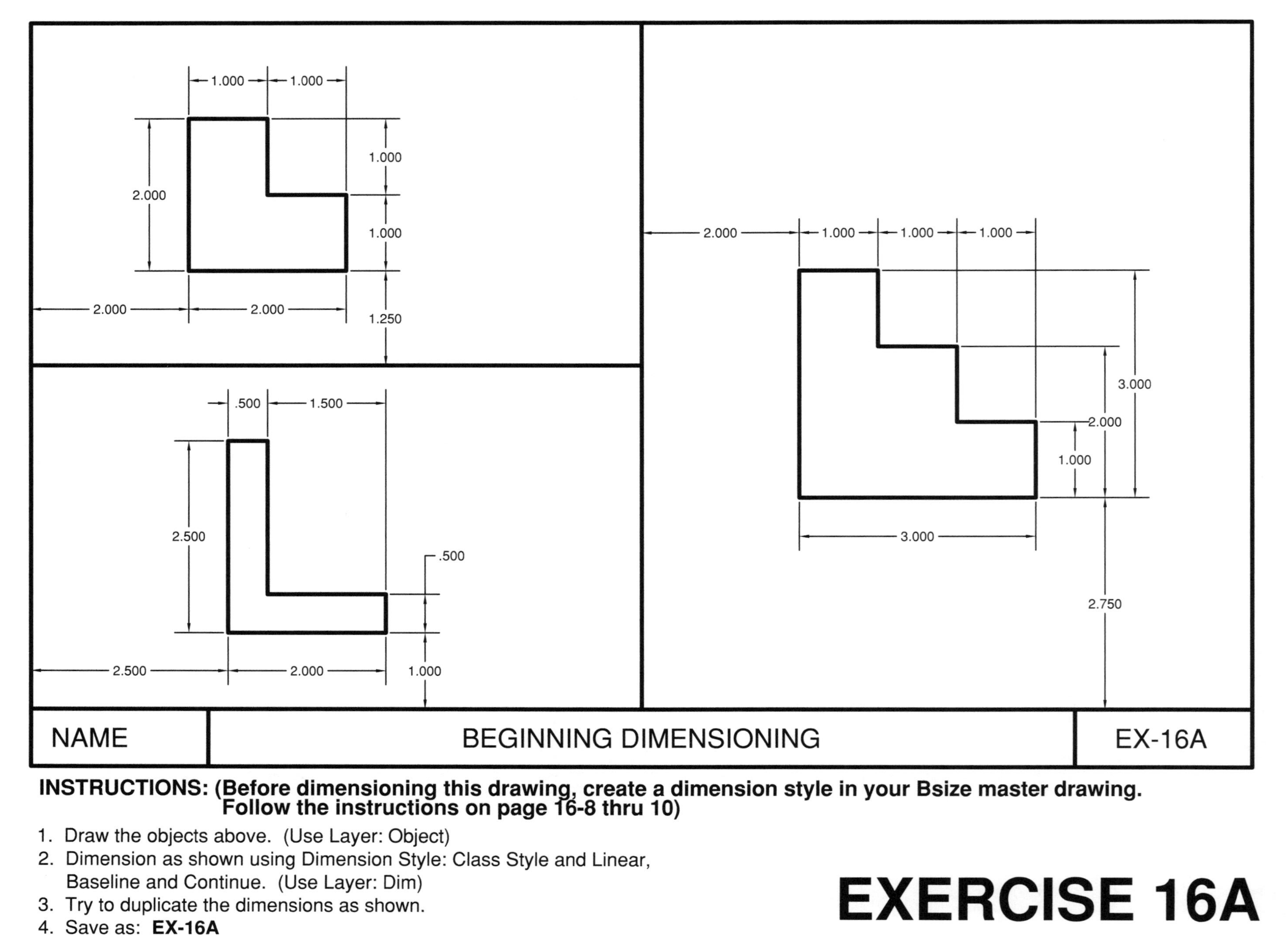

INSTRUCTIONS: (Before dimensioning this drawing, create a dimension style in your Bsize master drawing. Follow the instructions on page 16-8 thru 10)

1. Draw the objects above. (Use Layer: Object)
2. Dimension as shown using Dimension Style: Class Style and Linear, Baseline and Continue. (Use Layer: Dim)
3. Try to duplicate the dimensions as shown.
4. Save as: **EX-16A**

EXERCISE 16A

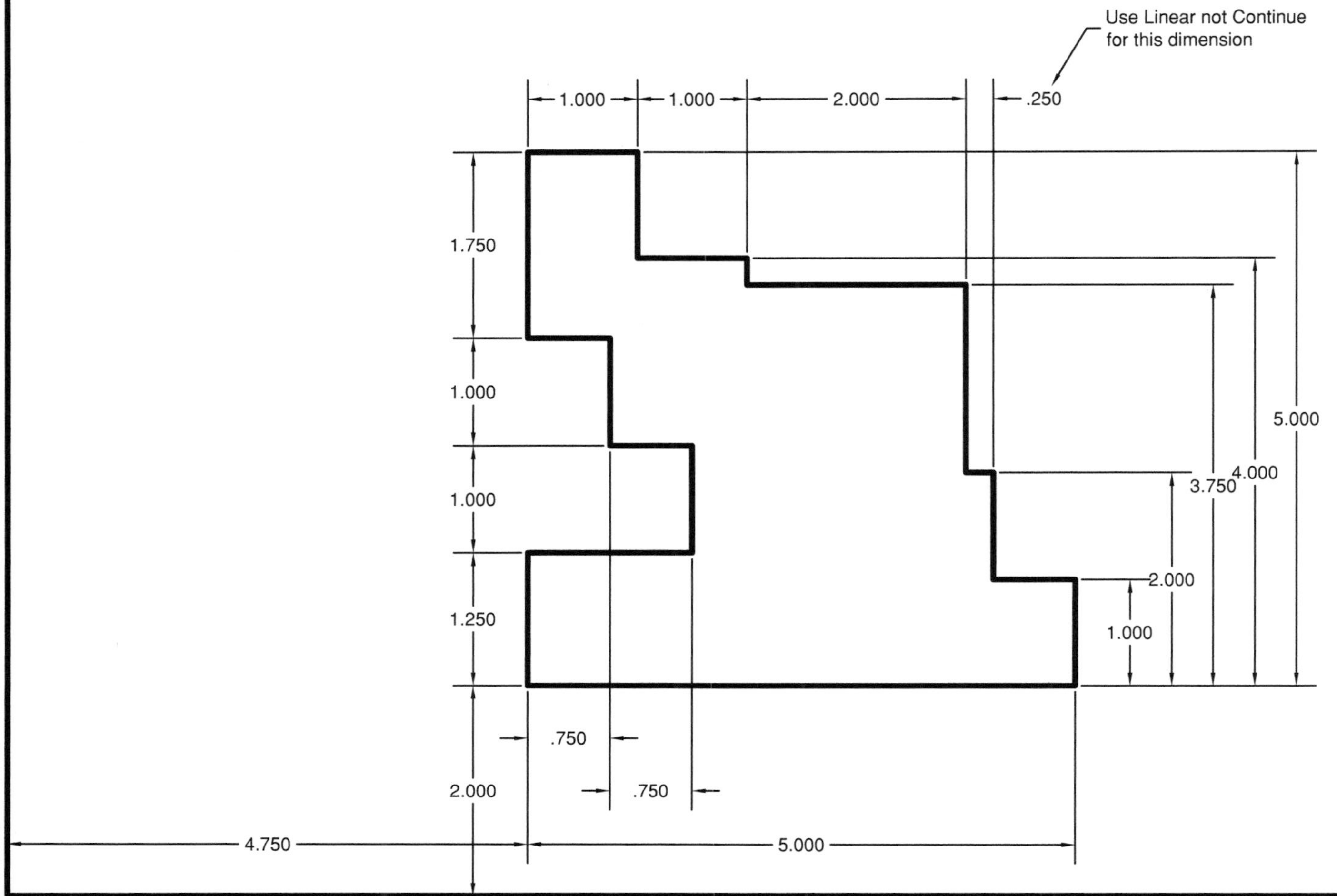

NAME	DIFFICULT DIMENSIONING	EX-16B

INSTRUCTIONS:

1. Draw the objects above. (Use Layer: Object)
2. Note: The OFFSET command would be more efficient than coordinate input.
3. Dimension as shown using Dimension Style: Class Style and Linear, Baseline and Continue. (Use Layer: Dim)
4. Try to duplicate the dimensions as shown.
5. Save as: **EX-16B**

EXERCISE 16B

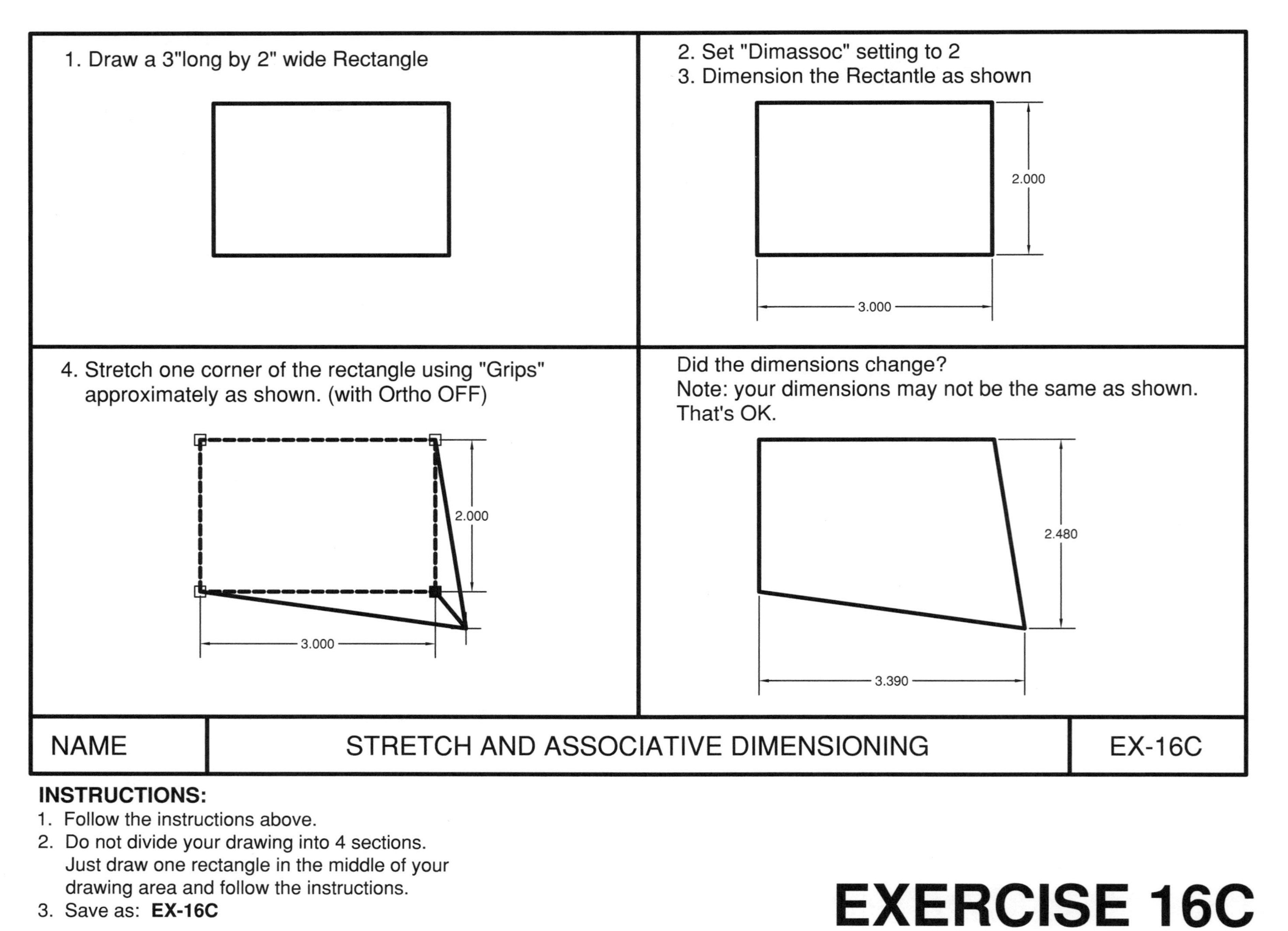
1. Draw a 3"long by 2" wide Rectangle
2. Set "Dimassoc" setting to 2
3. Dimension the Rectantle as shown
2.000
3.000
4. Stretch one corner of the rectangle using "Grips" approximately as shown. (with Ortho OFF)
2.000
3.000
Did the dimensions change?
Note: your dimensions may not be the same as shown.
That's OK.
2.480
3.390
NAME
STRETCH AND ASSOCIATIVE DIMENSIONING
EX-16C
INSTRUCTIONS:
1. Follow the instructions above.
2. Do not divide your drawing into 4 sections.
Just draw one rectangle in the middle of your drawing area and follow the instructions.
3. Save as: EX-16C
EXERCISE 16C

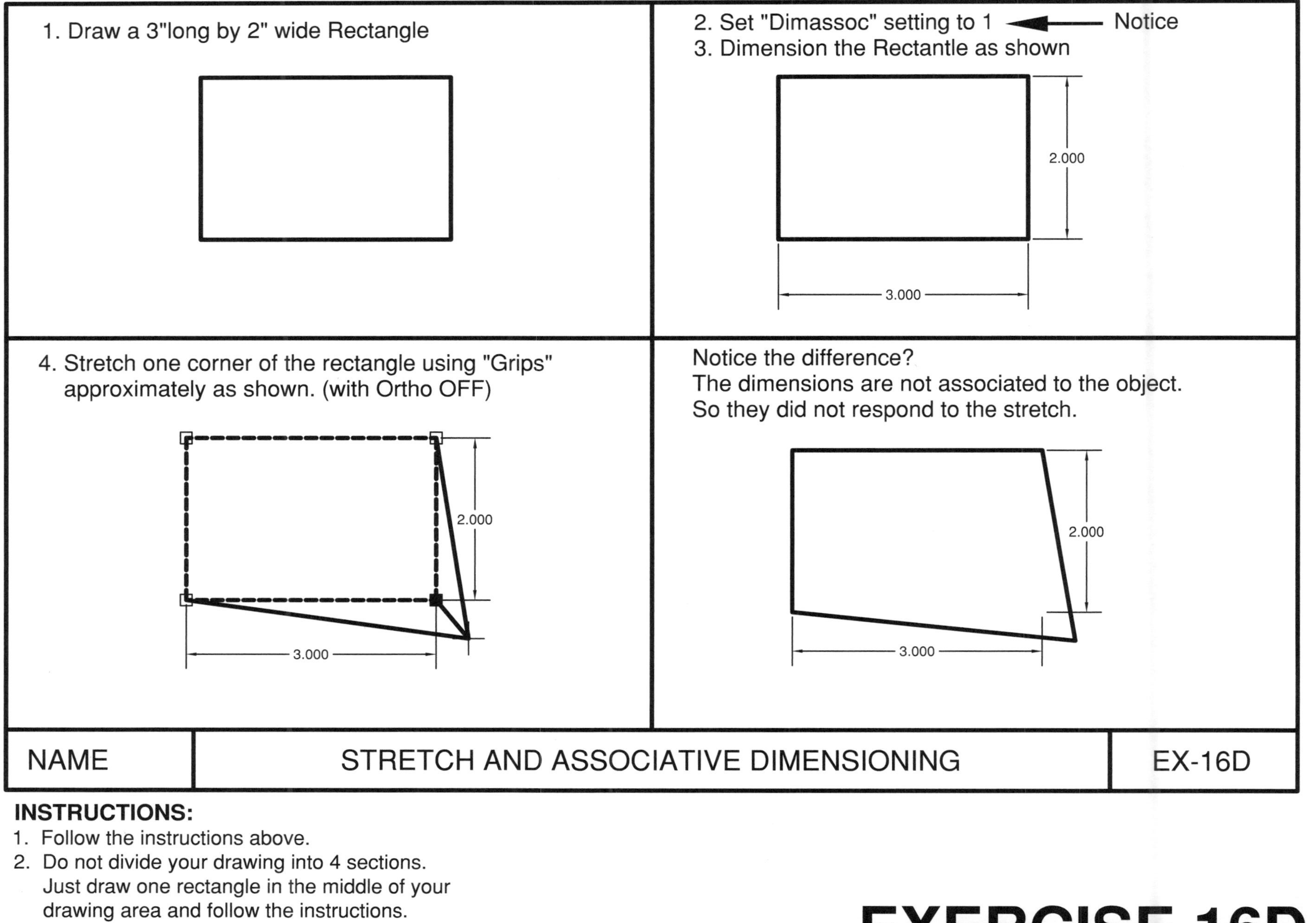

INSTRUCTIONS:

1. Follow the instructions above.
2. Do not divide your drawing into 4 sections. Just draw one rectangle in the middle of your drawing area and follow the instructions.
3. Save as: **EX-16D**

EXERCISE 16D

NOTES:

LEARNING OBJECTIVES

After completing this lesson, you will be able to:

1. Edit dimension text values
2. Edit the Dimension position.
3. Modify an entire Dimension Style.
4. Override a dimension style.
5. Edit a dimension using Properties Palette.

LESSON 17

EDITING DIMENSION TEXT VALUES

Sometimes you need to modify the dimension text. You may add a symbol, a note or even change the text of an existing dimension. There are 2 methods.

Example: Add the word "Max." to the existing dimension.

Before Editing **After Editing**

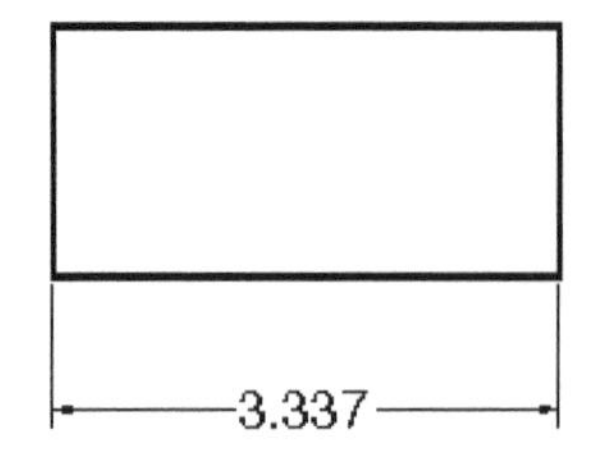

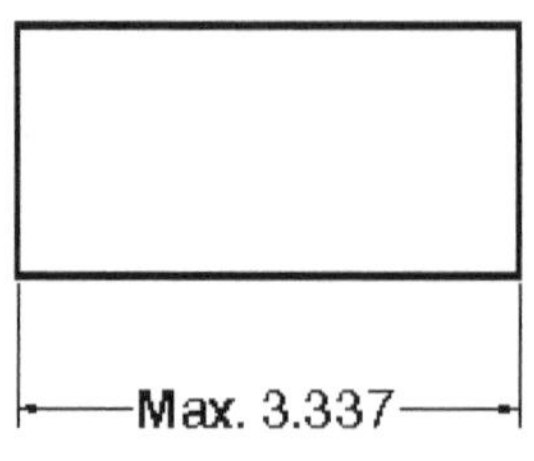

Method 1.

1. Double click on the dimension that you want to change.
   The Properties Palette will appear.
2. Scroll down to Text / Text override
   (Notice that the actual measurement is directly above it.)
3. Type the new text and press <enter>

Rotated Dir
Text
Fractional t... Horizontal
Text color ByBlock
Text height .200
Text offset .050
Text outsid... Off
Text pos hor Centered
Text pos vert Centered
Text style STANDARD
Text inside ... Off
Text positio... 1.643
Text positio... 6.383
Text rotation 0
Measurement 3.337
Text override Max. 3.337
Fit
Primary Units
Alternate Units
Tolerances
Specifies the text string of the dimension (overrides Measur...
PROPERTIES

Important:
If you use this method the dimension is no longer Associative.

Method 2.

1. Type on the Command line: **DDEDIT** <enter>
2. Select the dimension you want to edit. (Only one at a time)
3. The "Multiline Text Editor" will appear.
   If the dimension is Associative, it will display the dimension text as a symbol <>.
   Do not erase *this symbol. You can add text before or after the symbol.*
4. Select the OK button.

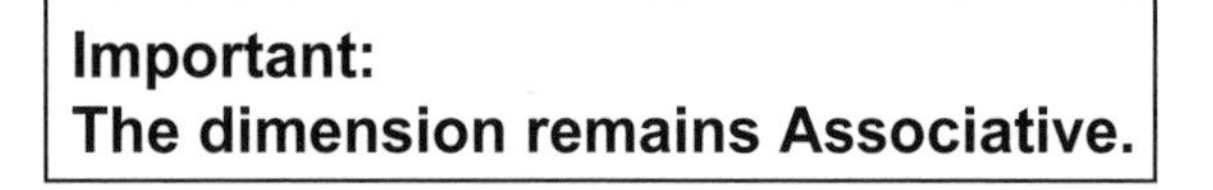

Important:
The dimension remains Associative.

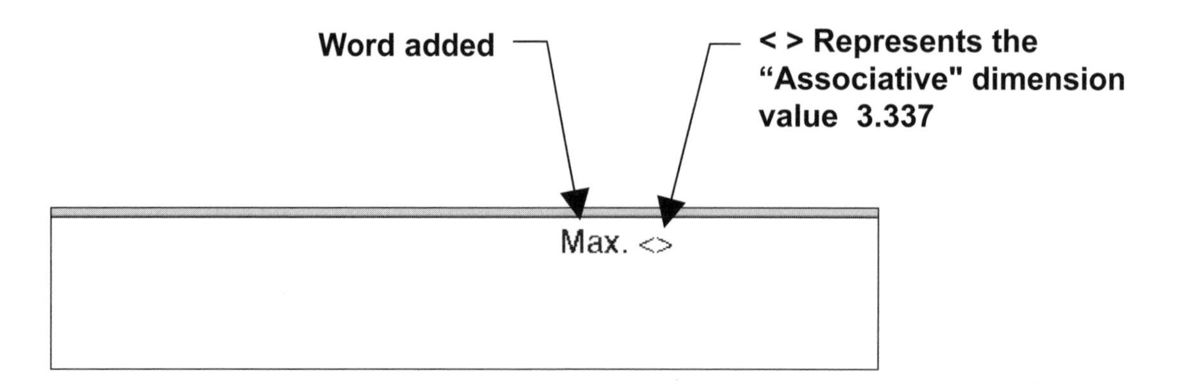

EDITING THE DIMENSION POSITION

Sometimes dimensions are too close and you would like to stagger the text or you need to move an entire dimension to a new location, such as the examples below.

Before Editing | **Slide the text down** | **Move an entire dimension**

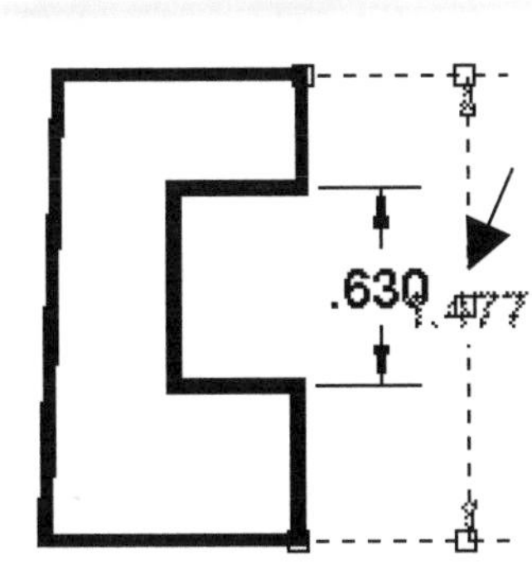

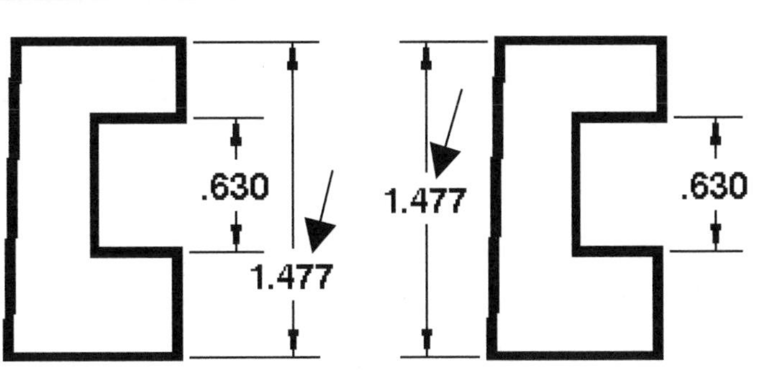

***Editing the position is easy*.**
1. Select the dimension that you want to change. (Grips will appear)
2. Select the middle grip. (It will become red and active, hot)
3. Move the cursor and the dimension will respond.
4. Press the left mouse button to place the dimension in the new location.

ADDITIONAL EDITING OPTIONS USING THE SHORTCUT MENU

1. Select the dimension that you want to change.
2. Press the Right Mouse button.
3. Select "Dimension Text Position" from the shortcut menu shown to the right.
4. A sub-menu appears with many editing options.

Experiment with these options. They will be very helpful in the future.

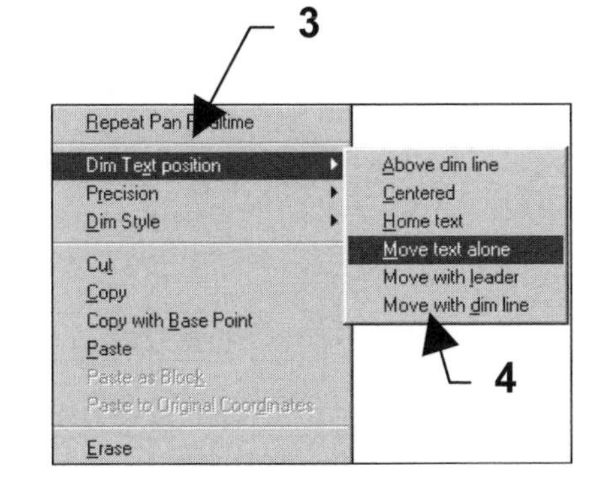

**Note: This command will not work with <u>Exploded</u> dimensions.
You will have to use the "<u>Stretch</u>" command to achieve the same results.**

MODIFY AN ENTIRE DIMENSION STYLE

After you have created a Dimension Style, you may find that you have changed your mind about some of the settings. You can easily change the entire Style by using the "Modify" button in the Dimension style Manager dialog box. This will not only change the Style for future use, but it will also update dimensions already in the drawing.

Note: if you do not want to update the dimensions already in the drawing, but want to make a change to the next dimension drawn, ***refer to Override, page 17-5***

1. Select the Dimension Style command. (Refer to page 16-7)

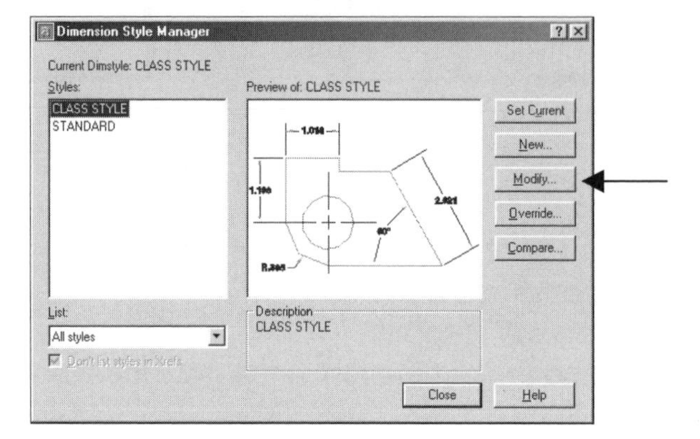

2. Select the Modify button from the Dimension Style Manager dialog box.
3. Make the desired changes to the settings.
4. Select the OK button.
5. Select the Close button.

Now look at your drawing. Have your dimensions updated?

Note: If some of the dimensions have not changed:
1. Select Dimension / Update
2. Select the dimension(s) you wish to update and press <enter>

Note: This feature will not work with Exploded dimensions.

<u>OVERRIDE</u> A DIMENSION STYLE

A dimension Override is a **temporary** change to the dimension settings.
An override **will not affect existing dimensions**. It will affect **new dimensions only**.

*Use this option when you want your **next dimension** just a little bit different but you don't want to create a whole new dimension style and you don't want the existing dimensions to change either.*

1. Select **Format / Dimension Style.**
2. Select the "**Style"** you want to override. (such as Class Style)
3. Select the **Override** button.
4. Make the desired changes to the settings.
5. Select the OK button.
6. Confirm the Override
   a. Look at the List of styles. Under the Style name, a sub heading of "Style overrides" should be displayed.
   b. The description box should display the style name and the override settings.

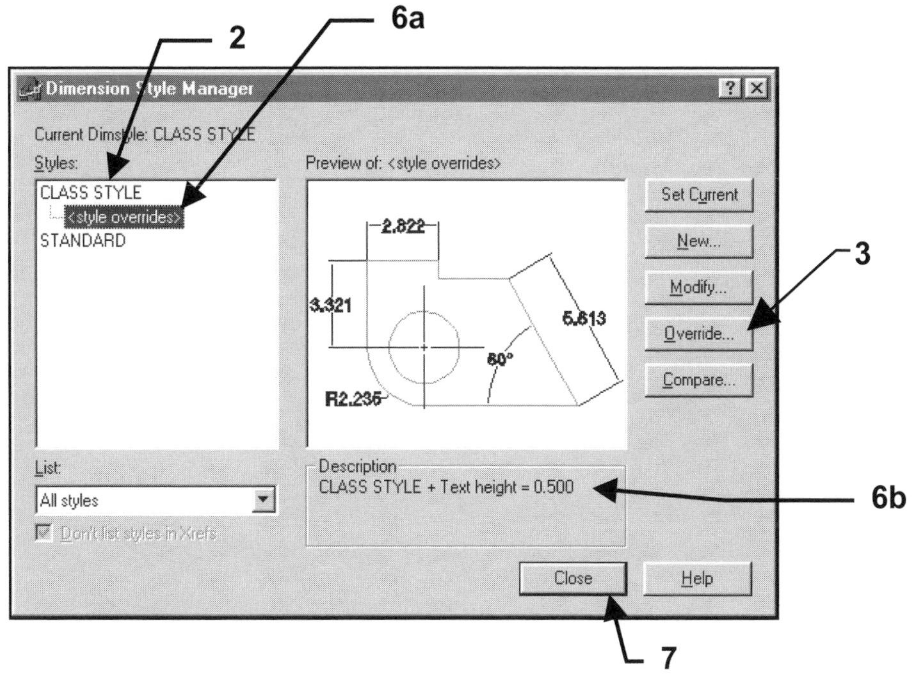

7. Select the Close button.

If you want to return to style "**Class Style"**, select "Class Style" and then select the "**Set Current"** button. Each time you select a different style, you must select the **Set Current** button to activate it.

EDITING A DIMENSION using PROPERTIES PALETTE

Sometimes you would like to modify the settings of an **individual existing** dimension. This can be achieved using the Modify Properties command.

1. Double click on the dimension that you wish to change.

   The Properties Palette will appear.

2. Select and change the desired settings.

3. Press <enter> . *(The change should have taken effect)*

4. Press the <esc> key **or** move the cursor onto the drawing area, right click and select "Deselect All" from the shortcut menu.

Note: This command will not work with Exploded dimensions.

A

6.000

6.000

B

NAME	OVERRIDE	EX-17A

INSTRUCTIONS:

1. Draw the 6" Long by 4" Wide Rectangle above. (Use Layer: Object)
2. Use dimension style "Class Style" for dimension A. (Use Layer: Dim)
3. Use OVERRIDE for dimension B. Change the setting for Text height to .500.
4. Save as: **EX-17A and plot.**

EXERCISE 17A

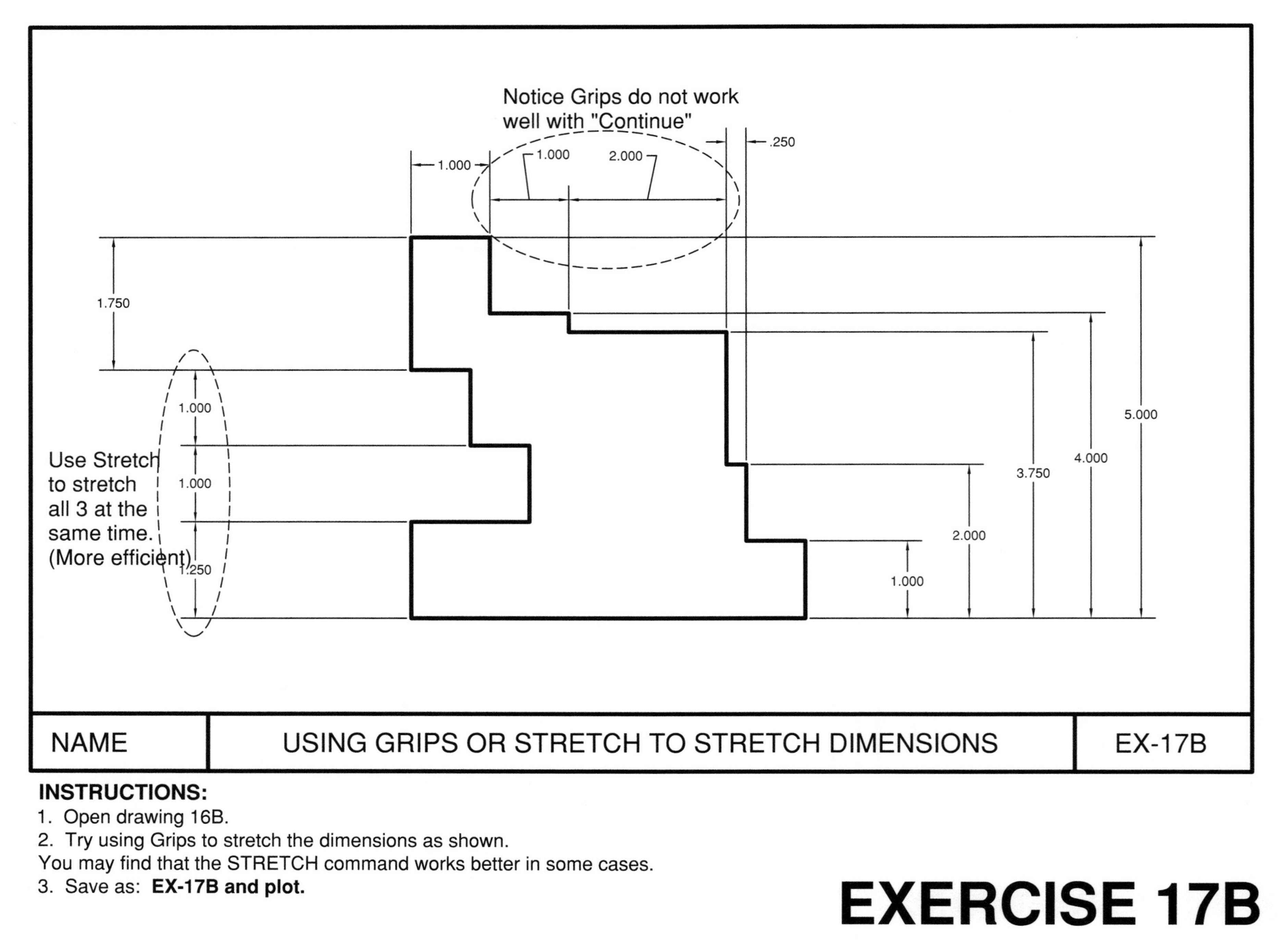

INSTRUCTIONS:

1. Open drawing 16B.
2. Try using Grips to stretch the dimensions as shown.

You may find that the STRETCH command works better in some cases.

3. Save as: **EX-17B and plot.**

EXERCISE 17B

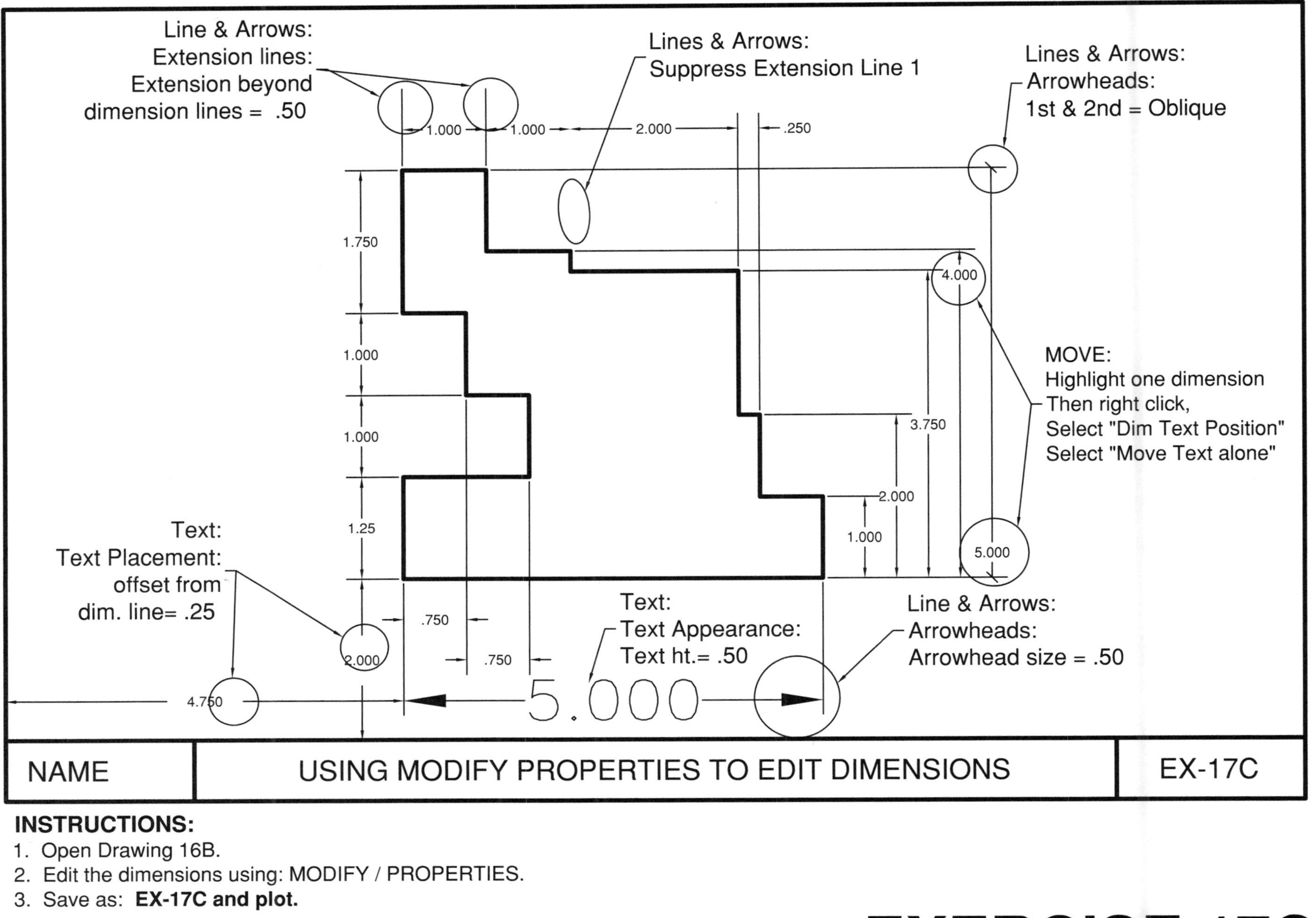

INSTRUCTIONS:

1. Open Drawing 16B.
2. Edit the dimensions using: MODIFY / PROPERTIES.
3. Save as: **EX-17C and plot.**

EXERCISE 17C

NOTES:

LEARNING OBJECTIVES

After completing this lesson, you will be able to:

1. Add Diameter, Radial and Angular dimensions to your drawing.
2. Draw Center Marks.
3. Control the size and appearance of the Center Marks.
4. Understand the need for Sub-Styles.
5. Create a Sub-Style.

LESSON 18

RADIAL DIMENSIONING

DIAMETER DIMENSIONING

The **DIAMETER** dimensioning command should be used when dimensioning circles and arcs of more than 180 degrees.

Center marks are automatically drawn as you use the diameter dimensioning command. If the circle already has a center mark or you do not want a center mark, set the center mark setting to **NONE** (Dimension Style / Lines and Arrows tab) before using Diameter dimensioning.

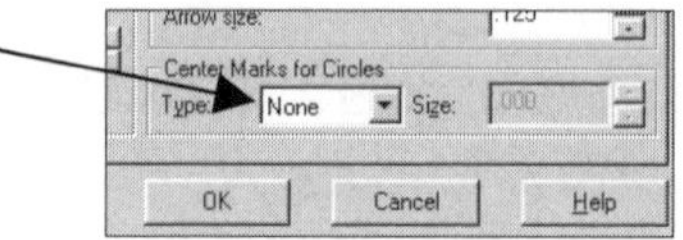

1. Select the **DIAMETER** command.
   Command: _dimdiameter
2. Select arc or circle: ***click on the arc or circle (P1), location is not important, do not use object snap.***
   Dimension text = ***the diameter will be displayed here***
3. Specify dimension line location or [Mtext/Text/Angle]: ***place dimension text location (P2)***

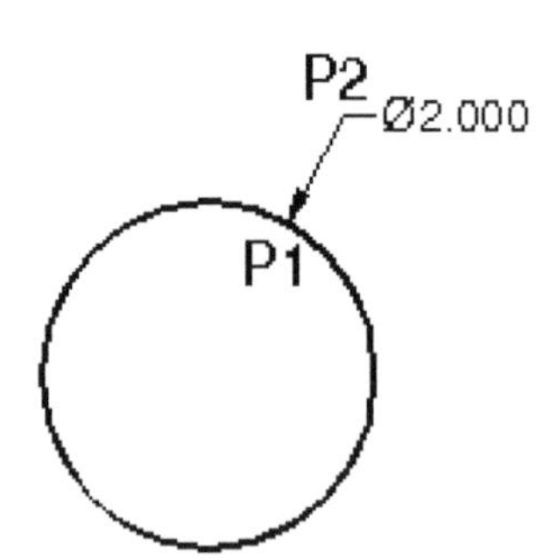

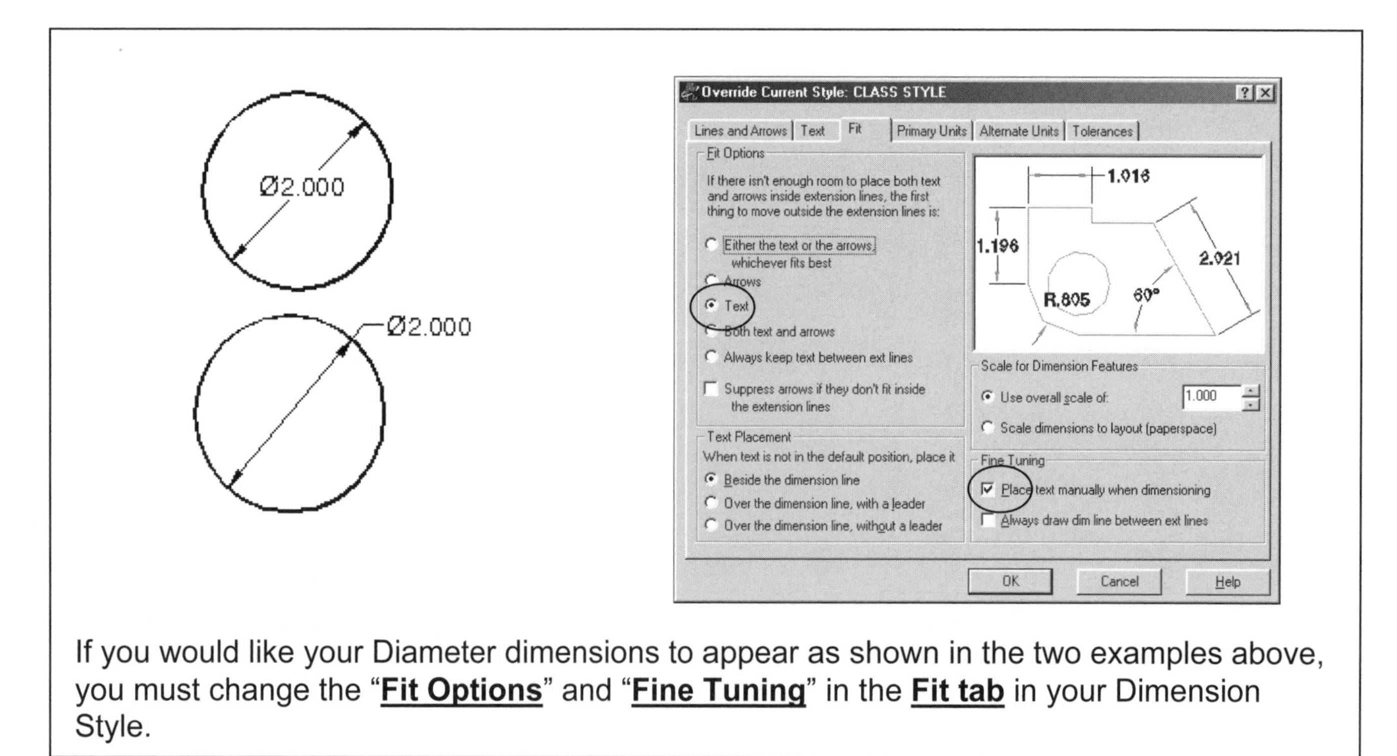

If you would like your Diameter dimensions to appear as shown in the two examples above, you must change the "**Fit Options**" and "**Fine Tuning**" in the **Fit tab** in your Dimension Style.

RADIAL DIMENSIONING (continued)

RADIUS DIMENSIONING

The **RADIUS** dimensioning command should be used when dimensioning arcs of <u>LESS than 180 degrees.</u>

<u>Center Marks</u> are automatically drawn as you use the **RADIUS** dimensioning command. If the circle already has a center mark, set the center mark to **<u>NONE</u>** (in the Dimension Style) before using RADIUS dimensioning. Or create a "sub-style" as shown on page 18-6.

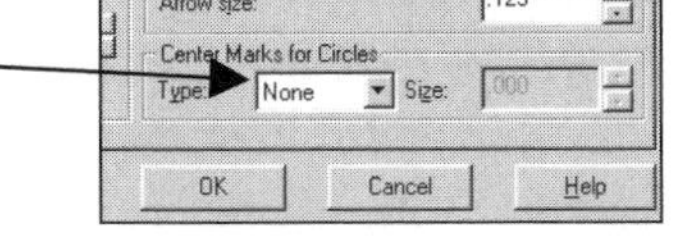

1. Select the **RADIUS** command.
   Command: _dimradius
2. Select arc or circle: ***click on the arc (P1), location is not important, do not use snap.***
   Dimension text = ***the radius will be displayed here***
3. Specify dimension line location or [Mtext/Text/Angle]: ***place dim text location (P2)***

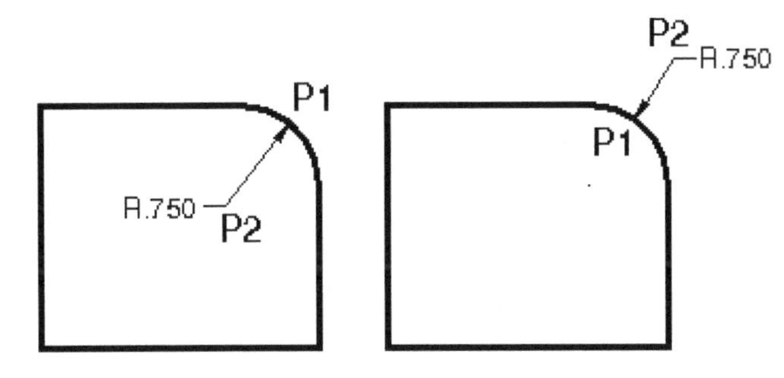

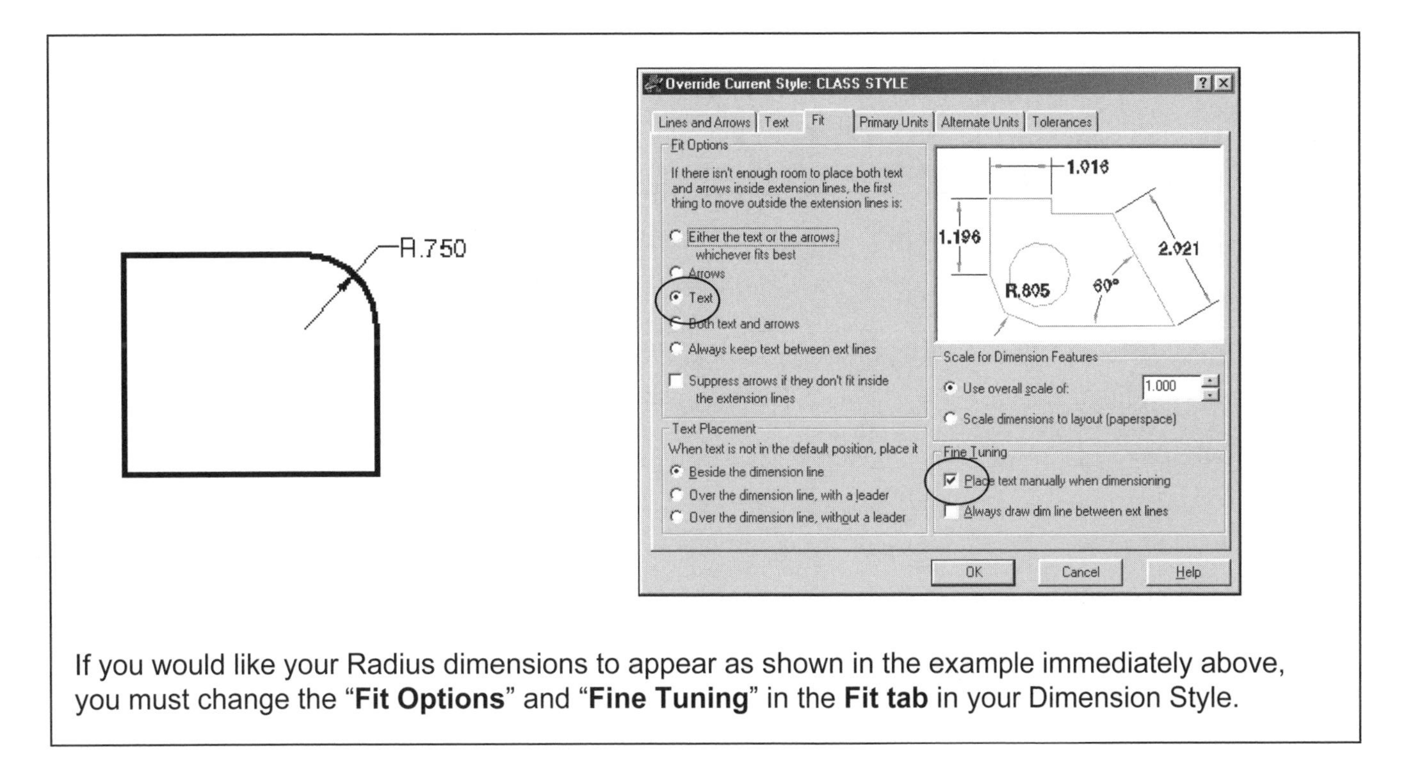

If you would like your Radius dimensions to appear as shown in the example immediately above, you must change the "**Fit Options**" and "**Fine Tuning**" in the **Fit tab** in your Dimension Style.

ANGULAR DIMENSIONING

The **ANGULAR** dimension command is used to create an angular dimension between two lines that form an angle. All that is necessary is the selection of the two lines and the location for the dimension text.

The **degree symbol** is automatically added as the dimension is created.

1. Select the **ANGULAR** command.
   Command: _dimangular
2. Select arc, circle, line, or <specify vertex>: ***click on the first line that forms the angle (P1) location is not important, do not use snap.***
3. Select second line: ***click on the second line that forms the angle (P2)***
4. Specify dimension arc line location or [Mtext/Text/Angle]: ***place dimension text location***
   Dimension text = ***angle will be displayed here***

Any of the 4 angular dimensions shown below can be created by clicking on the 2 lines (P1 and P2) that form the angle.

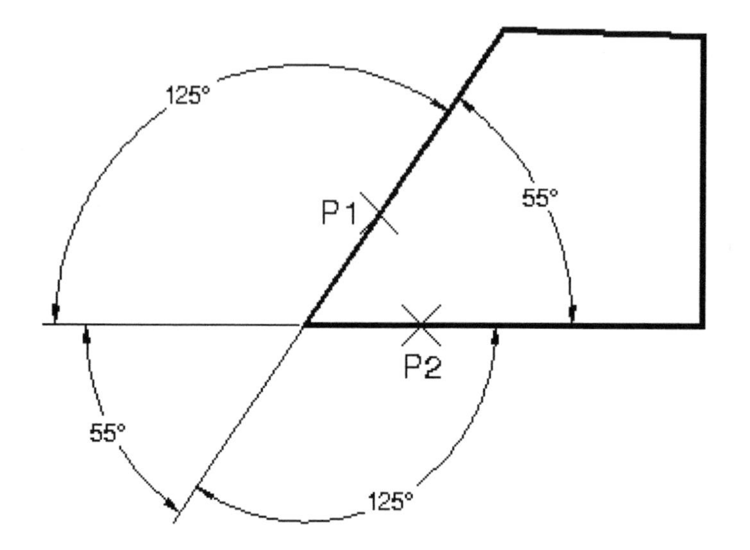

CENTERMARK

CENTERMARKS can ONLY be drawn with circular objects like Circles and Arcs. You set the size and type.

The Center Mark has three types, **None, Mark** and **Line** as shown below.

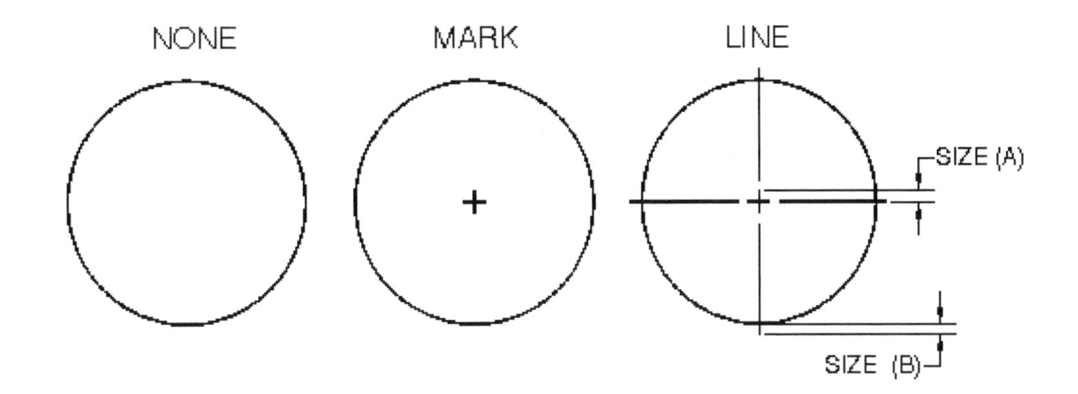

What does "SIZE" mean?

The **size** setting determines both, (A) the length of half of the intersection line and (B) the length extending beyond the circle. (See above)

Where do you set the CENTERMARK "TYPE" and "SIZE"

1. Select the ***Dimension Style*** command.
2. Select: ***New, Modify, or Override.***
3. Select: ***LINES and ARROWS*** *tab*
4. Select the ***Center mark type***
5. Set the **Size**.

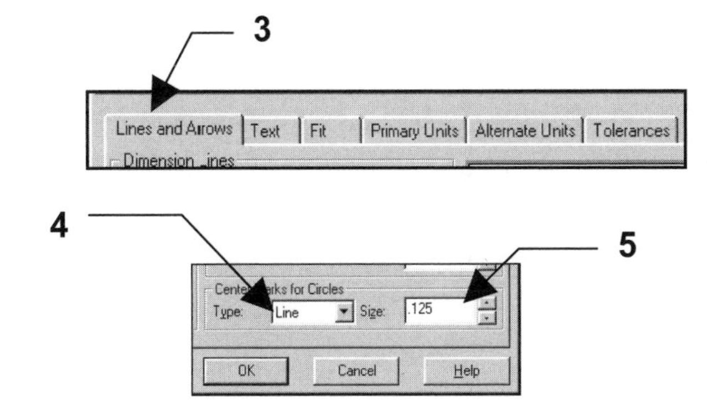

To draw a CENTER MARK

1. Select the **Center mark** command using one of the following:

   TYPING = DIMCENTER or DCE
   PULLDOWN = DIMENSION / CENTERMARK
   TOOLBAR = DIMENSION

2. Select arc or circle: ***select the arc or circle with the cursor.***

CREATING A DIMENSION SUB-STYLE

Sometimes when using a dimension style, you would like the Linear, Angular, Diameter and Radius dimensions to have different settings. But you want them to use the same dimension style. To achieve this, you must create a “sub-style”.
Sub-styles have also been called “children” of the “Parent” dimension style. As a result, they form a family.
A Sub-style is permanent, unlike the Override command, which is temporary.

LET’S CREATE A SUB-STYLE FOR RADIUS.

We will set the center mark to None for the Radius command only.
The Diameter command center mark will not change.

1. Open your **BSIZE** drawing.
2. Select the **DIMENSION / STYLE** command.
3. Select **“Class Style”** from the Style List.
4. Select the **NEW** button.
5. Change the “Use for” to:
   Radius Dimensions
6. Select **Continue**
7. Select the **Lines and Arrows** tab.
8. Change the “Center Marks for Circles” Type: **NONE**
9. Select the **OK** button

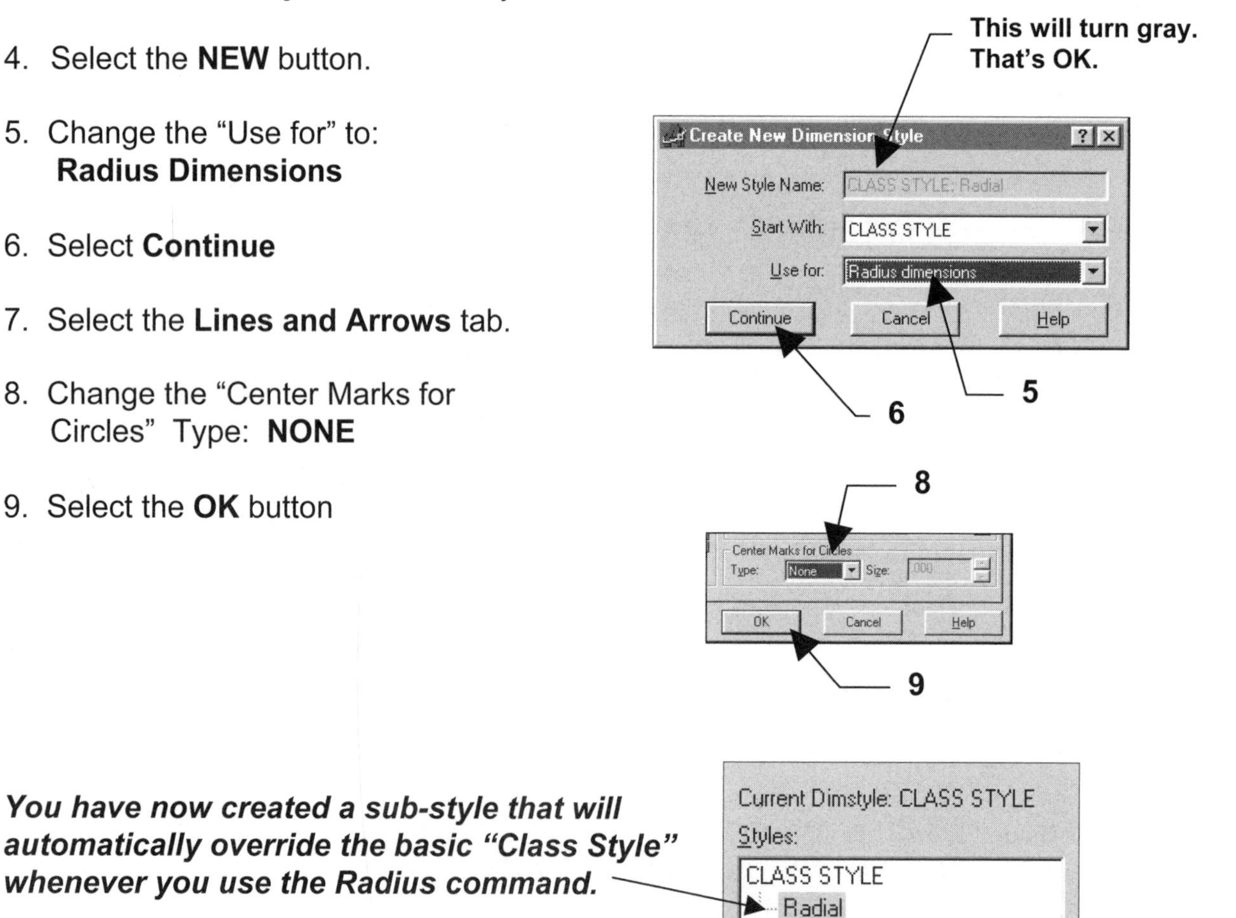

You have now created a sub-style that will automatically override the basic “Class Style” whenever you use the Radius command.

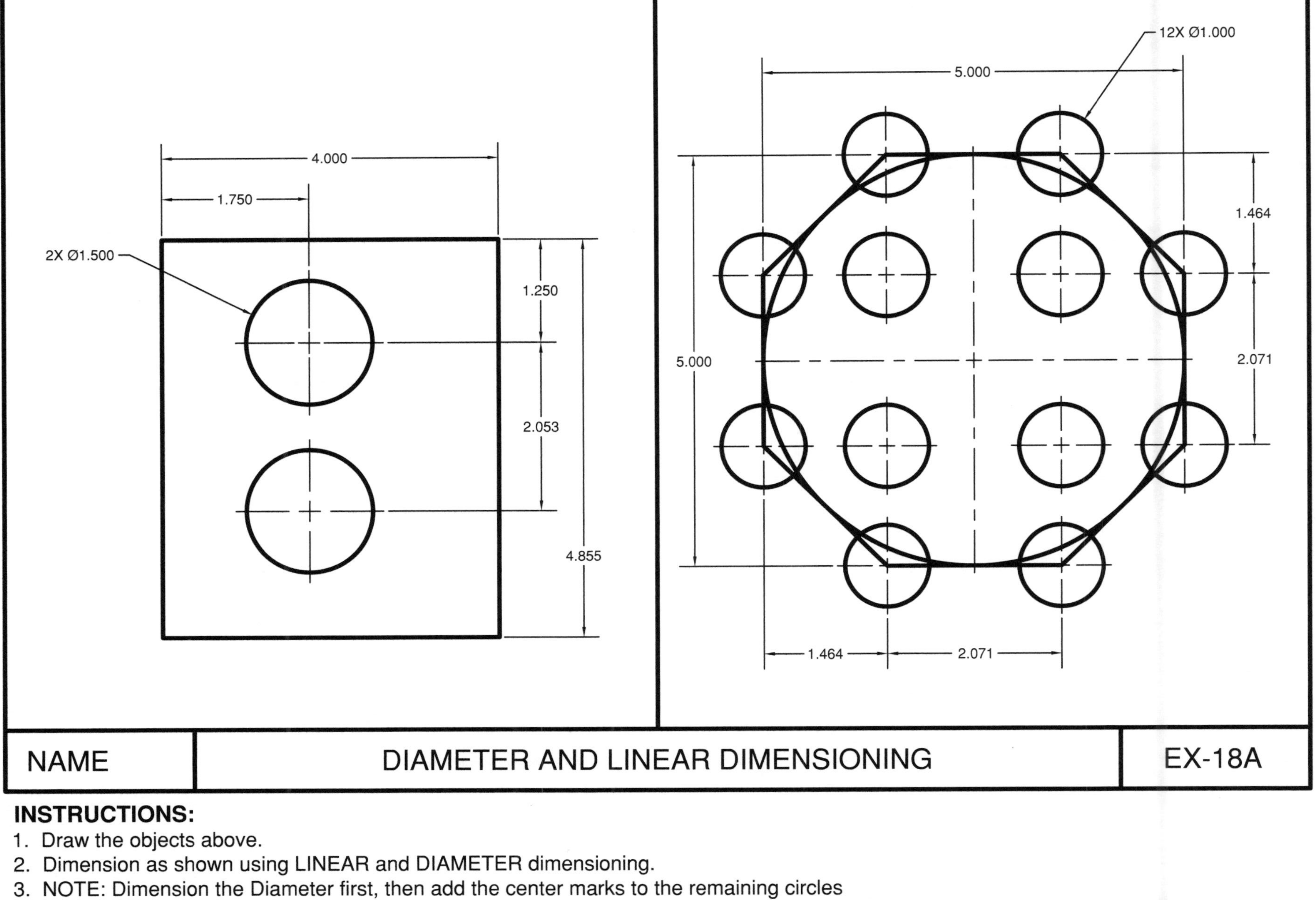

INSTRUCTIONS:

1. Draw the objects above.
2. Dimension as shown using LINEAR and DIAMETER dimensioning.
3. NOTE: Dimension the Diameter first, then add the center marks to the remaining circles
   Draw the Linear dimensions last so you can snap to the endpoint of the center marks.
4. Save as: **EX-18A and plot.**

EXERCISE 18A

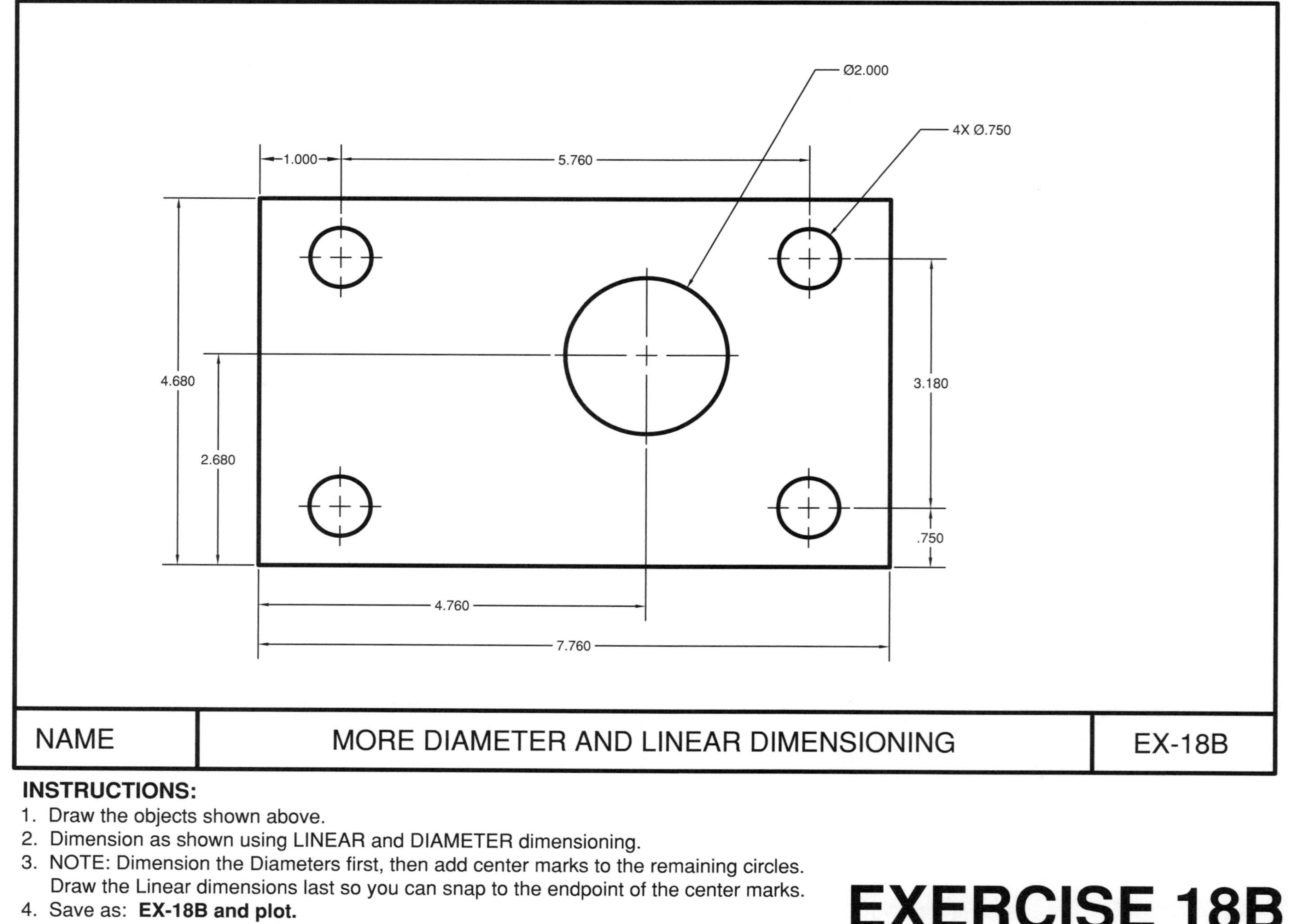

INSTRUCTIONS:

1. Draw the objects shown above.
2. Dimension as shown using LINEAR and DIAMETER dimensioning.
3. NOTE: Dimension the Diameters first, then add center marks to the remaining circles. Draw the Linear dimensions last so you can snap to the endpoint of the center marks.
4. Save as: **EX-18B and plot.**

EXERCISE 18B

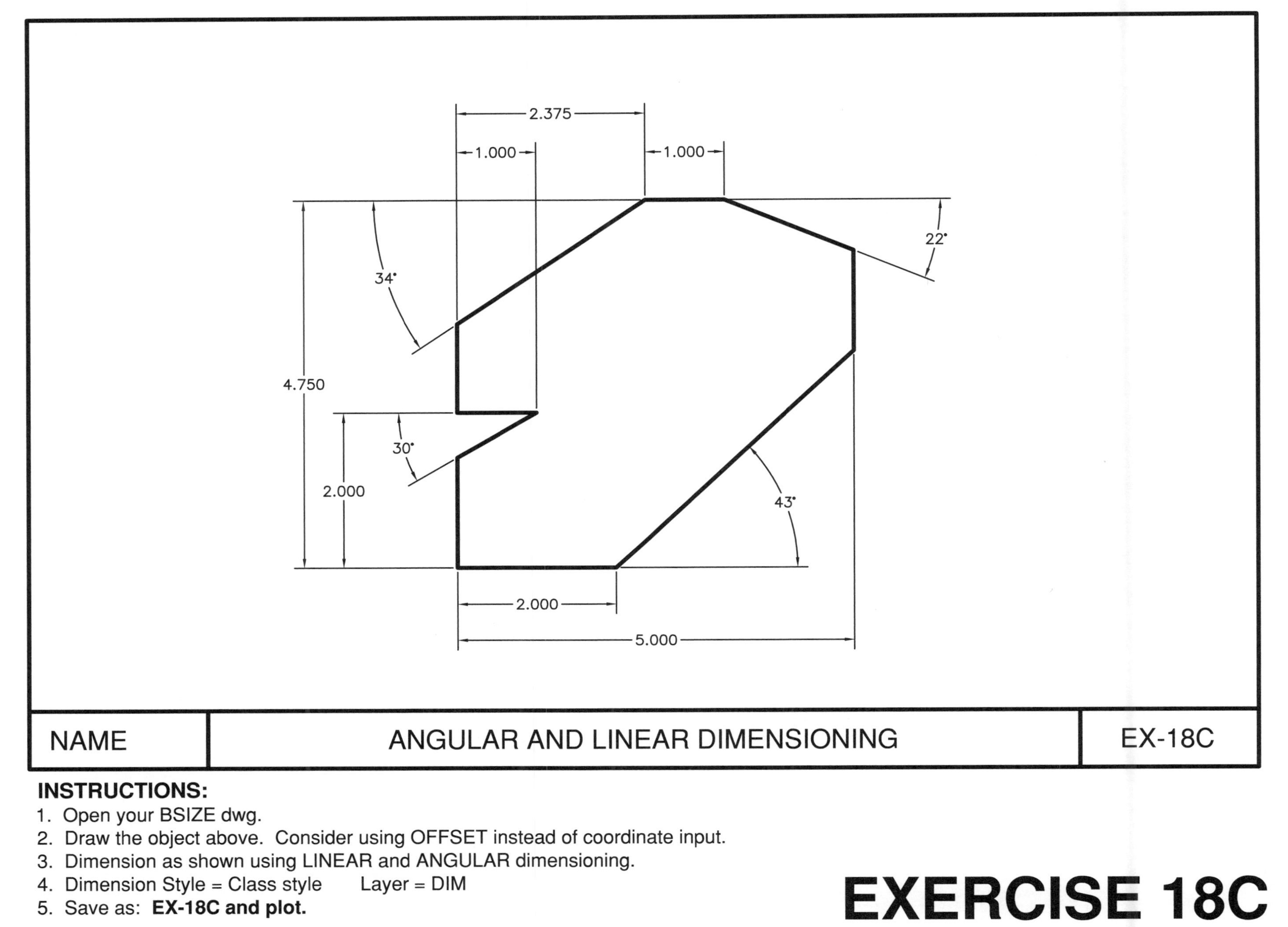

INSTRUCTIONS:

1. Open your BSIZE dwg.
2. Draw the object above. Consider using OFFSET instead of coordinate input.
3. Dimension as shown using LINEAR and ANGULAR dimensioning.
4. Dimension Style = Class style   Layer = DIM
5. Save as: **EX-18C and plot.**

EXERCISE 18C

NOTES:

LEARNING OBJECTIVES

After completing this lesson, you will be able to:

1. Dimension objects that are on an angle.
2. Draw a Leader.
3. Add symbols such as: diameter, plus or minus and degree to text.
4. Pre-assign a prefix or suffix to a dimension.

LESSON 19

ALIGNED DIMENSIONING

The **ALIGNED** dimension command aligns the dimension with the angle of the object that you are dimensioning. The process is the same as Linear dimensioning. It requires two extension line origins and placement of text location. (Example below)

1. Select the **ALIGNED** command using one of the following:

   TYPE = DIMALIGNED or DIMALI or DAL
   PULLDOWN = DIMENSION / ALIGNED
   TOOLBAR = DIMENSION

   Command: _dimaligned
2. Specify first extension line origin or <select object>: ***select the first extension line origin (P1)***
3. Specify second extension line origin: ***select the second extension line origin (P2)***

4. Specify dimension line location or [Mtext/Text/Angle]: ***place dimension text location***

   Dimension text = ***the dimension value will appear here***

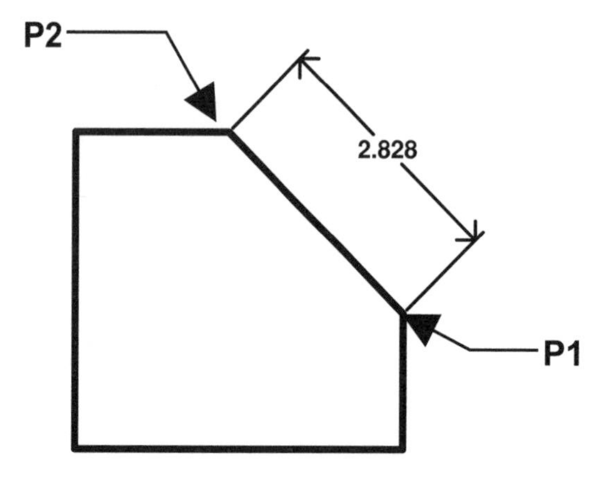

LEADER

The **LEADER** command is primarily used to add a **note** to an object. A Leader's appearance is very similar to a Radial dimension **but** a Leader **should not** to be used for Radial dimensioning. A Leader allows you to type a note at the end of the hook. (Note: Leader and Qleader are the same command)

If associative dimensioning is turned on with DIMASSOC, the leader start point can be associated with a location on an object. If the object is relocated, the arrowhead remains attached to the object and the leader line stretches, but the text or feature control frame remains in place.

1. Select the **LEADER** command using one of the following.

    TYPE = LE or QLEADER
    PULLDOWN = DIMENSION / LEADER
    TOOLBAR = DIMENSION

See "Settings" below

Command: _qleader

2. Specify first leader point, or [Settings]<Settings>: ***select the location for the arrowhead (P1)***
3. Specify next point: ***select next location (P2)***
4. Specify next point: ***select another point or press <enter> to input text***
5. Specify text width <.000>: ***press <enter>***
6. Enter first line of annotation text <Mtext>: ***type your desired text here***
7. Enter next line of annotation text: ***type more text or press <enter> to stop***

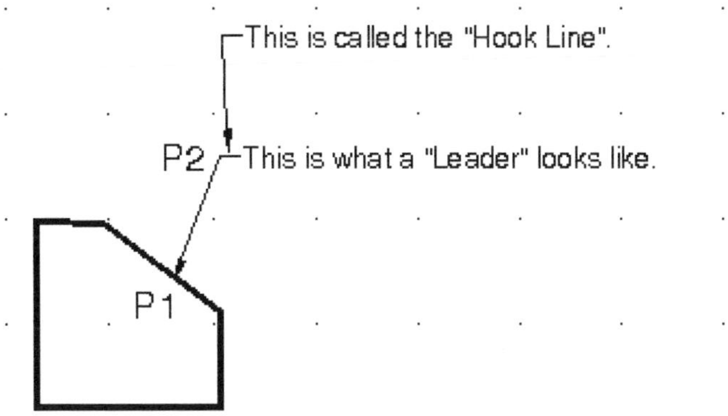

This is called the "Hook Line".
P2 This is what a "Leader" looks like.
P1

HOOK LINE
The hook line is automatically added to the last line segment (P2) if the leader line is 15 degrees or more from horizontal. The length of the Hook Line is controlled by the arrow length setting in the dimension style.

SETTINGS:
If you would like to make changes to the appearance of the Leader, select **"Settings"** before you place the first point (arrowhead location). Selecting **"Settings"** will display a dialog box with many options. (These options are discussed in the Advanced Workbook.)

SPECIAL TEXT CHARACTERS

Characters such as the ***degree symbol, diameter symbol*** and the ***plus / minus symbols*** are created by typing **%%** and then the appropriate "code" letter.
For example: entering 350**%%D** will create: **350°.** The **"D"** is the **"code"** letter.

SYMBOL	CODE
∅ Diameter	%%C
° Degree	%%D
± Plus / Minus	%%P

SINGLE LINE TEXT

If you are using "Single Line Text", type the code in the sentence. While you are typing, the code will appear, in the sentence, on the command line. But when you are finished typing, and press <enter>, the symbol will appear.

MULTILINE TEXT

The code displays the same as single line text if you type your text in the text editor. But multiline text offers another method for inserting special character "symbols".

1. While you are typing in the multiline text editor, instead of typing the code, right click and the shortcut menu will appear.

2. Select "Symbol".

3. Select the symbol desired and the code or the actual symbol will automatically appear in the sentence. (*This is not a great time saver but it does relieve you of having to memorize frequently used codes)*

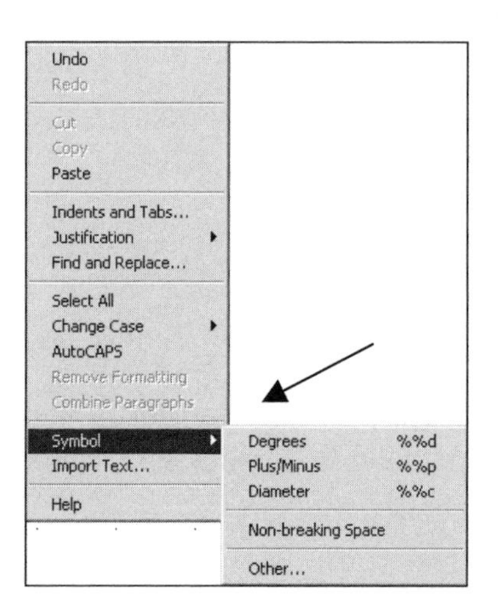

AutoCAD has many special characters but these are the most frequently used. Refer to the AutoCAD users guide for a list of the special characters

PREFIX and SUFFIX

The **PREFIX** (before) and the **SUFFIX** (after) allows you to preset text to be inserted automatically as you dimension.

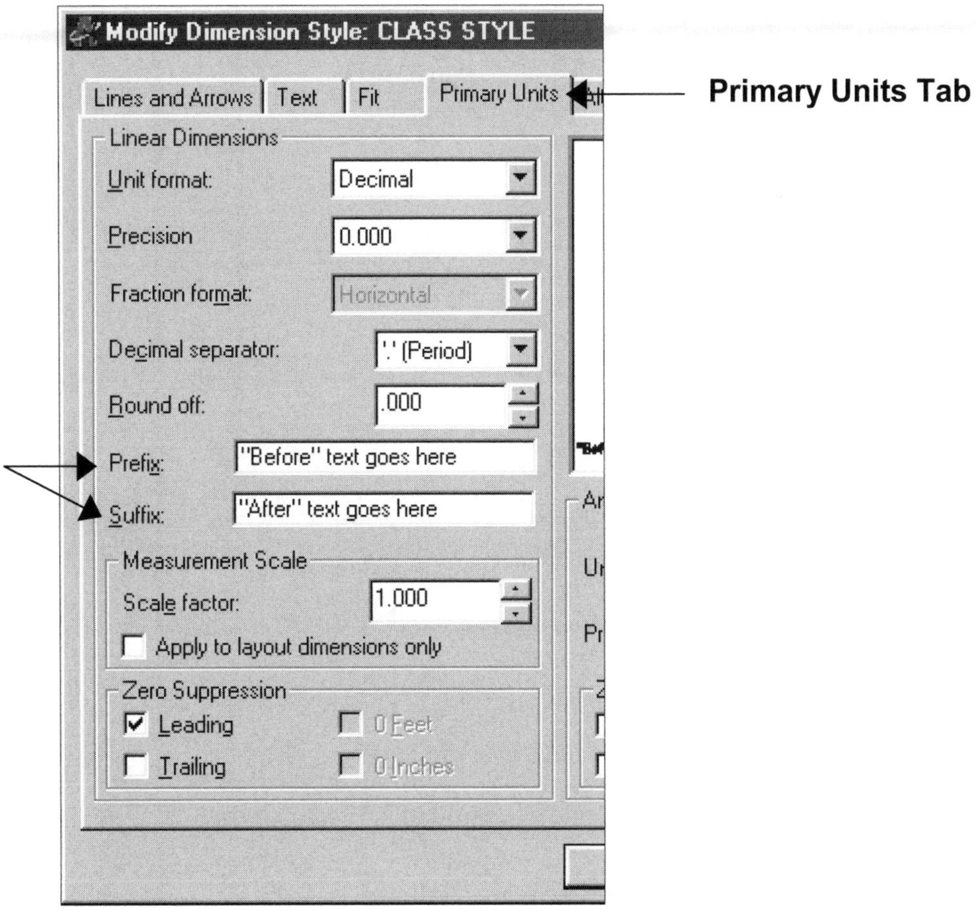

Primary Units Tab

Examples:

If you wanted all of your dimensions to end with ***Ref***, you would type **<space>Ref** in the suffix box.

If you wanted all of your dimensions to begin with ***2 places***, you would type **2 places** in the prefix box.

Note: If you enter text in the Prefix box when drawing Radial dimensions, the **"R"** for radius or the **diameter symbol**, will not be automatically drawn. You must add the symbol code, also, to the prefix box.

Example: If you would like a diameter dimension to have 2X as a Prefix, you must type the following, in the Prefix box: **2X <space> %%C**.

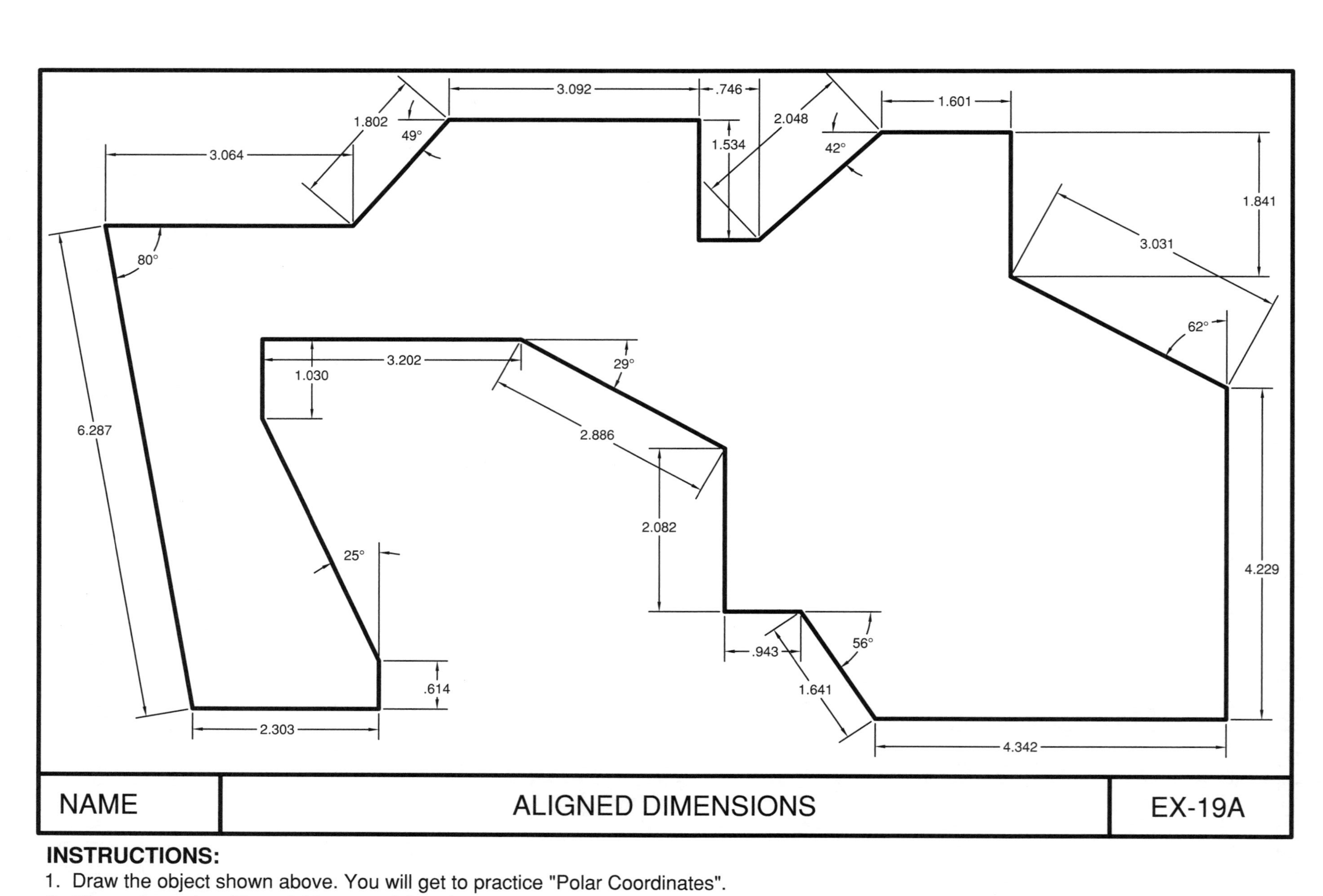

INSTRUCTIONS:

1. Draw the object shown above. You will get to practice "Polar Coordinates".
2. Dimension using Aligned, Linear and Angular.
3. Use Dimension Style "Class Style".
4. Save as 19A and Plot.

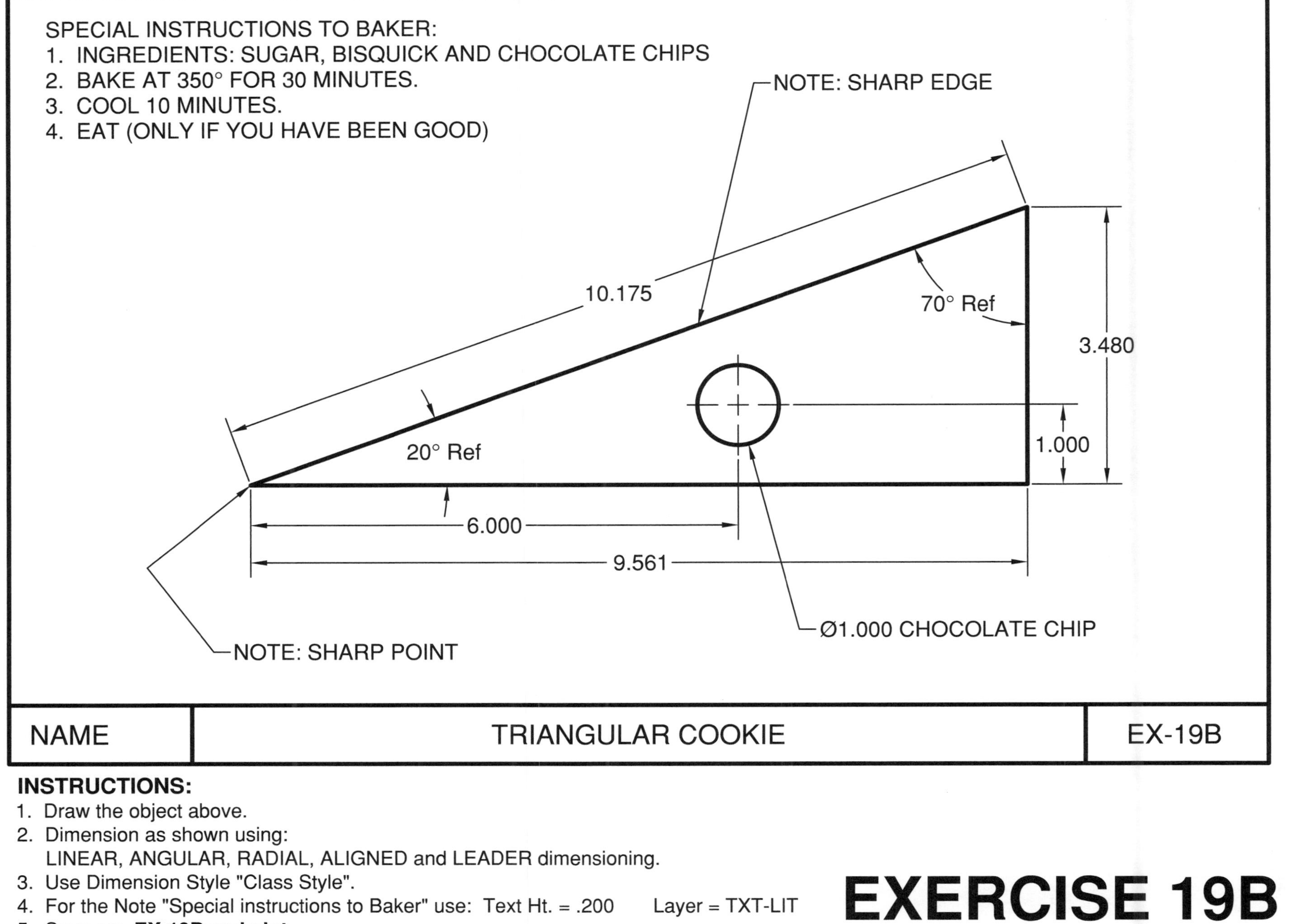

INSTRUCTIONS:

1. Draw the object above.
2. Dimension as shown using:
   LINEAR, ANGULAR, RADIAL, ALIGNED and LEADER dimensioning.
3. Use Dimension Style "Class Style".
4. For the Note "Special instructions to Baker" use: Text Ht. = .200 Layer = TXT-LIT
5. Save as: **EX-19B and plot.**

EXERCISE 19B

NOTES:

LEARNING OBJECTIVES

After completing this lesson, you will be able to:

1. Use Multiple Automatic dimensioning.
2. Edit Multiple dimensions.

LESSON 20

QUICK DIMENSION

Sorry LT users, this command is not available to you. Skip to Lesson 21.

Quick Dimension creates multiple dimensions with one command. Quick Dimension can create Continuous, Baseline, Radius and Diameter dimensions. *(Ordinate will be discussed in the Advanced workbook)*

1. Select the **Quick Dimension** command using one of the following:

TYPE = QDIM
PULLDOWN = Dimension/Quick Dimension
TOOLBAR = DIMENSION

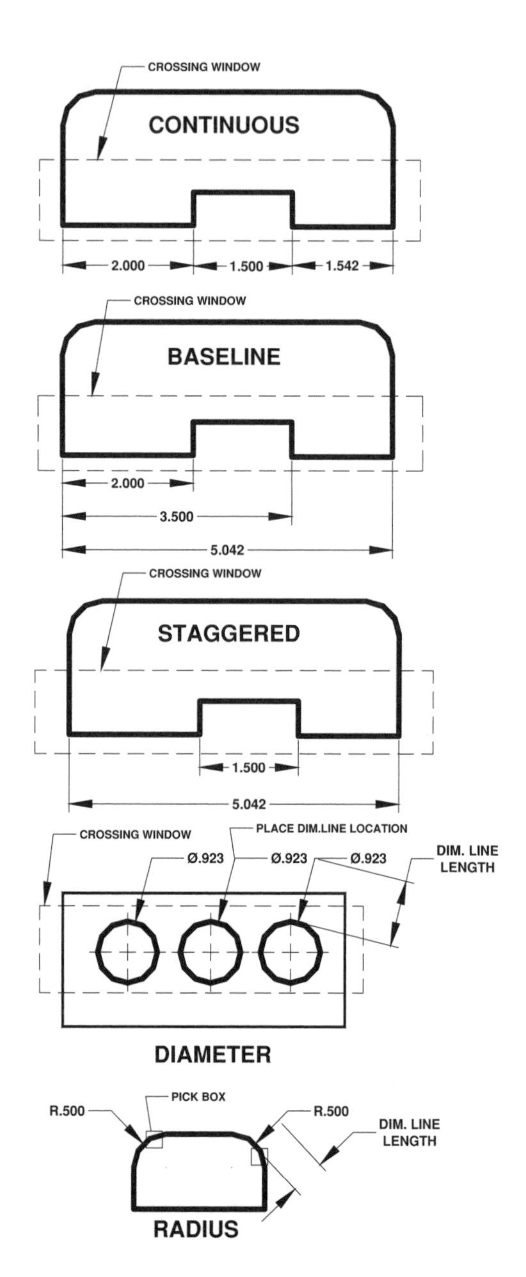

CONTINUOUS
2. Select the objects to be dimensioned with a crossing window or pick each object
3. Press <enter> to stop
4. Select **"C"** <enter> for ***Continuous.***
5. Select the location of the dimension line.

BASELINE
2. Select the objects to be dimensioned with a crossing window or pick each object.
3. Press <enter> to stop
4. Select **"B"** <enter> for ***Baseline.***
5. Select the location of the dimension line.

STAGGERED
2. Select the objects to be dimensioned with a crossing window or pick each object.
3. Press <enter> to stop
4. Select **"S"** <enter> for ***Staggered***.
5. Select the location of the dimension line.

DIAMETER
2. Select the objects to be dimensioned with a crossing window or pick each object (Qdim will automatically filter out any linear dims)
3. Press <enter> to stop
4. Select **"D"** <enter> for ***Diameter***
5. Select the location of the dimension line.

(Dimension line length is determined by the "Baseline Spacing" setting)

RADIUS
2. Select the objects to be dimensioned with a crossing window or pick each object. ***(Quick Dimension will automatically filter out any linear dimensions)***
3. Press <enter> to stop selecting objects
4. Select **"R"** <enter> for ***Radius***.

The dimensions are automatically placed, you do not select the location for the dimension line. *Dimension line length is determined by the "Baseline Spacing" setting)*

EDITING MULTIPLE DIMENSIONS

You can edit existing multiple dimensions using the QDIM / Edit command. The Qdim, edit command will edit all multiple dimensions, no matter whether they were created originally with Qdim or not. All multiple: linear, baseline and continue dimensions respond to this editing command.

1. Select the QDIM command.
2. Select the dimensions to edit.
3. Press <enter> to stop selecting.
4. Select "E" for edit.

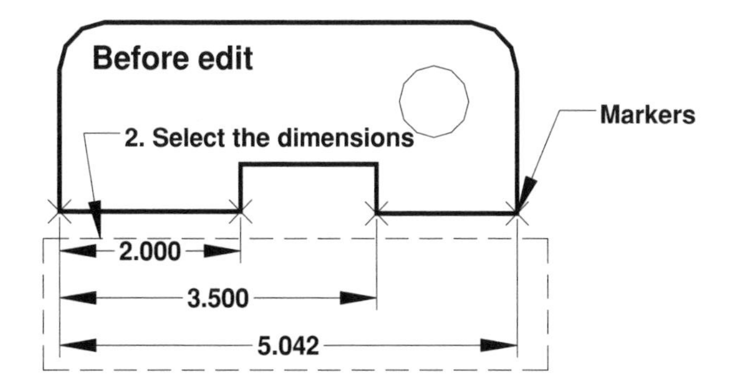

Small ***markers*** will appear at the extension line origins. These markers are called **"dimension points".** You can **Add** or **Remove** dimensions by selecting these markers.

REMOVE

AutoCAD assumes that you want to Remove dimensions, so this is the default setting.

5. Indicate dimension point to remove, or [Add/eXit] <eXit>:.
6. Click on the extension line markers you want to remove.
7. Press "X" to stop selecting.
8. Press "ESC" to stop.

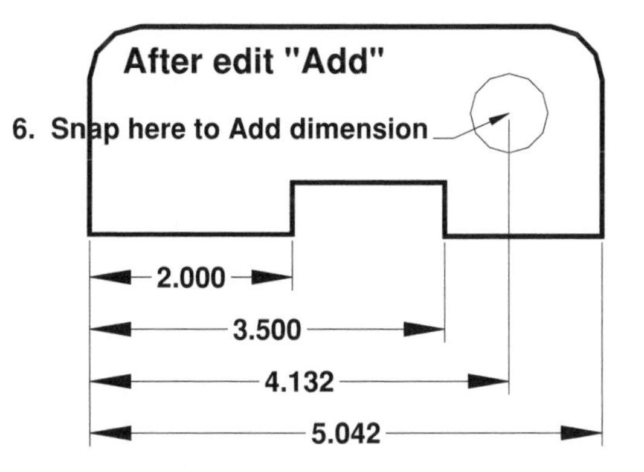

ADD

5. Select "A" <enter> for Add
6. Using Object Snap, select the extension line origins of the dimension you want to Add.
7. Press "X" to stop selecting.
8. Place the dimension line location.

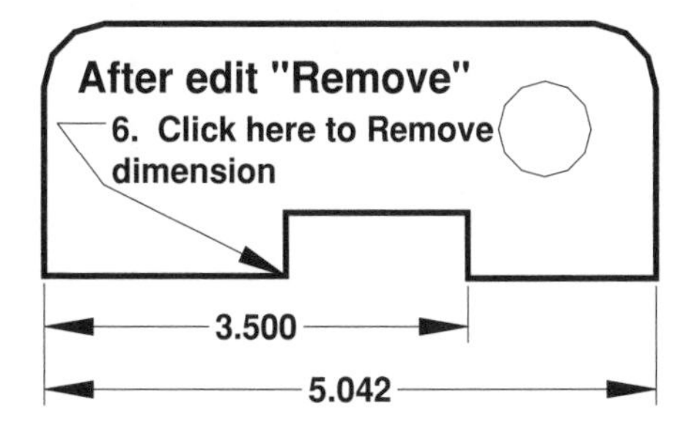

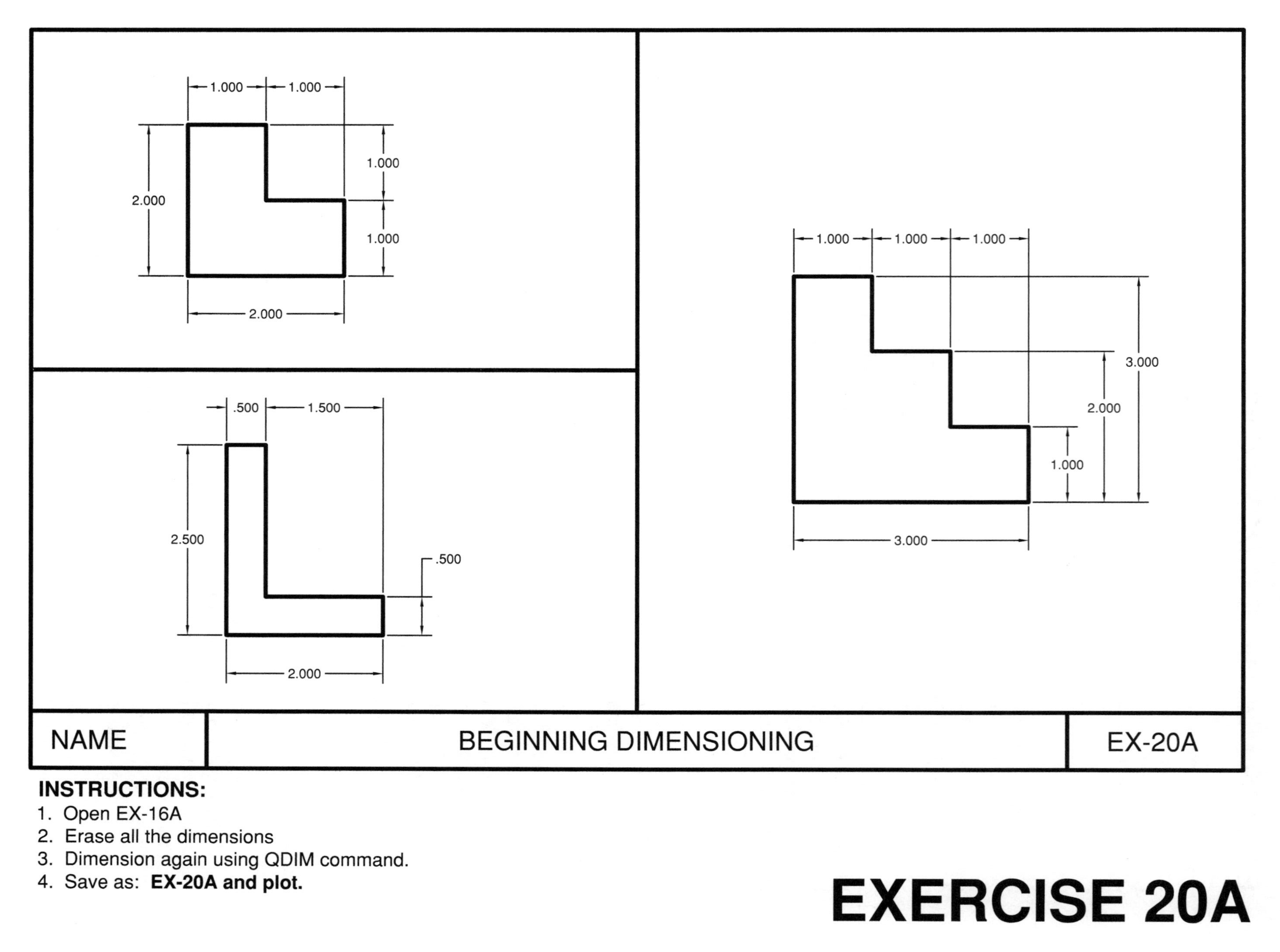

INSTRUCTIONS:
1. Open EX-16A
2. Erase all the dimensions
3. Dimension again using QDIM command.
4. Save as: **EX-20A and plot.**

EXERCISE 20A

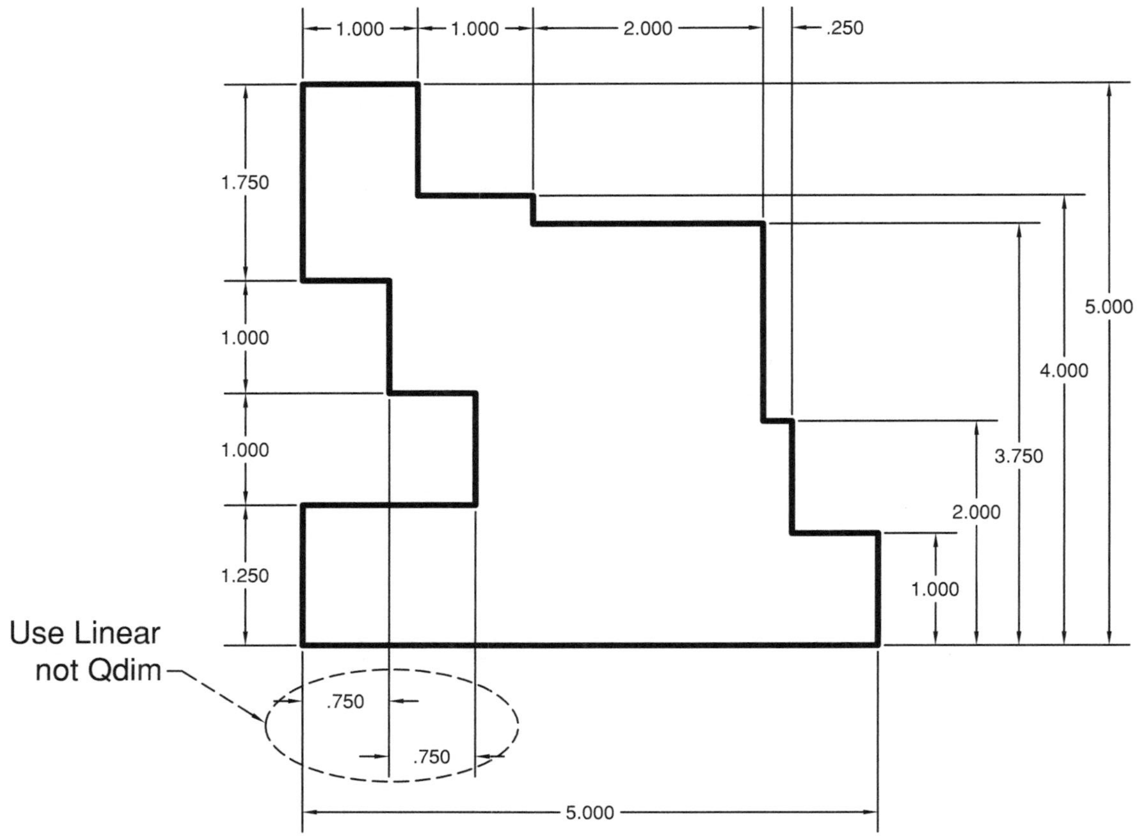

1.000 1.000 2.000 .250

1.750

1.000

1.000

1.250

5.000 4.000 3.750 2.000 1.000

Use Linear not Qdim

.750 .750

5.000

NAME	DIFFICULT DIMENSIONING	EX-20B

INSTRUCTIONS:

1. Open EX-16B.
2. Erase all of the Dimensions.
3. Dimension again using QDIM.
4. Save as: **EX-20B and plot.**

EXERCISE 20B

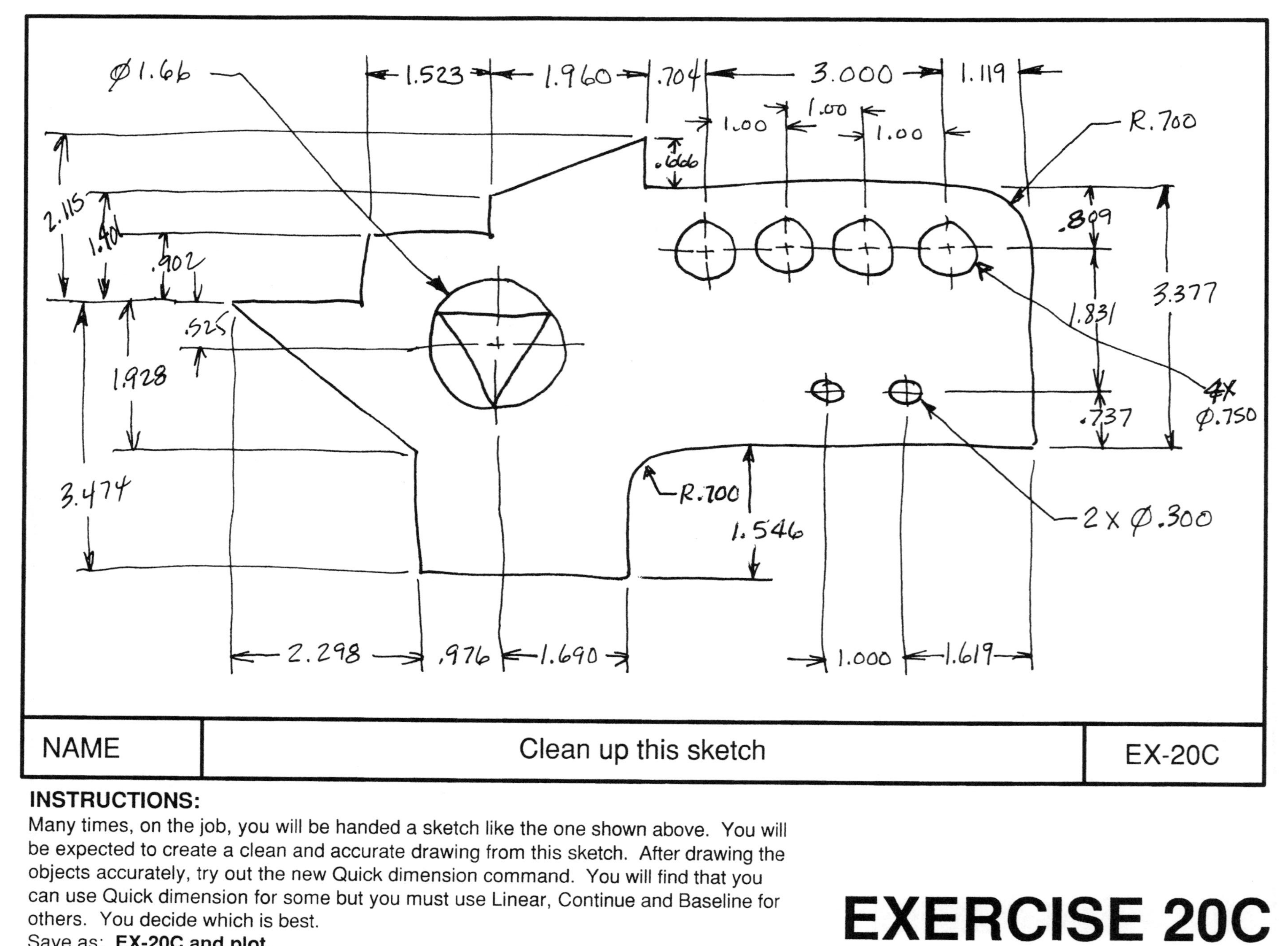

INSTRUCTIONS:

Many times, on the job, you will be handed a sketch like the one shown above. You will be expected to create a clean and accurate drawing from this sketch. After drawing the objects accurately, try out the new Quick dimension command. You will find that you can use Quick dimension for some but you must use Linear, Continue and Baseline for others. You decide which is best.

Save as: **EX-20C and plot.**

EXERCISE 20C

LEARNING OBJECTIVES

After completing this lesson, you will be able to:

1. Change an object's properties to match the properties of another object.
2. Create a Revision Cloud
3. Cover part of the drawing with a blank patch.

LESSON 21

MATCH PROPERTIES

Match Properties is used to "paint" the properties of one object to another. This is a simple and useful command. You first select the object that has the desired properties (the source object) and then select the object you want to "paint" the properties to (destination object).

Only one "source object" can be selected but its properties can be painted to any number of "destination objects".

1. Select the Match Properties command using one of the following:

   TYPE = MATCHPROP or MA
   PULLDOWN = MODIFY / MATCH PROPERTIES
   TOOLBAR = STANDARD

   Command: matchprop
2. Select source object: ***select the object with the desired properties to match***

3. Select destination object(s) or [Settings]: ***select the object(s) you want to receive the matching properties.***

4. Select destination object(s) or [Settings]: ***select more objects or <enter> to stop.***

Note: If you do not want to match all of the properties, right click and select "Settings" from the short cut menu, before selecting the destination object. Uncheck all the properties you do not want to match and select the OK button. Then select the destination object(s).

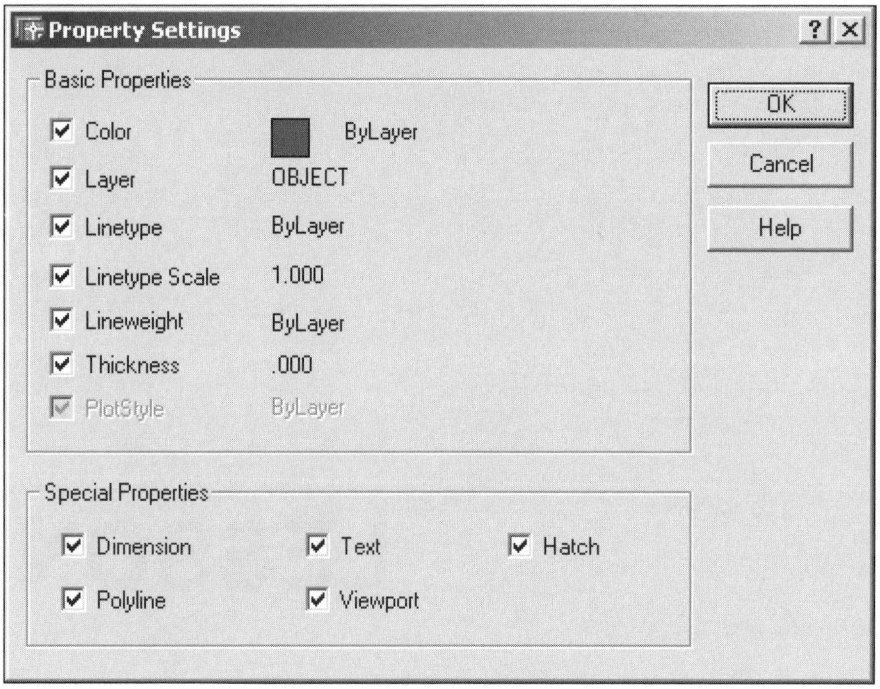

Property Settings
Basic Properties
Color ByLayer
Layer OBJECT
Linetype ByLayer
Linetype Scale 1.000
Lineweight ByLayer
Thickness .000
PlotStyle ByLayer
OK
Cancel
Help
Special Properties
Dimension
Text
Hatch
Polyline
Viewport

CREATING A REVISION CLOUD

When you make a revision to a drawing it is sometimes helpful to highlight the revision for someone viewing the drawing. A common method to highlight the area is to draw a "Revision Cloud" around the revised area. This can be accomplished easily with the "Revision Cloud" command.

The Revision Cloud command creates a series of sequential arcs to form a cloud-shaped object. You set the minimum and maximum arc lengths. (Maximum arc length cannot exceed three times the minimum arc length.) If you set the minimum and maximum different lengths the arcs will vary in size and will display an irregular appearance.

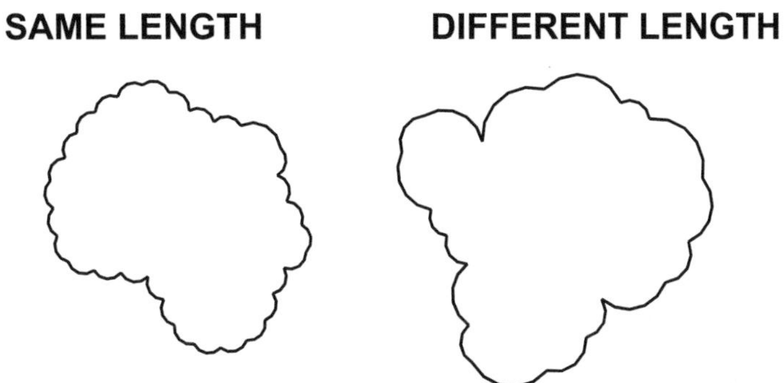

To draw a Revision Cloud you specify the start point with a left click but then gently drag the cursor to form the outline. AutoCAD automatically draws the arcs. When the cursor gets very close to the start point, AutoCAD snaps the last arc to the first arc and closes the shape.

1. Select the Revision Cloud command using one of the following:

   TYPE = REVCLOUD
   PULLDOWN = DRAW / REVISION CLOUD
   PULLDOWN (LT) = TOOLS / REVISION CLOUD
   TOOLBAR = DRAW

   Command: _revcloud
   Minimum arc length: .500000   Maximum arc length: .500000

2. Specify start point or [Arc length/Object] <Object>: ***Select "Arc length"***

3. Specify minimum length of arc <.500000>: ***Specify the minimum arc length***

4. Specify maximum length of arc <.500000>: ***Specify the maximum arc length***

5. Specify start point or [Object] <Object>: ***Place cursor at start location and click.***

6. Guide crosshairs along cloud path...***Move the cursor to create the cloud outline.***

7. Revision cloud finished. ***When the cursor approaches the start point, the cloud closes automatically.***

CONVERT A CLOSED OBJECT INTO A REV CLOUD

You can convert a closed object, such as a circle, ellipse, rectangle or closed polyline to a revision cloud. The original object is deleted when it is converted.
(If you want the original object to remain, set the variable "delobj" to "0". The default setting is "1".)

1. Draw a closed object such as a circle.
2. Select the Revision Cloud command using one of the following:

   TYPE = REVCLOUD
   PULLDOWN = DRAW / REVISION CLOUD
   PULLDOWN (LT) = TOOLS / REVISION CLOUD
   TOOLBAR = DRAW

   Command: _revcloud
   Minimum arc length: .500000 Maximum arc length: .500000

3. Specify start point or [Arc length/Object] <Object>: ***Select "Arc length"***
4. Specify minimum length of arc <.500000>: ***Specify the minimum arc length***
5. Specify maximum length of arc <.500000>: ***Specify the maximum arc length***
6. Specify start point or [Object] <Object>: ***Select "Object".***
7. Select object: ***Select the object to convert***
8. Select object: Reverse direction [Yes/No] <No>: ***Select Yes or No***
   Revision cloud finished.

REVERSE DIRECTION

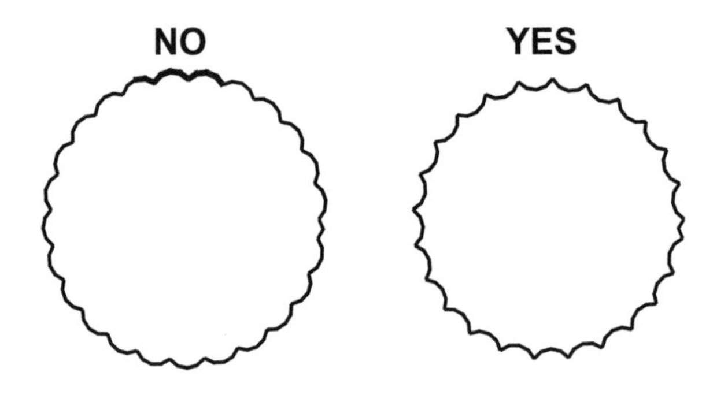

NOTE:
1. The Match Properties command will not match the arc length from the source cloud to the destination cloud.
2. Make sure that you can see the entire area to be outlined with REVCLOUD before you begin the command. REVCLOUD is not designed to support transparent and real-time panning and zooming.

WIPEOUT

The Wipeout command creates a blank area that covers existing objects. The area has a background that matches the background of the drawing area. This area is bounded by the wipeout frame, which you can turn on or off.

1. Select the Wipeout command using one of the following:

   TYPE = WIPEOUT
   PULLDOWN = DRAW / WIPEOUT
   TOOLBAR = NONE

2. Command: _wipeout Specify first point or [Frames/Polyline] <Polyline>: ***specify the first point of the shape (P1)***
3. Specify next point: ***specify the next point (P2)***
4. Specify next point or [Undo]: ***specify the next point (P3)***
5. Specify next point or [Undo]: ***specify the next point (P4)***
6. Specify next point or [Close/Undo]: ***specify the next point or <enter> to close***

BEFORE WIPEOUT **AFTER WIPEOUT**

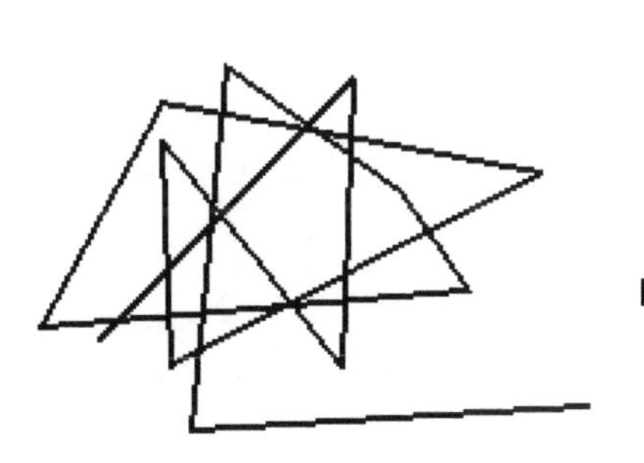

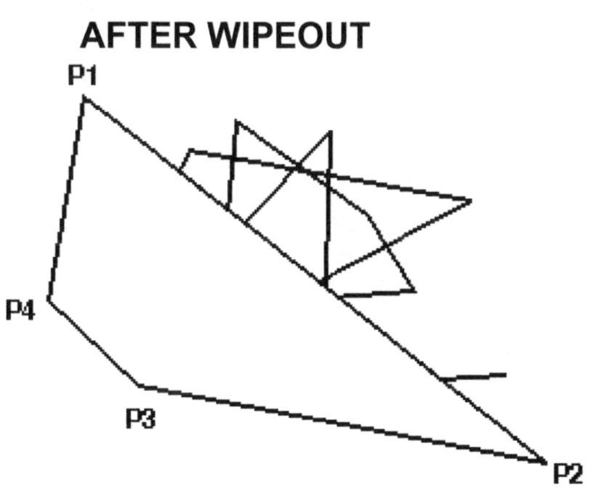

TURNING FRAMES ON OR OFF

1. Select the Wipeout command.
2. Select the "Frames" option.
3. Enter ON or OFF.

ON **OFF**

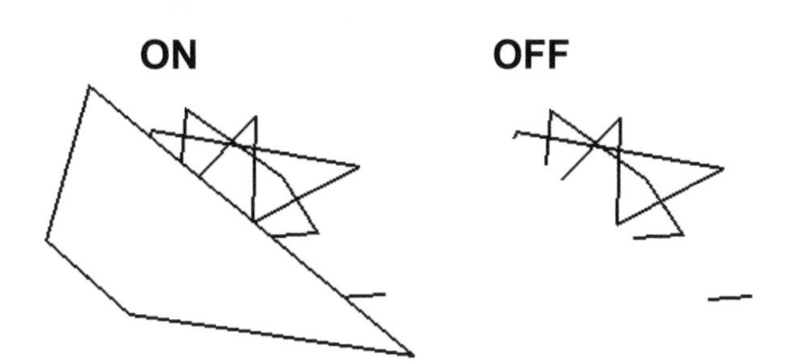

Note: If you want to move the objects and the wipeout area, move them together not separately. If you do move the objects and the wipeout, the objects under the wipeout area may reappear. Select **View / Regen** and they will disappear again.

EXERCISE 21A

INSTRUCTIONS:

1. Open **EX 5F.dwg**.
2. Using **Match Properties,** change the properties of the Polygon and Arcs to the same properties of the Donut.
3. Using Wipeout, block out the bottom donut approximately as shown.
4. **Save** this drawing as: **EX21A and Plot.**

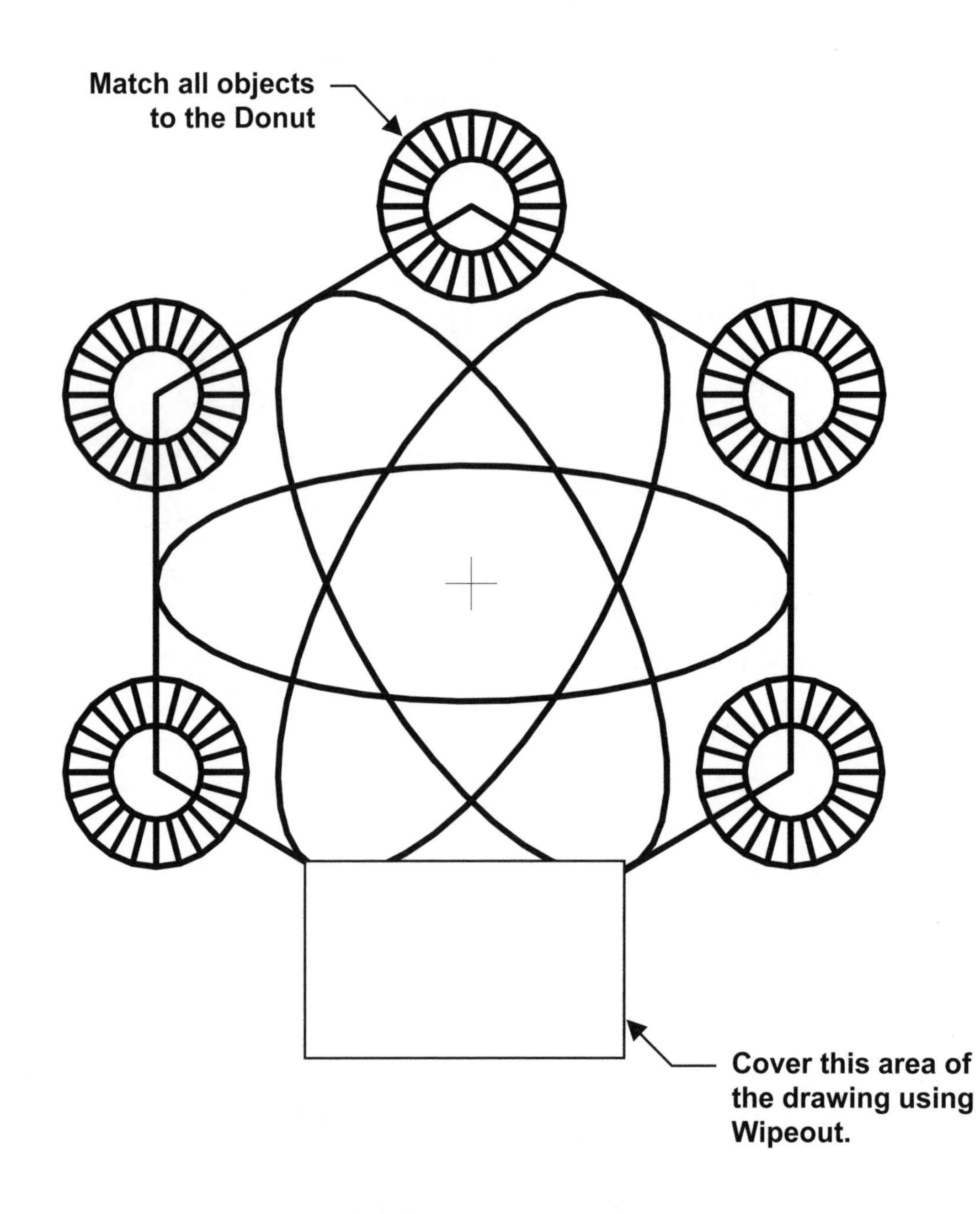

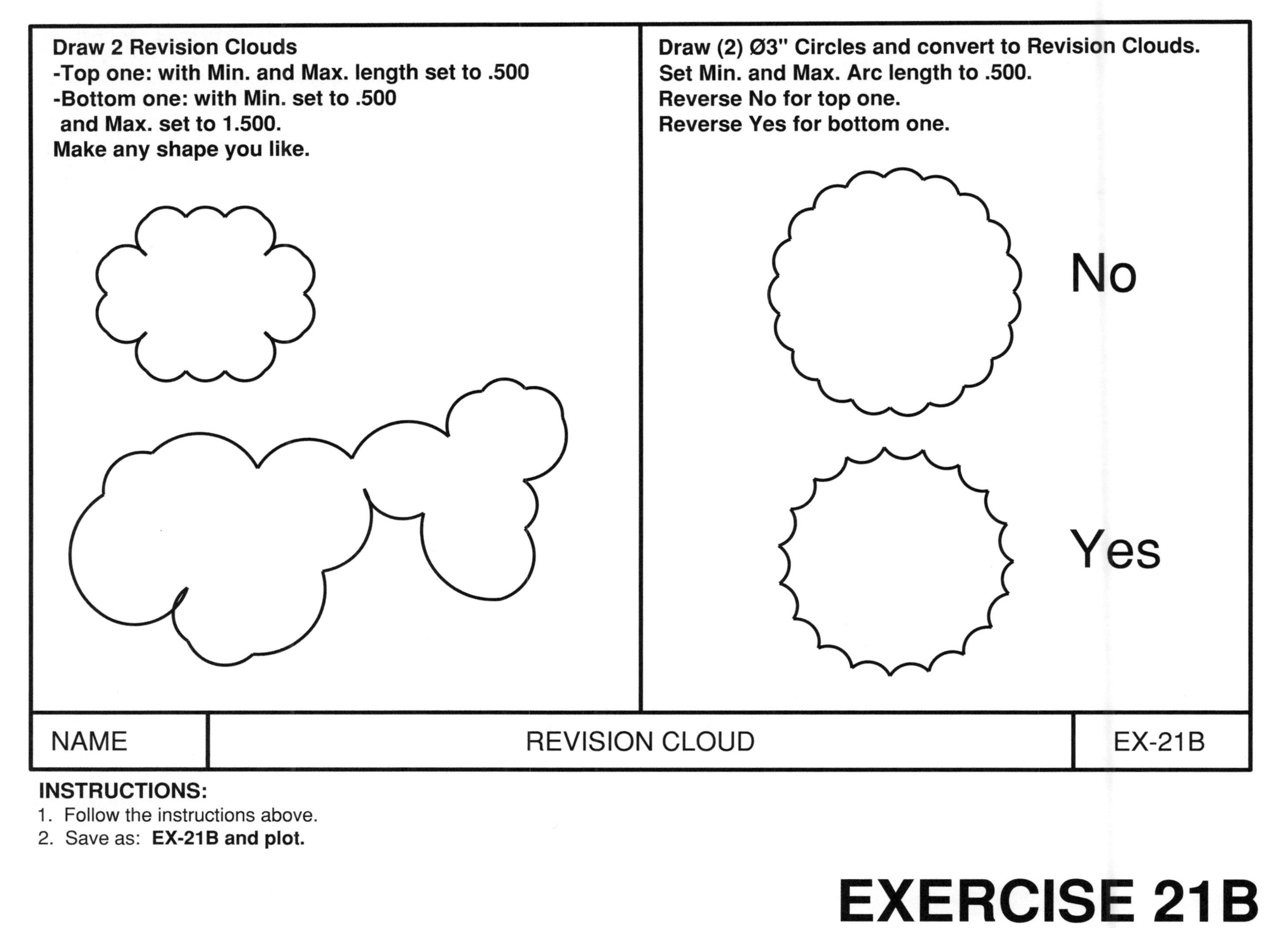

INSTRUCTIONS:

1. Follow the instructions above.
2. Save as: **EX-21B and plot.**

EXERCISE 21B

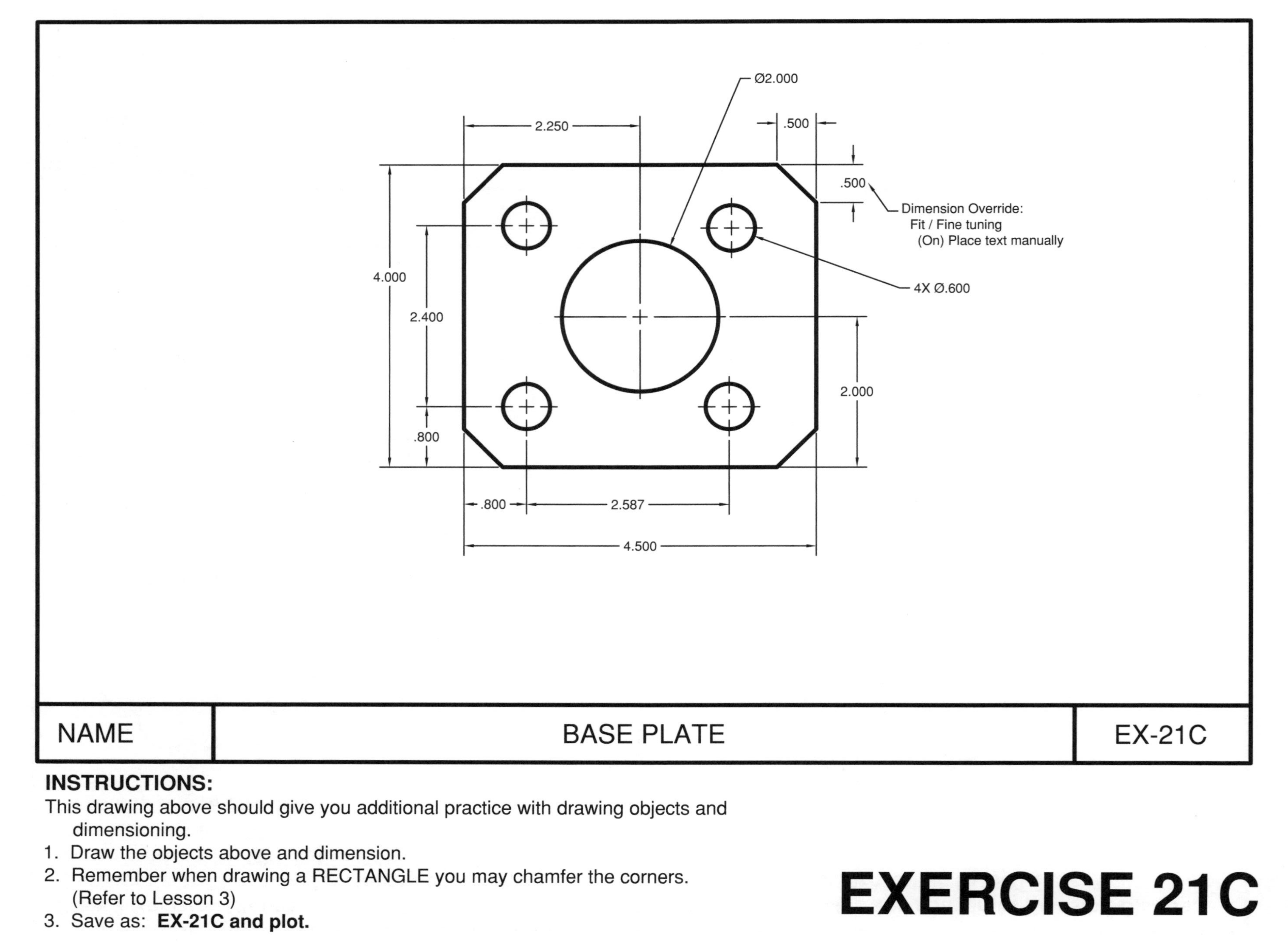

INSTRUCTIONS:

This drawing above should give you additional practice with drawing objects and dimensioning.

1. Draw the objects above and dimension.
2. Remember when drawing a RECTANGLE you may chamfer the corners. (Refer to Lesson 3)
3. Save as: **EX-21C and plot.**

EXERCISE 21C

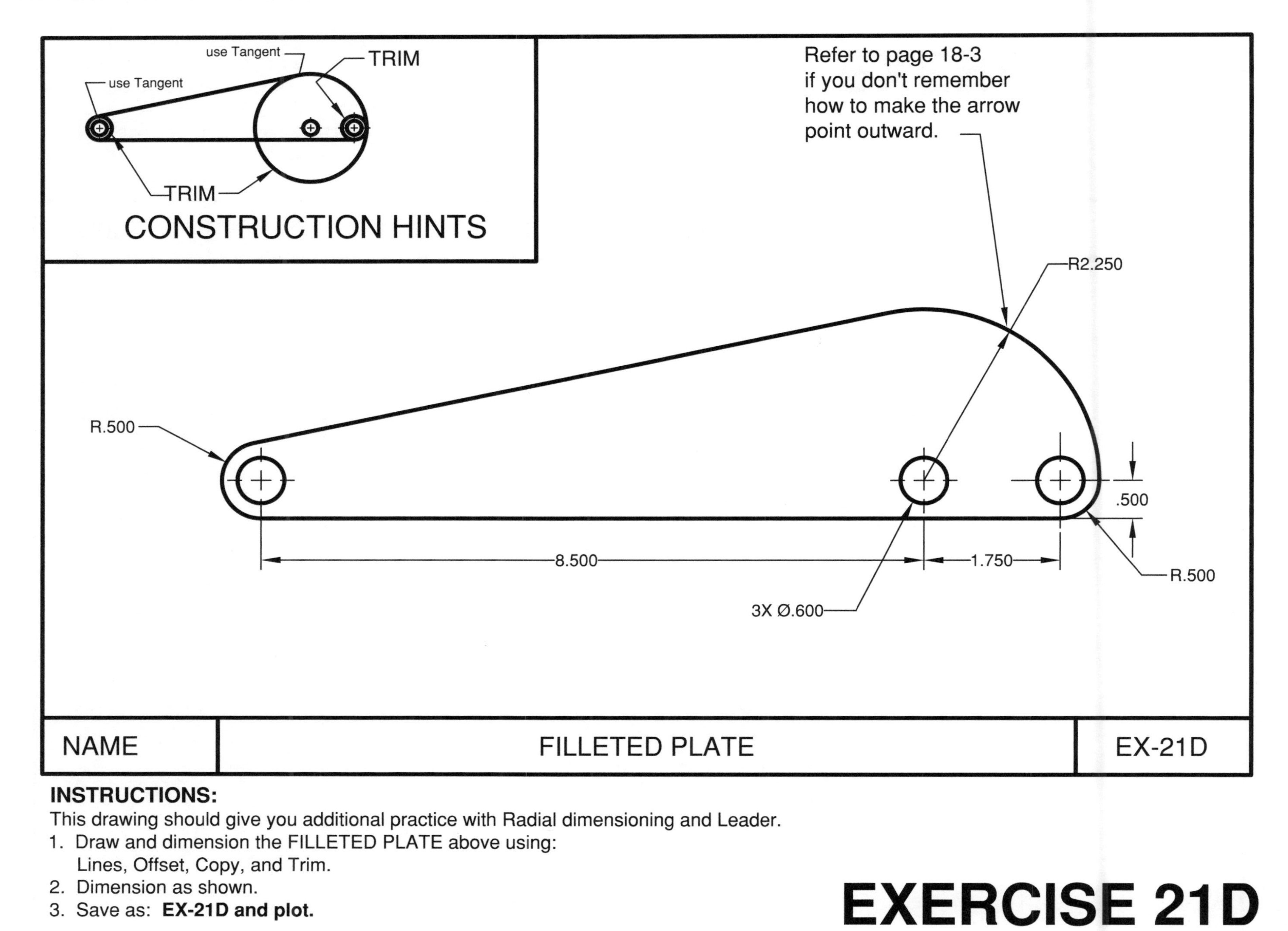

INSTRUCTIONS:
This drawing should give you additional practice with Radial dimensioning and Leader.
1. Draw and dimension the FILLETED PLATE above using:
   Lines, Offset, Copy, and Trim.
2. Dimension as shown.
3. Save as: **EX-21D and plot.**

EXERCISE 21D

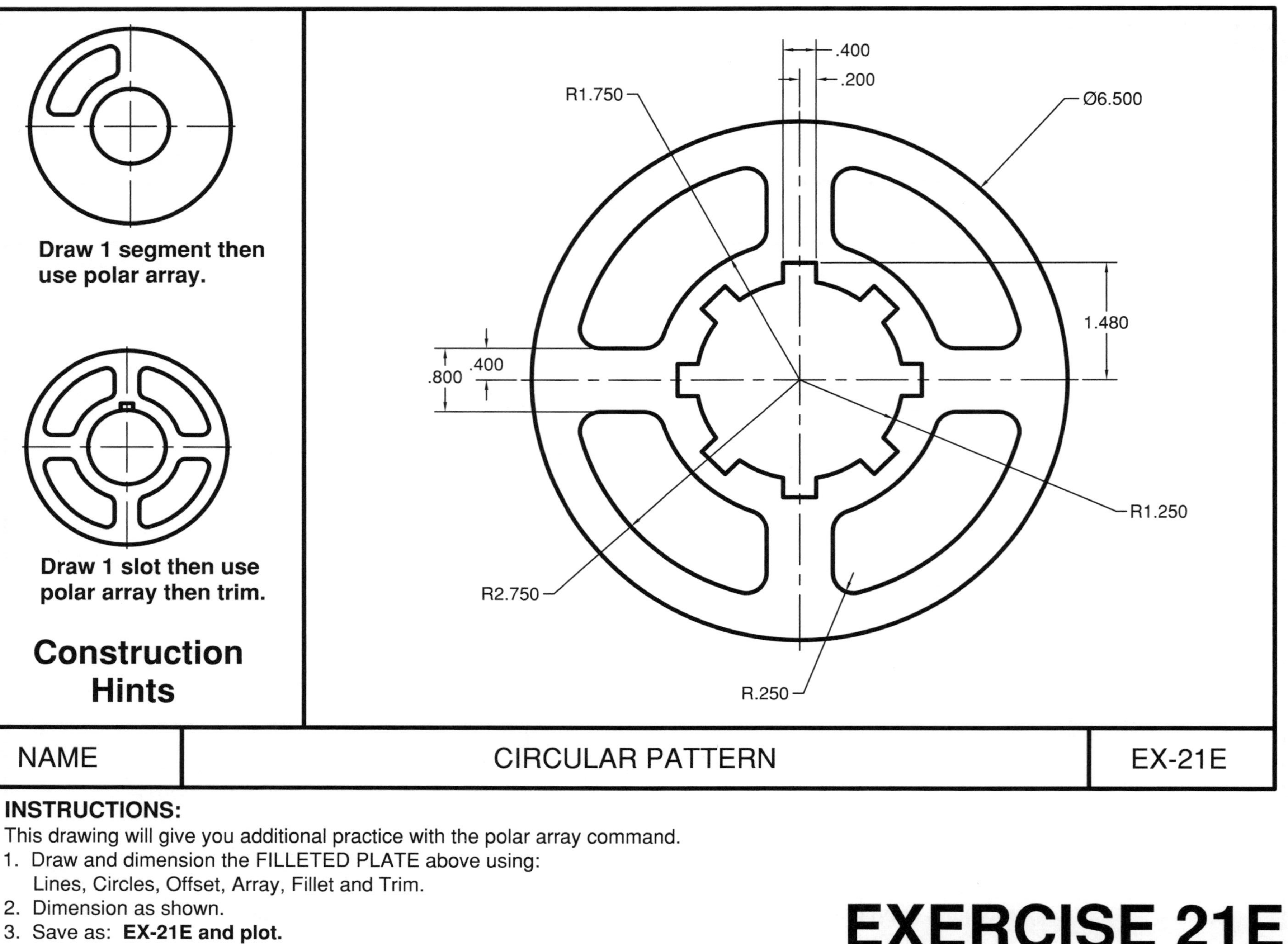

INSTRUCTIONS:

This drawing will give you additional practice with the polar array command.

1. Draw and dimension the FILLETED PLATE above using:
   Lines, Circles, Offset, Array, Fillet and Trim.
2. Dimension as shown.
3. Save as: **EX-21E and plot.**

EXERCISE 21E

LEARNING OBJECTIVES

After completing this lesson, you will be able to:

1. Draw an Arc using 10 different methods.

LESSON 22

ARC

TYPING = A
PULLDOWNS = DRAW / ARC
TOOLBARS =DRAW

There are 10 ways to draw an ARC in AutoCAD. Not all of the ARCS options are easy to create so you may find it is often easier to **trim a Circle** or use the **Fillet** command.

On the job, you will probably only use 2 of these methods. Which 2 depends on the application.

An **ARC** is a segment of a circle and must be less than 360 degrees.

Most ARCS are drawn counter-clockwise but you will notice, in the examples on the following pages, that some may be drawn clockwise by entering a negative input.

Examples of each of the ARC options are shown in EXERCISES 22A through 22C. Also, refer to the "Help" menu for additional example of the use of the Arc command.

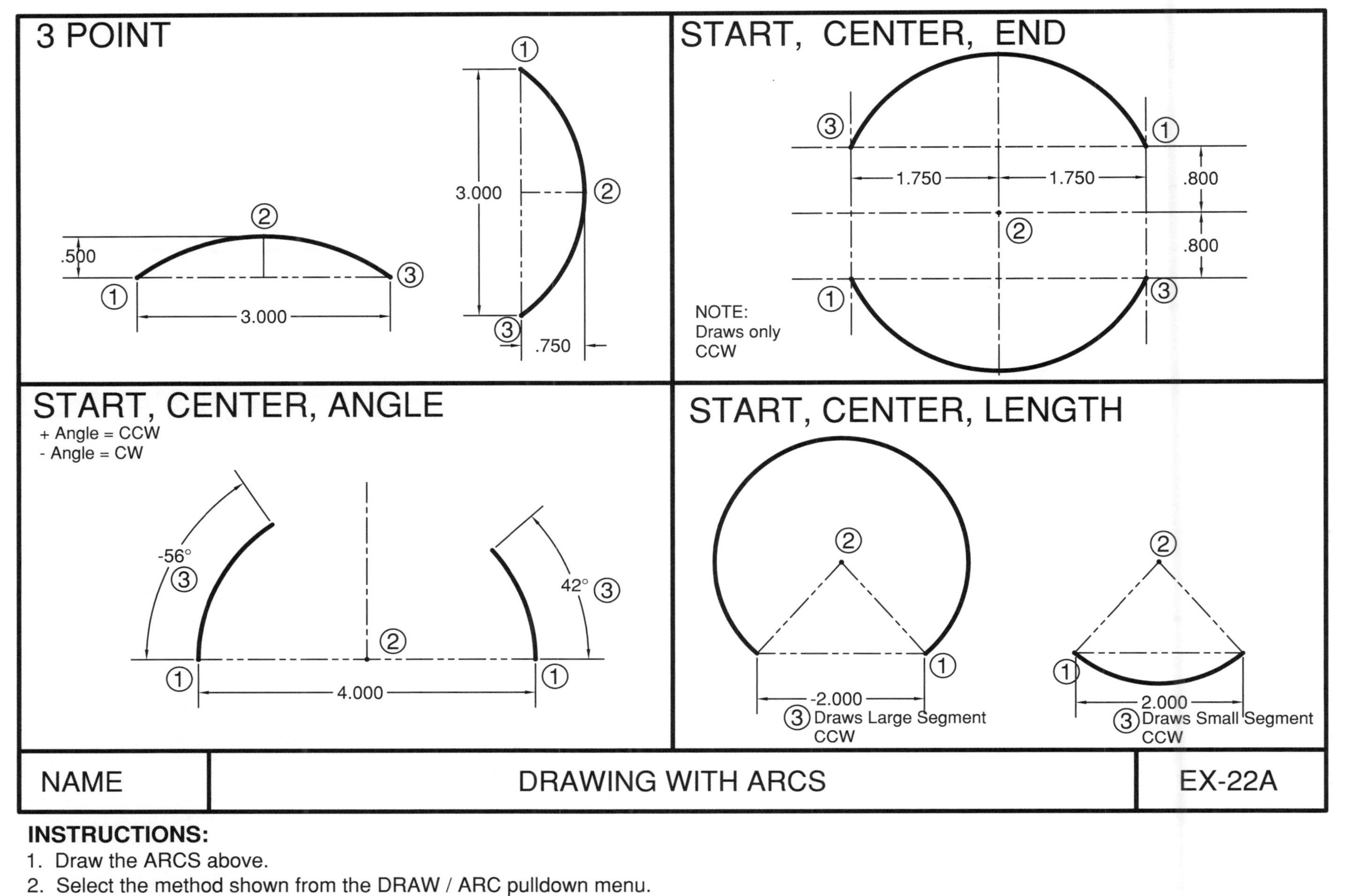

INSTRUCTIONS:

1. Draw the ARCS above.
2. Select the method shown from the DRAW / ARC pulldown menu.

Note: Yours may not appear exactly as the example...that's OK.

3. Save as: **EX-22A.**

EXERCISE 22A

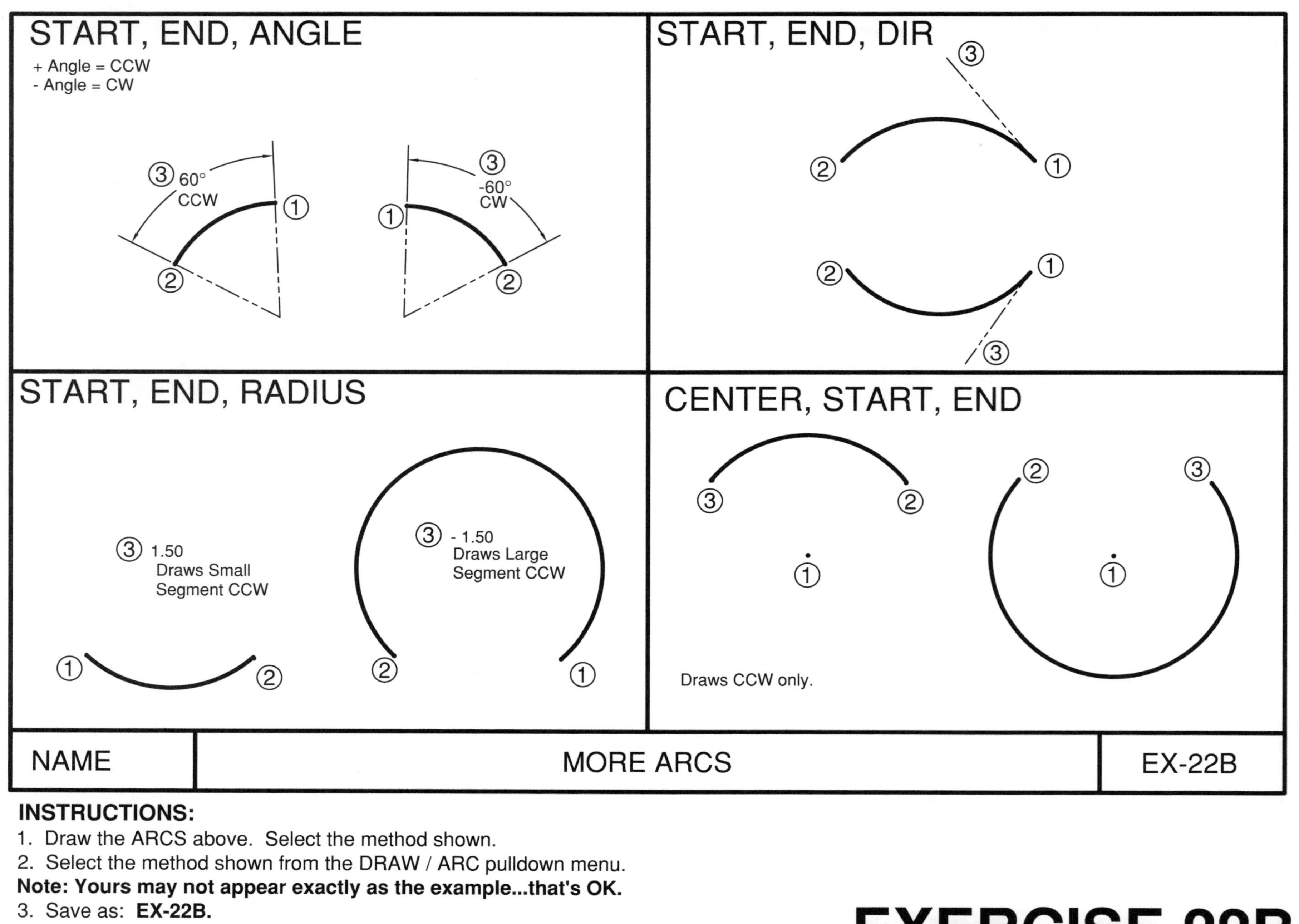

INSTRUCTIONS:

1. Draw the ARCS above. Select the method shown.
2. Select the method shown from the DRAW / ARC pulldown menu.

Note: Yours may not appear exactly as the example...that's OK.

3. Save as: **EX-22B.**

EXERCISE 22B

EXERCISE 22C

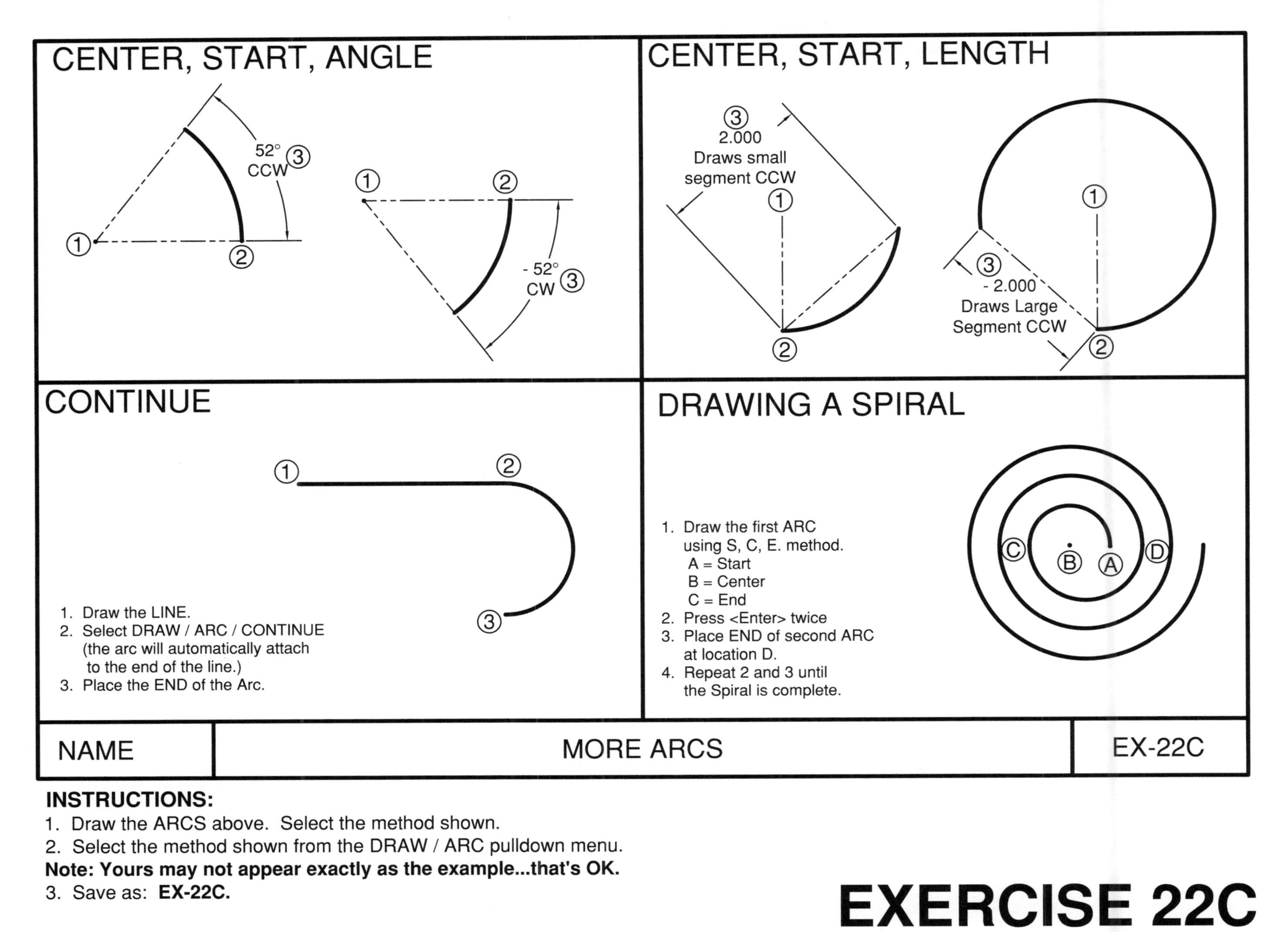

INSTRUCTIONS:

1. Draw the ARCS above. Select the method shown.
2. Select the method shown from the DRAW / ARC pulldown menu.

Note: Yours may not appear exactly as the example...that's OK.

3. Save as: **EX-22C.**

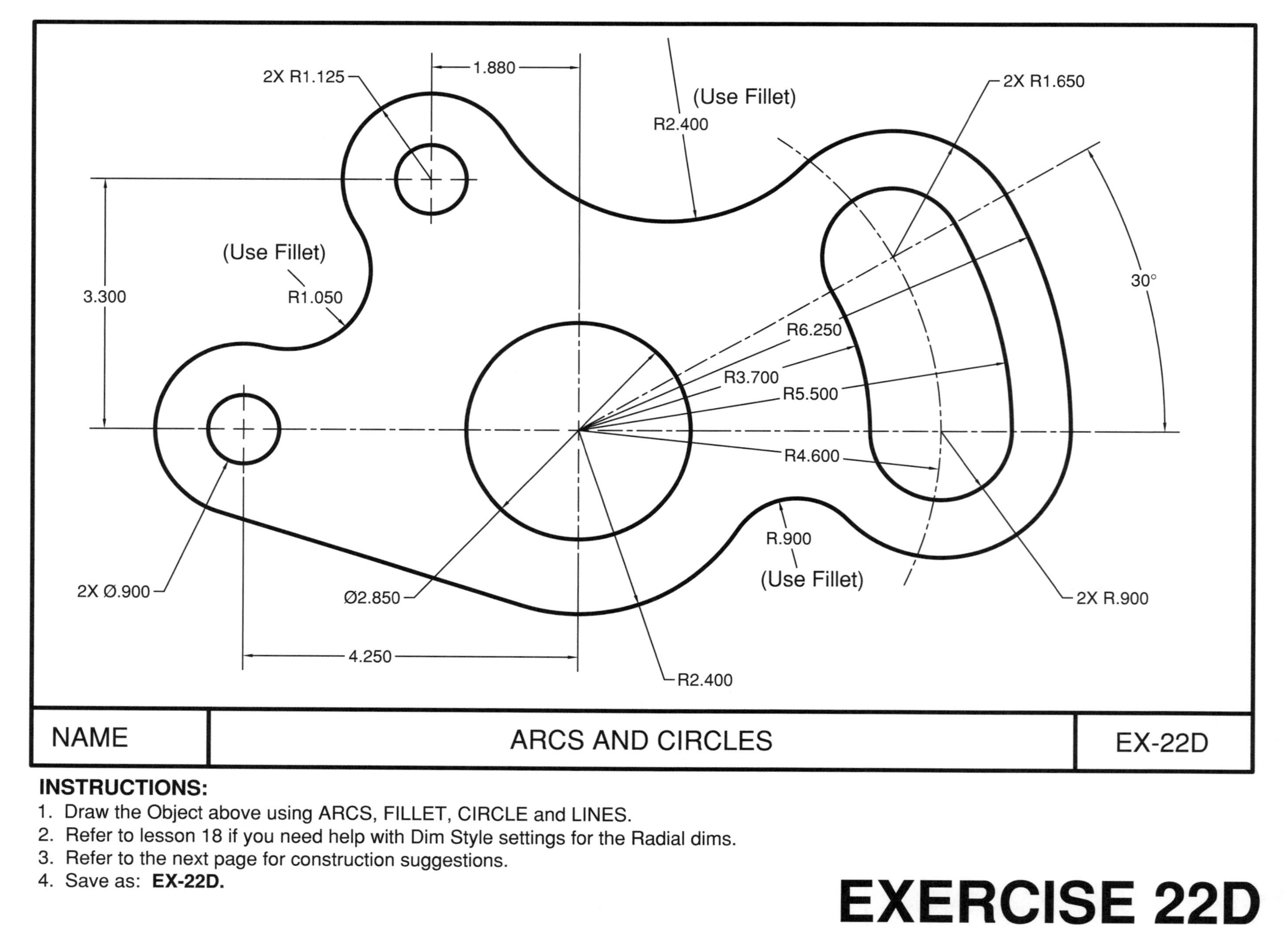

INSTRUCTIONS:

1. Draw the Object above using ARCS, FILLET, CIRCLE and LINES.
2. Refer to lesson 18 if you need help with Dim Style settings for the Radial dims.
3. Refer to the next page for construction suggestions.
4. Save as: **EX-22D.**

EXERCISE 22D

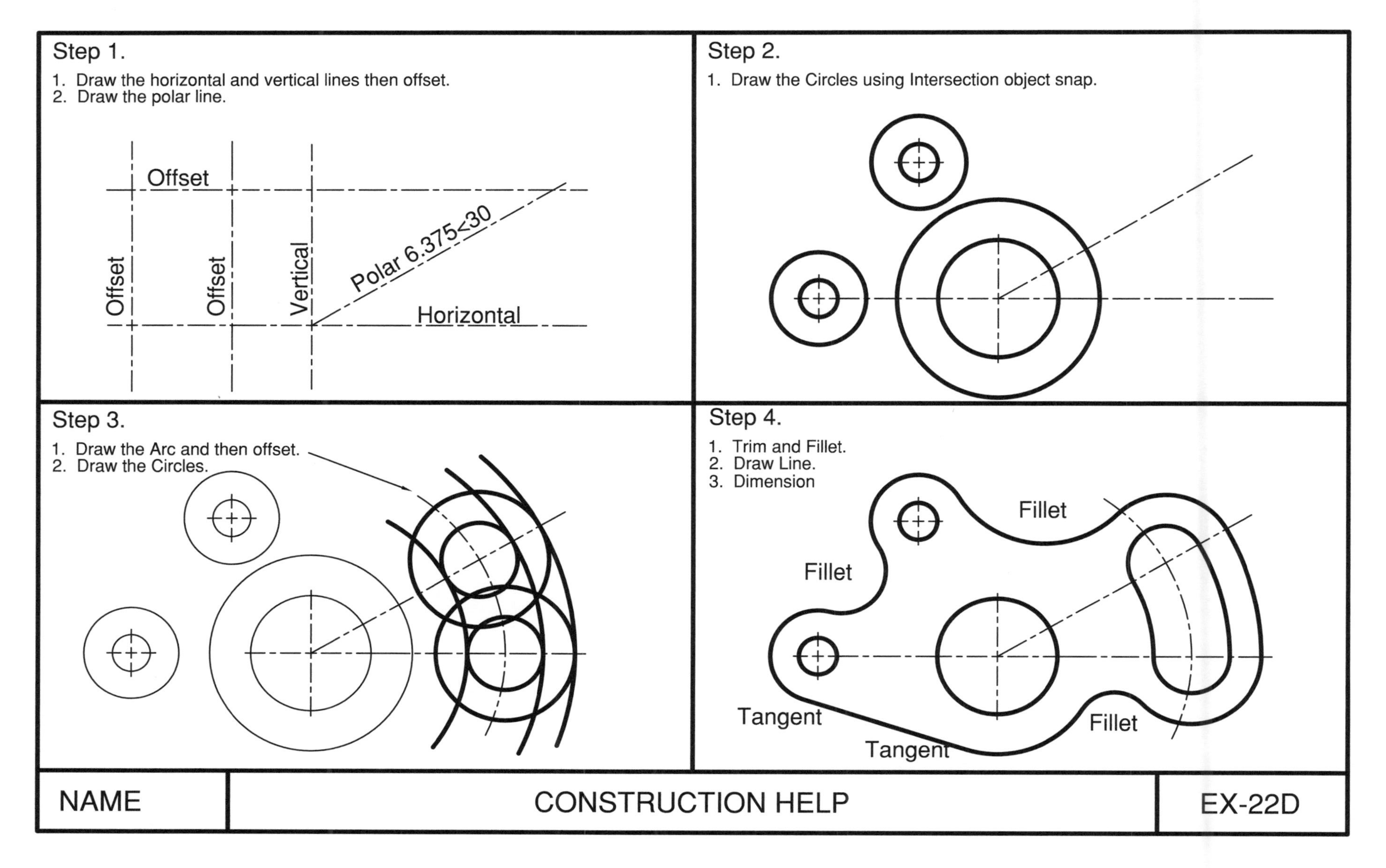

EXERCISE 22D

NOTES:

LEARNING OBJECTIVES

After completing this lesson, you will be able to:

1. Understand what is a Polyline.
2. Draw a Polyline and Polyarc.
3. Assign widths to polylines.
4. Set the Fill mode to On or Off.

LESSON 23

POLYLINES

A **POLYLINE** is very similar to a LINE. It is created in the same way a line is drawn. It requires first and second endpoints. But a POLYLINE has additional features, as follows:

1. A **POLYLINE** is ONE object, even though it may have many segments.
2. You may specify a specific width to each segment.
3. You may specify a different width to the start and end of a polyline segment.

THE FOLLOWING ARE EXAMPLES OF POLYLINES WITH WIDTHS ASSIGNED.

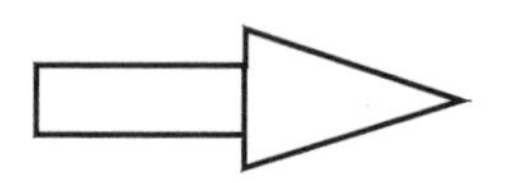

Fill Mode = Off

Fill Mode = On

To turn FILL MODE on or off.

1. Command: ***type*** **FILL *<enter>***
2. Enter mode [On / Off] <ON>: ***type ON or Off <enter>***
3. Command: ***type REGEN*** <enter> ***or select: View / Regen***

NOTE: If you explode a POLYLINE it loses its width and turns into a regular line.

THE FOLLOWING IS AN EXAMPLE OF DRAWING A POLYLINE WITH "WIDTH"

1. Select the POLYLINE command using one of the following:

   TYPE = PL
   PULLDOWN = DRAW / POLYLINE
   TOOLBAR = DRAW

   Command: _pline
2. Specify start point: ***place the first endpoint of the line***
   Current line-width is 0.000
3. Specify next point or [Arc/Close/Halfwidth/Length/Undo/Width]: ***type w*** <enter>
4. Specify starting width <0.000>: ***type the desired width*** <enter>
5. Specify ending width <0.000>: ***type the desired width*** <enter>
6. Specify next point or [Arc/Close/Halfwidth/Length/Undo/Width]: ***place the next endpoint***
7. Specify next point or [Arc/Close/Halfwidth/Length/Undo/Width]: ***place the next endpoint***
8. Specify next point or [Arc/Close/Halfwidth/Length/Undo/Width]: ***place the next endpoint***
9. Specify next point or [Arc/Close/Halfwidth/Length/Undo/Width]: ***type C <enter>***

OPTIONS:

WIDTH
Specify the starting and ending width.

You can create a tapered polyline by specifying different starting and ending widths.

HALFWIDTH
The same as Width except the starting and ending halfwidth specifies half the width rather than the entire width.

ARC
This option allows you to create a circular polyline less than 360 degrees.

CLOSE
The close option is the same as in the Line command. Close attaches the last segment to the first segment.

LENGTH
This option allows you to draw a polyline at the same angle as the last polyline drawn. This option is very similar to the OFFSET command. You specify the first endpoint and the length. The new polyline will automatically be drawn at the same angle as the previous polyline.

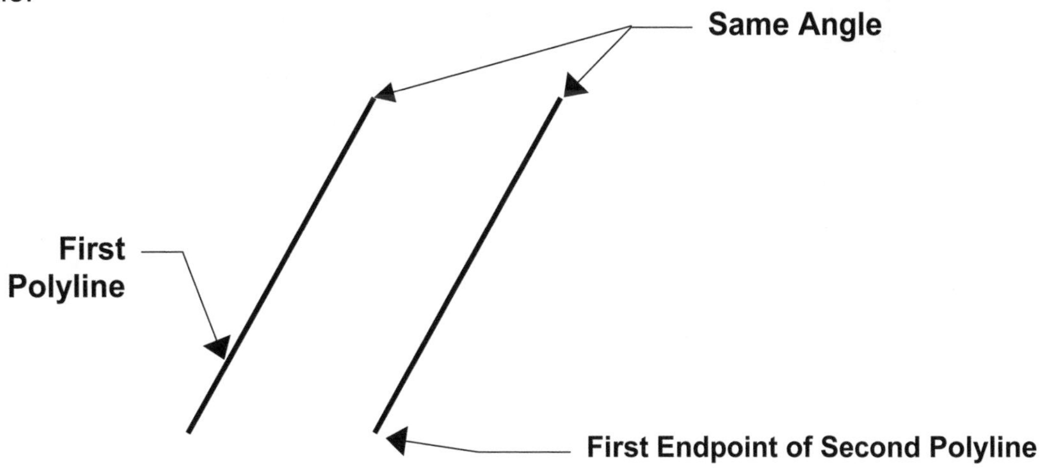

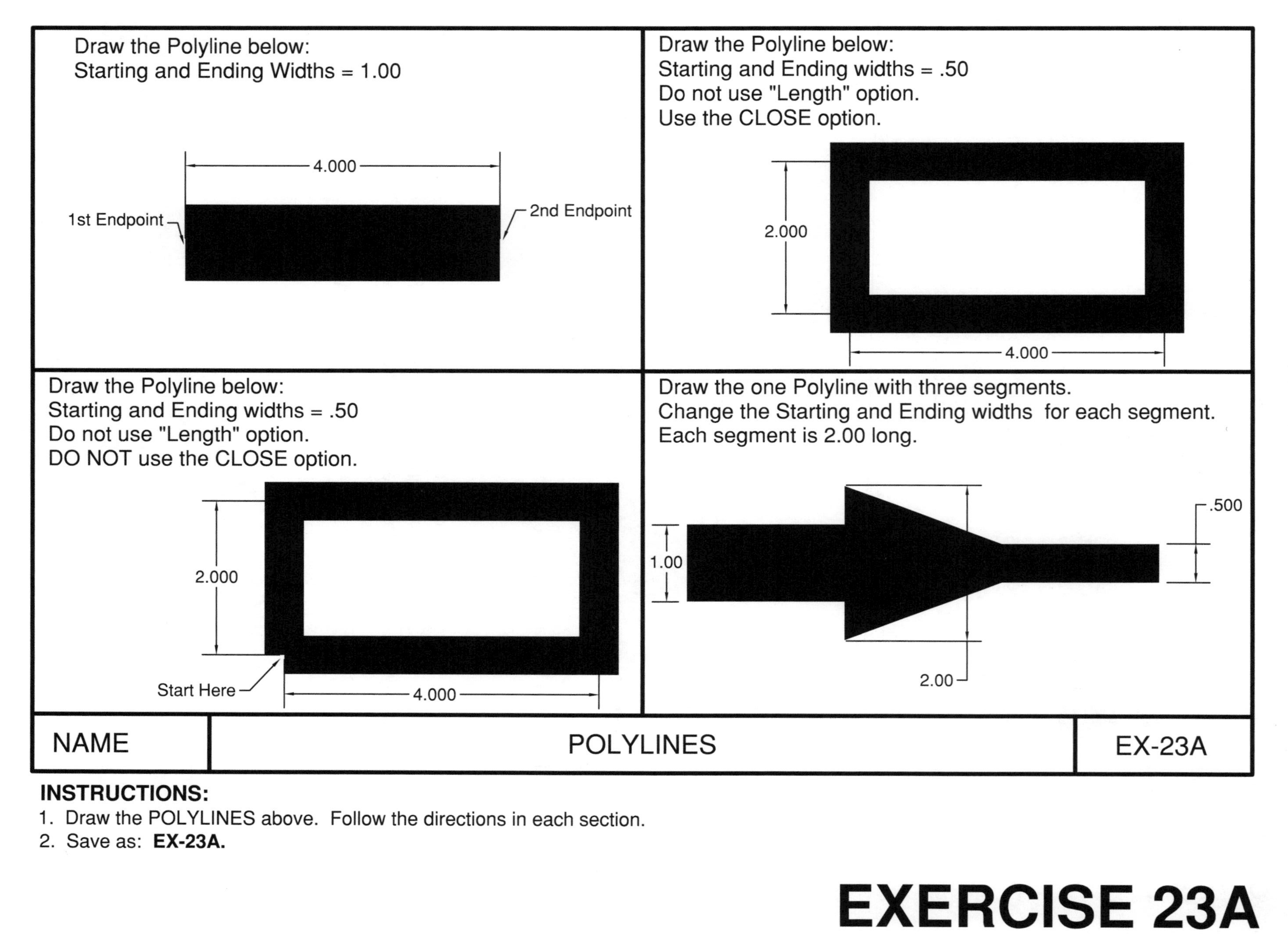

INSTRUCTIONS:

1. Draw the POLYLINES above. Follow the directions in each section.
2. Save as: **EX-23A.**

EXERCISE 23A

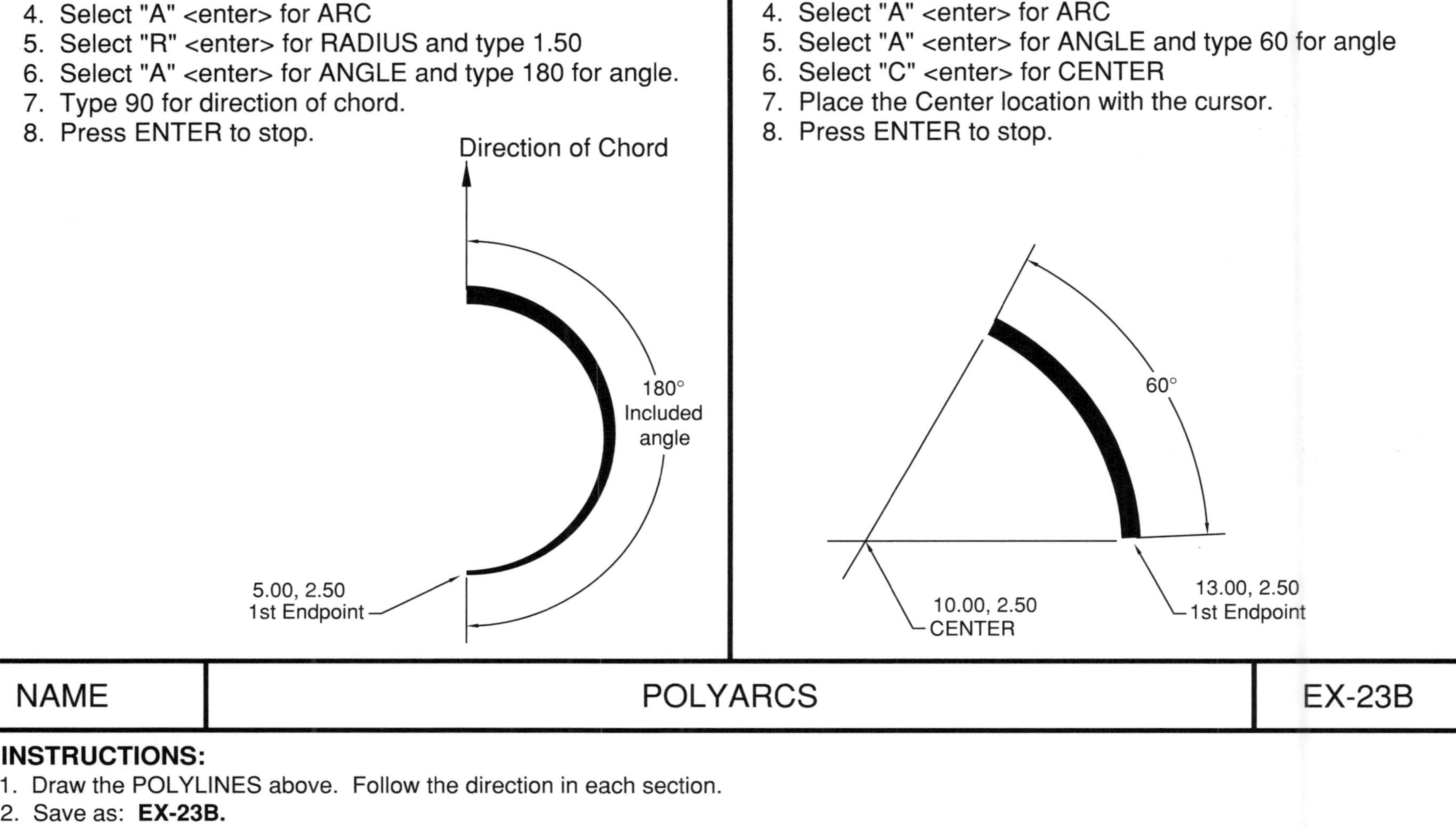

INSTRUCTIONS:

1. Draw the POLYLINES above. Follow the direction in each section.
2. Save as: **EX-23B.**

EXERCISE 23B

NOTES:

LEARNING OBJECTIVES

After completing this lesson, you will be able to:

1. Edit the width of a Polyline.
2. Join Polylines.
3. Convert a Polyline to a spline.
4. Convert a basic Line into a Polyline.

LESSON 24

EDITING POLYLINES

The **POLYEDIT** command allows you to make changes to a polyline's option, such as the width. You can also change a regular line into a polyline and JOIN the segments.

1. Select the **POLYEDIT** command using one of the following:

    TYPE = PE
    PULLDOWN = MODIFY / OBJECT/ POLYLINE
    TOOLBAR =MODIFY II

Note: AutoCAD 2002 allows you to modify "Multiple" polylines simultaneously.

2. PEDIT Select polyline or [Multiple]: ***select the polyline to be edited or "M"***
3. Enter an option [Close/Join/Width/Edit vertex/Fit/Spline/Decurve/Ltypegen/Undo]: ***select an Option (descriptions of each are listed below.)***

Note: If you select a line that is **NOT a POLYLINE**, the prompt will ask if you would like to turn it into a POLYLINE.

OPTIONS: (Step by step instructions in the following exercises)

CLOSE (Refer to Exercise 24A)
CLOSE connects the last segment with the first segment of an Open polyline. AutoCAD considers a polyline open unless you use the "Close" option to connect the segments originally.

OPEN (Refer to Exercise 24A)
OPEN removes the closing segment, but only if the CLOSE option was used to close the polyline originally.

JOIN (Refer to Exercise 24D)
The JOIN option allows you to join individual polyline segments into one polyline. The segments must have matching endpoints.

WIDTH (Refer to Exercise 24B)
The WIDTH option allows you to change the width of the polyline. But the entire polyline will have the same width.

EDIT VERTEX (Refer to Exercise 24C)
This option allows you to change the starting and ending width of each segment individually.

SPLINE (Refer to Exercise 24D)
This option allows you to change straight polylines to curves.

DECURVE
This option removes the SPLINE curves and returns the polyline to its original straight line segments.

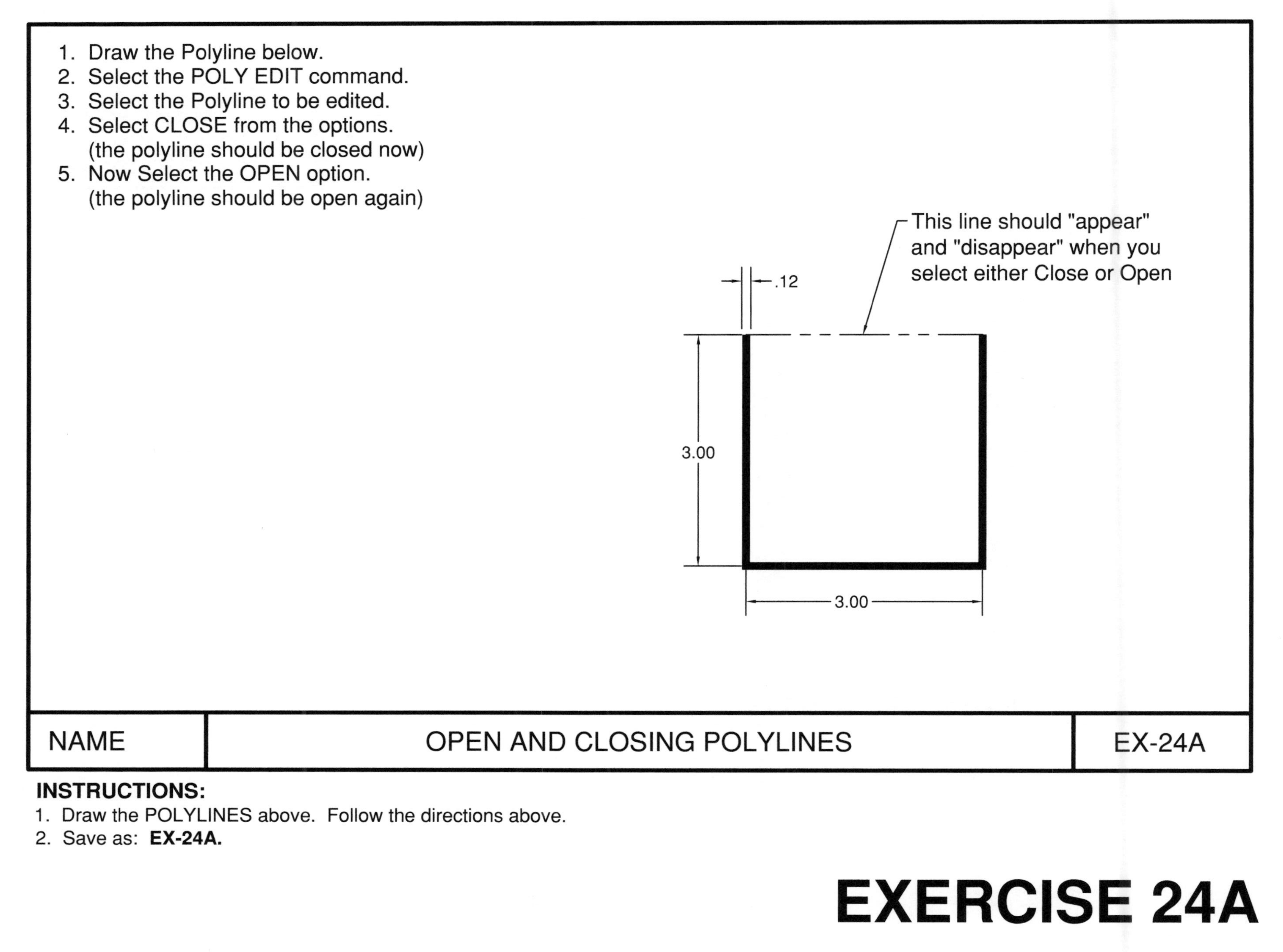

INSTRUCTIONS:

1. Draw the POLYLINES above. Follow the directions above.
2. Save as: **EX-24A.**

EXERCISE 24A

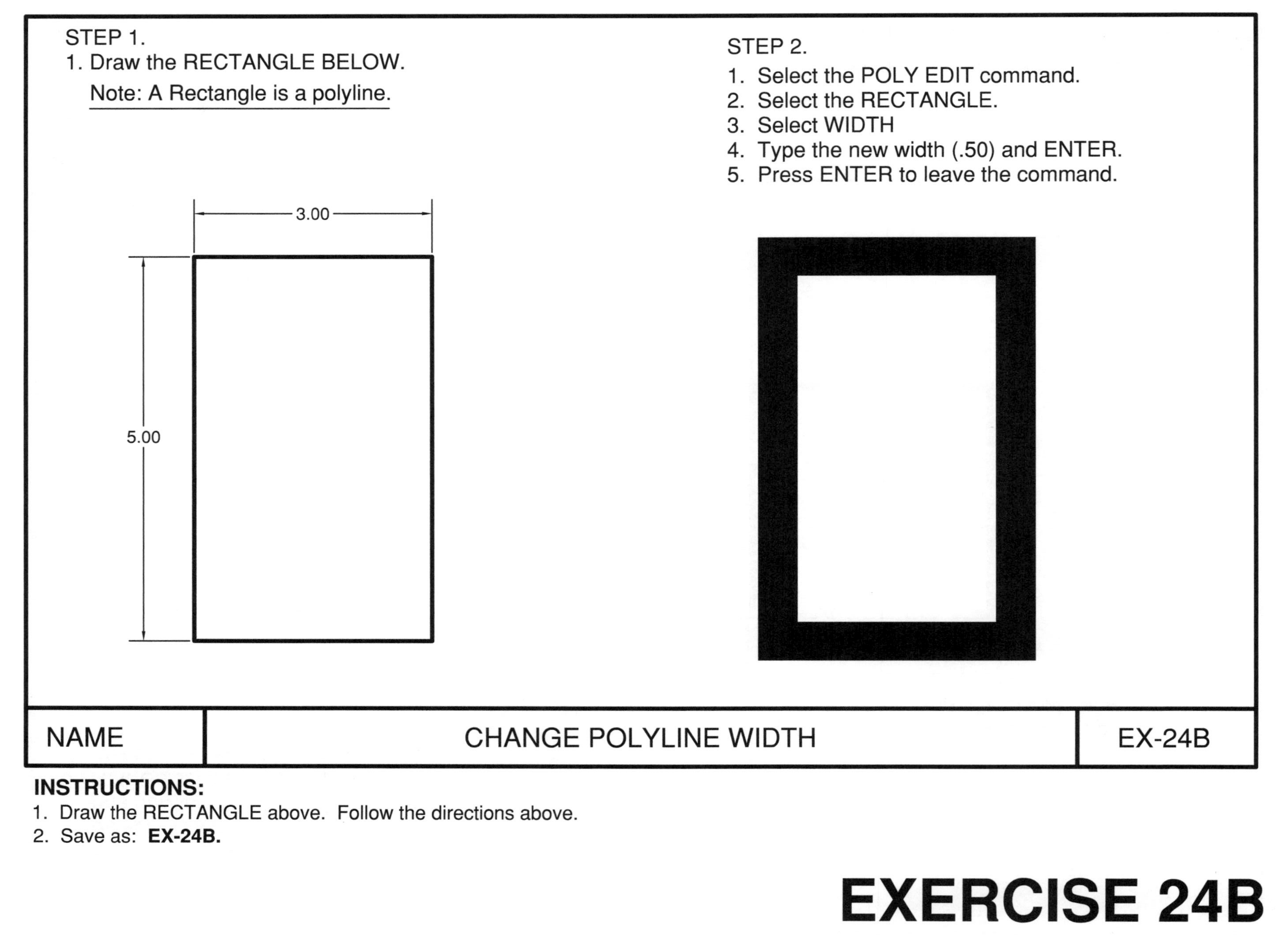

INSTRUCTIONS:

1. Draw the RECTANGLE above. Follow the directions above.
2. Save as: **EX-24B.**

EXERCISE 24B

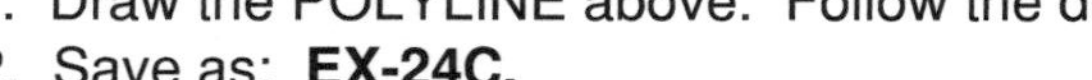

EXERCISE 24C

STEP 1. Draw 1 polyline made of 3 connecting (joined) segments, each 2 inches long

STEP 2.

1. Select the POLY EDIT command.
2. Select the first segment.
3. Select the EDIT VERTEX option.

(Notice the "X" marking the starting point.)

4. Select the WIDTH option
5. Type the STARTING (1.00) and ENDING (.50) widths.
6. Select NEXT.

(Notice the "X" moved to the "next" segment starting point)

7. Select the WIDTH option.
8. Type the STARTING (.50) and ENDING (.50) widths.
9. Select NEXT. ("X" moved again)
10. Select the WIDTH option.
11. Type the STARTING (2.00) and ENDING (0)widths.
12. Press <enter> to stop.

STEP 1

2.00

2.00

2.00

First Segment Starts Here

STEP 2

2.00

.50

1.00

NAME	CHANGE WIDTHS WITH EDIT VERTEX	EX-24C

INSTRUCTIONS:

1. Draw the POLYLINE above. Follow the directions above.
2. Save as: **EX-24C.**

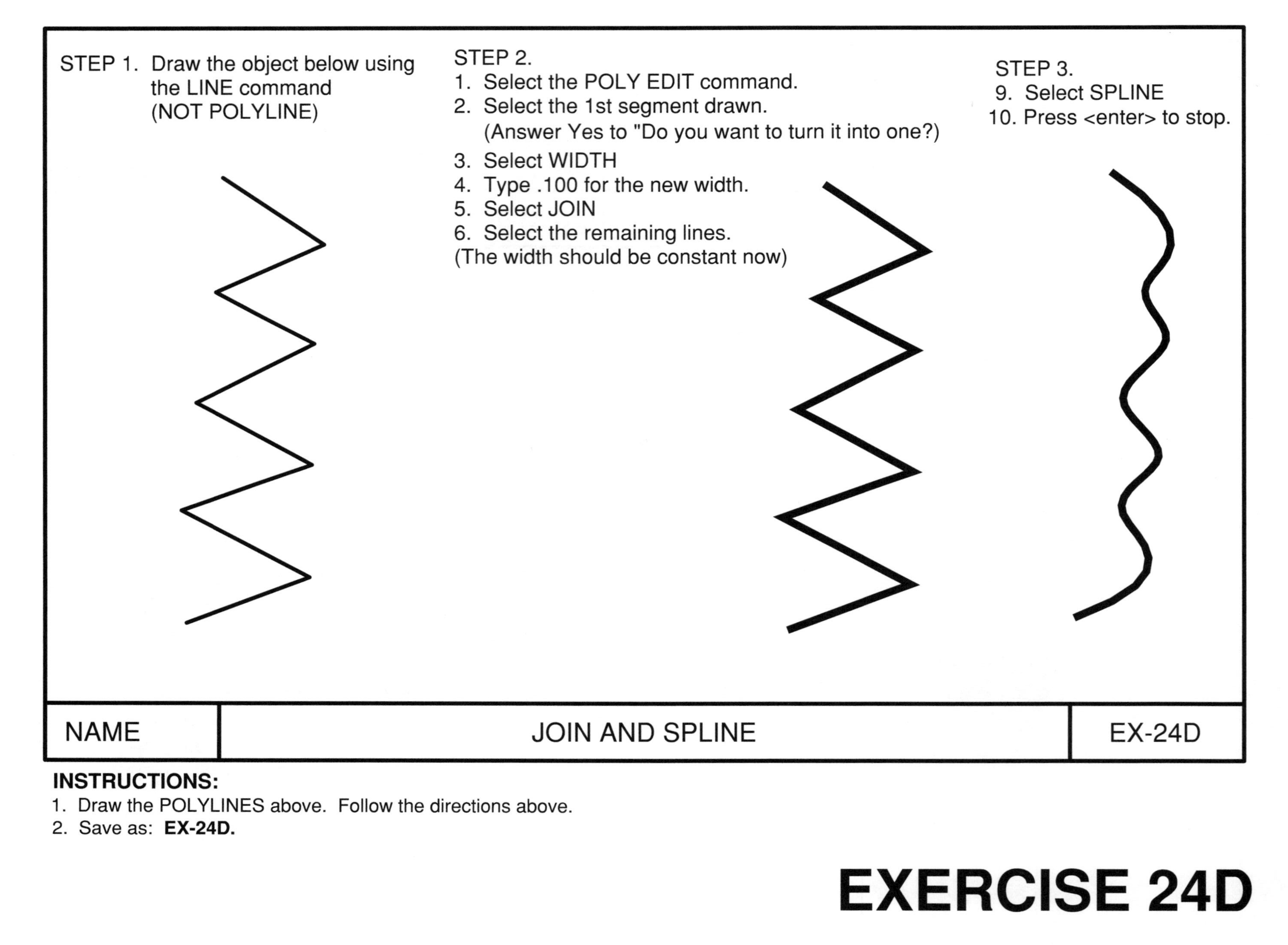

INSTRUCTIONS:

1. Draw the POLYLINES above. Follow the directions above.
2. Save as: **EX-24D.**

EXERCISE 24D

LEARNING OBJECTIVES

After completing this lesson, you will be able to:

1. Create a new Text Style.
2. Change and Existing Text Style.
3. Change the Point Style.
4. Place a Point at designated intervals on an object.

LESSON 25

CREATING NEW TEXT STYLES

AutoCAD provides you with only one Text Style named “Standard”. You may want to create a new text style with a different font and effects. Steps 1 through 8 below will guide you through the process.

1. Select the TEXT STYLE command using one of the following:

   TYPE = STYLE or ST
   PULLDOWN = FORMAT / TEXT STYLE
   TOOLBAR = FORMAT

   The TEXT STYLE dialog box below should appear.

The information in this dialog box is a description of the Text Style highlighted in the Style Name box.

2. Select the **NEW** button.

3. Type the new style name in **STYLE NAME** box. Then select the **OK** button.

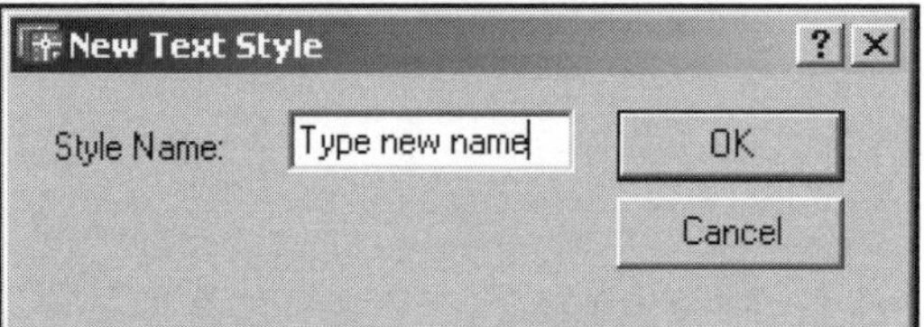

Text Styles can have a maximum of 31 characters, including letters, numbers, dashes, underlines and dollar signs. You can use Upper or Lower case.

4. Select the **FONT**.

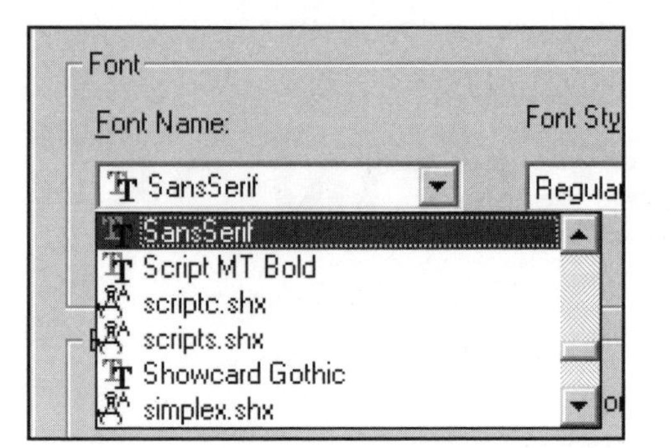

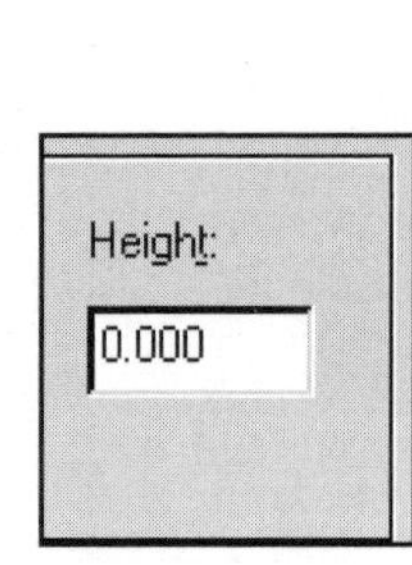

5. Enter the value of the Height. **(If the value is 0, AutoCAD will always prompt you for a height. If you enter a number the new text style will have a fixed height and AutoCAD will not prompt you for the height)**

6. Assign **EFFECTS.**

Effects
Upside down
Backwards
Vertical
Width Factor: 1.000
Oblique Angle: 0

UPSIDE-DOWN
Each letter will be created upside-down in the order in which it was typed.
(Note: this is different from rotating text 180 degrees.)

BACKWARDS
The letters will be created backwards as typed.

VERTICAL
Each letter will be inserted directly under the other. Only **.shx** fonts can be used.
VERTICAL text will not display in the **PREVIEW** box.

OBLIQUING ANGLE
Creates letter with a slant, like italic. An angle of 0 creates a vertical letter. A positive angle will slant the letter forward. A negative angle will slant the letter backward.

WIDTH FACTOR
This effect compresses or extends the width of each character. A value less than 1 compresses. A value greater than 1 extends each character.

7. **PREVIEW** your settings.

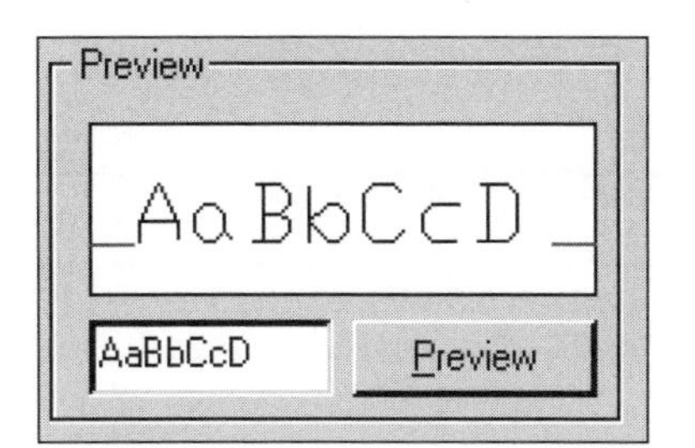

8. Select the **APPLY** button and then the **CLOSE** button.

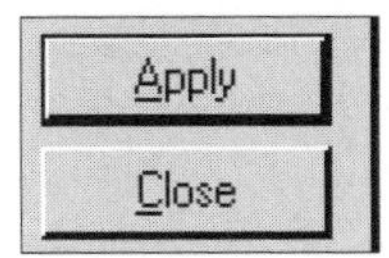

CREATING YOUR NEW TEXT STYLE IS NOW COMPLETE.

CHANGING TEXT STYLES

RENAMING

1. Select the **TEXT STYLE** command.
   The Text Style Dialog box appears.

2. Select the style you want to rename.

3. Click on **RENAME** button.

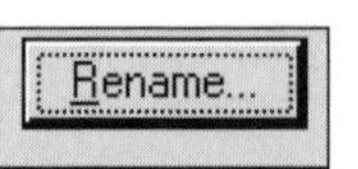

4. Type the New name then click on the **OK** button.

5. Click on the **CLOSE** button.

DELETING

1. Select the **TEXT STYLE** command.
   The Text Style Dialog box appears.

2. Select the style you want to DELETE.

3. Click on the **DELETE** button.

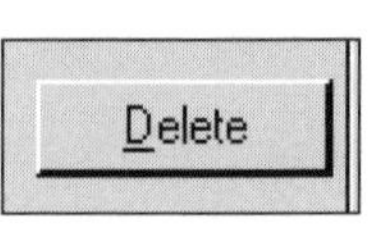

4. Warning appears, select **Yes or No**.

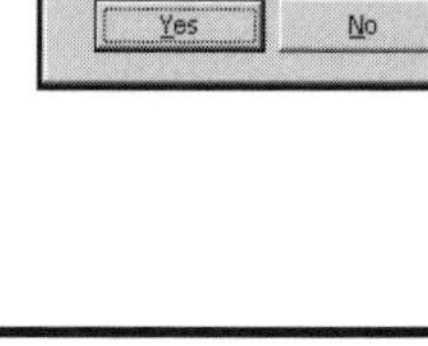

5. Click on the **CLOSE** button.

CHANGING EFFECTS

1. Select the **TEXT STYLE** command.
   The Text Style Dialog box appears.

2. Make the changes.

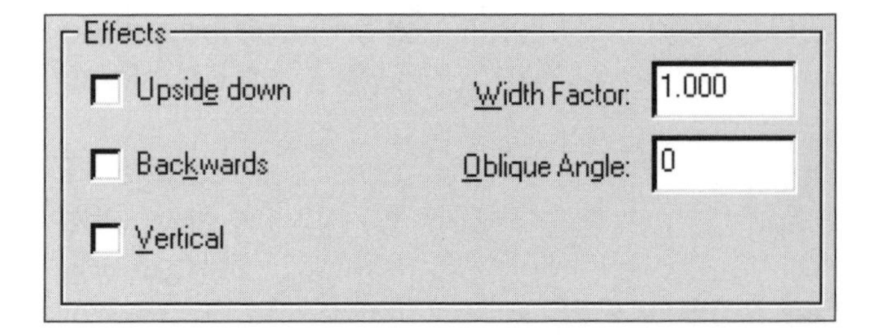

3. Click on the **APPLY** button and then **CLOSE.**

DIVIDE

The DIVIDE command divides an object mathematically by the NUMBER of segments you designate. It then places a POINT (object) at each interval on the object.

Note: the object selected is **NOT** broken into segments. The POINTS are simply drawn **ON** the object.

First select the **POINT STYLE** to be placed on the object. ***Refer to lesson 5***

Next select the **DIVIDE** command using one of the following:

TYPE = DIV
PULL DOWN = DRAW / POINT / DIVIDE
TOOLBAR = DRAW

Select object to divide: ***select the object to divide.***
Enter the number of segments or [Block]: ***type the number of segments <enter>***

EXAMPLE:

This LINE has been DIVIDED into 4 EQUAL lengths.
But remember, the line is not broken into segments.
The Points are simply drawn ON the object.

MEASURE

The **MEASURE** command is very similar to the **DIVIDE** command because point objects are drawn at intervals on an object. However, the **MEASURE** command allows you to designate the **LENGTH** of the segments rather than the number of segments.

Note: the object selected is **NOT** broken into segments. The **POINTS** are simply drawn **ON** the object.

First select the **POINT STYLE** to be placed on the object. ***Refer to lesson 5***

Next select the **MEASURE** command using one of the following:

TYPE = ME
PULL DOWN = DRAW / POINT / MEASURE
TOOLBAR = DRAW

Command: _measure
Select object to measure: ***select the object to measure.***
(Note: this selection point is also where the MEASUREment will start.)
Specify length of segment or [Block]: ***type the length of one segment <enter>***

EXAMPLE:

Select Object here to start the measurement from the left.

The MEASUREment was started at the left endpoint, and ended just short of the right end of the line. The remainder is less than the measurement length designated.

EXERCISE 25A

INSTRUCTIONS:

1. Open your "**BSIZE**" drawing.
2. Create the 3 text styles listed below.
3. Save as EX-25A (The text is not visible, but it will be after completing EX-25B)

Follow the instruction **"CREATING NEW TEXT STYLES"** on page 25-2.

Name:	Style1
Font:	Romand.shx
Height:	.50
Upside down:	No
Backwards:	No
Vertical:	Yes
Width factor: 1	
Obliquing angle:	30

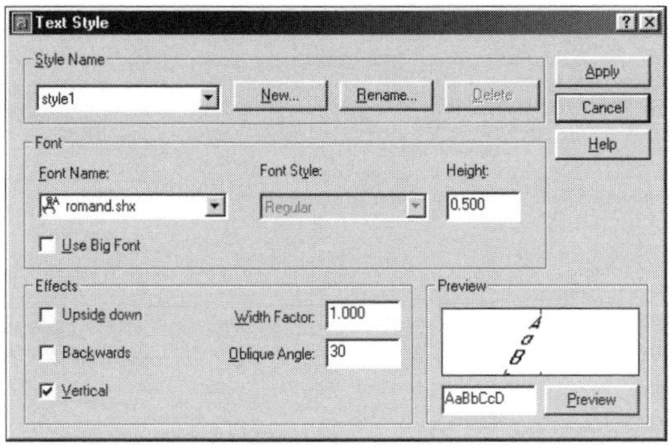

Name:	Style2
Font:	gothice.shx
Height:	1.00
Upside down:	No
Backwards:	Yes
Vertical:	No
Width factor: .75	
Obliquing angle:	0

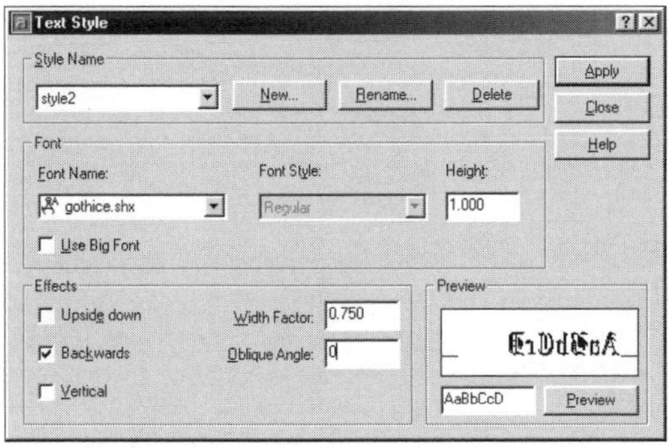

Name:	Style3
Font:	Italict.shx
Height:	2.00
Upside down:	Yes
Backwards:	No
Vertical:	No
Width factor: .75	
Obliquing angle:	0

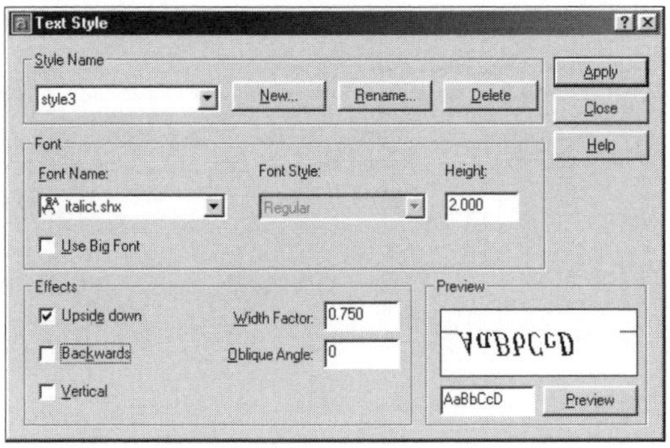

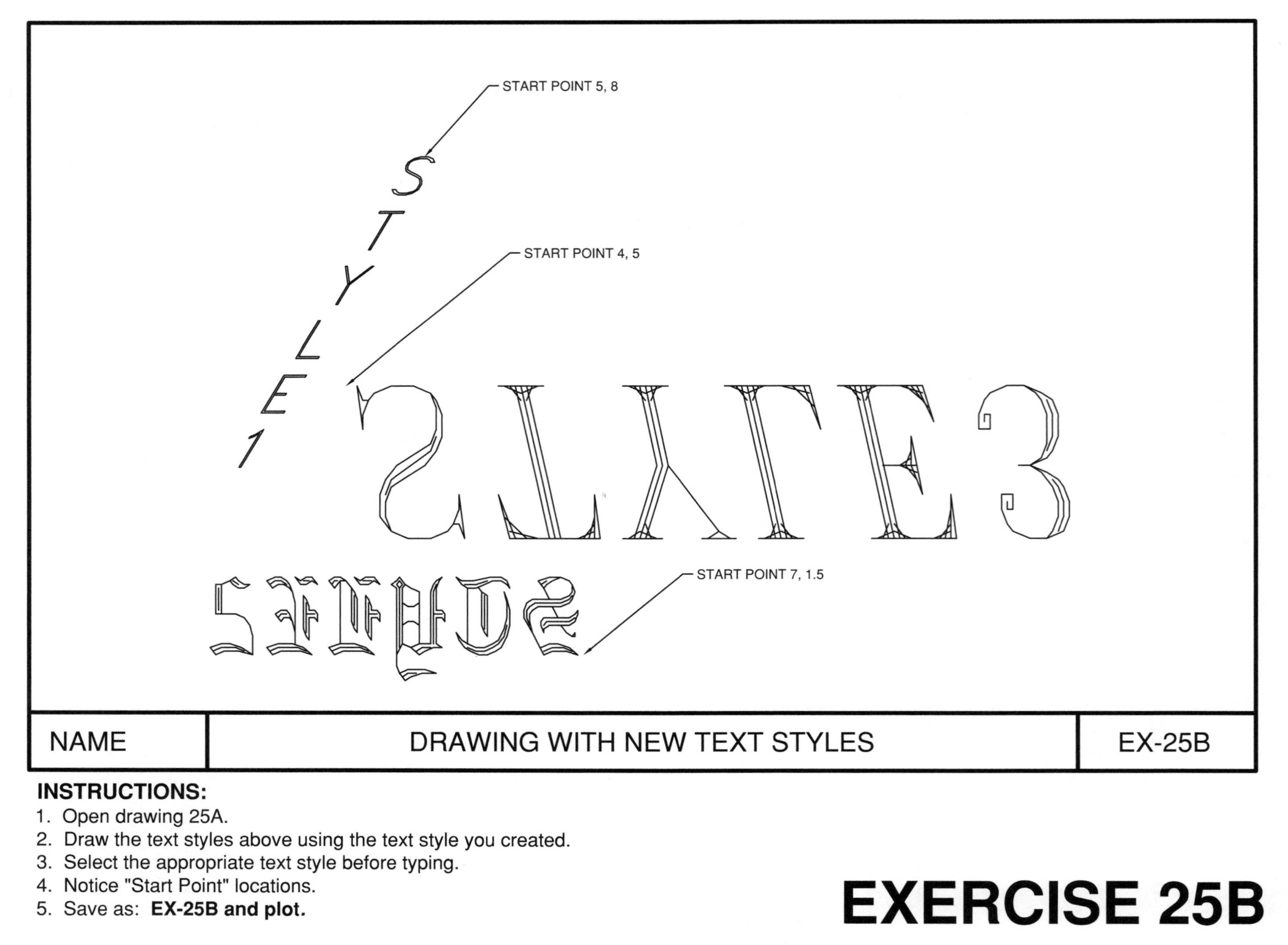

INSTRUCTIONS:

1. Open drawing 25A.
2. Draw the text styles above using the text style you created.
3. Select the appropriate text style before typing.
4. Notice "Start Point" locations.
5. Save as: **EX-25B and plot.**

EXERCISE 25B

INSTRUCTIONS:

1. Open drawing 25B.
2. Using MODIFY / PROPERTIES change the 3 text styles to Text Style "Standard".
3. Note: you must change one at a time and "deselect" between style changes.
   If you do not deselect between style changes, you will not acheive the same results.
4. Save as: **EX-25C and plot.**

EXERCISE 25C

EXERCISE 25D

INSTRUCTIONS:

1. Open File **25C.**
2. Delete text styles **1,2 and 3.**

<u>There are 2 methods. Try both.</u>

METHOD 1.

1. Select **FORMAT / TEXT STYLE**
2. Select the "Standard" text style.
3. Select the CLOSE button. (This makes Standard the current text style)
4. Now return to **FORMAT / TEXT STYLE**
5. Select the style you want to **delete**.
6. Click on the **DELETE** button.

AutoCAD will not allow you to delete an active or current text style. That is why number 2, above, said to select Standard and close.

METHOD 2.

1. Select **File / Drawing Utilities / Purge**
2. Select the **+ sign** beside Text Styles
3. Select the Style 1, 2 and 3
4. Select **Purge All** button.
5. Select **Close**.

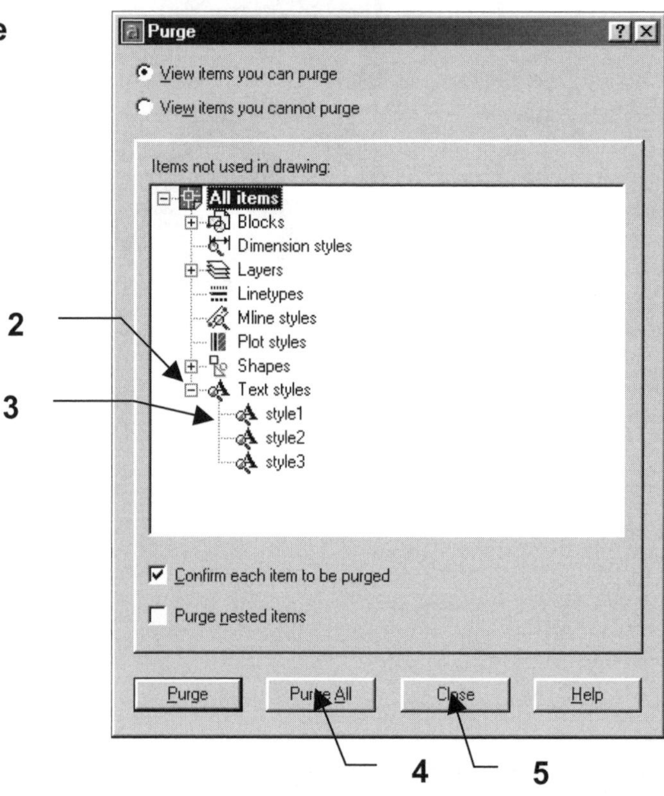

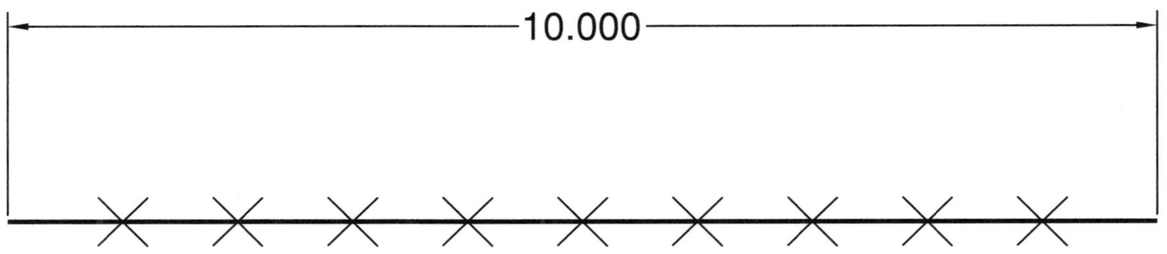

10.000

NAME	DIVIDE	EX-25E

INSTRUCTIONS:

1. Draw a LINE 10 inches long.
2. Set the POINT STYLE to **X** using: FORMAT / POINT STYLE.
3. DIVIDE the line into 10 equal divisions using: **a**. DRAW / POINT / DIVIDE.
   b. *Select the object to divide*. **c**. *Type 10 for number of segments.*
4. Save as: **EX-25E and plot.**

EXERCISE 25E

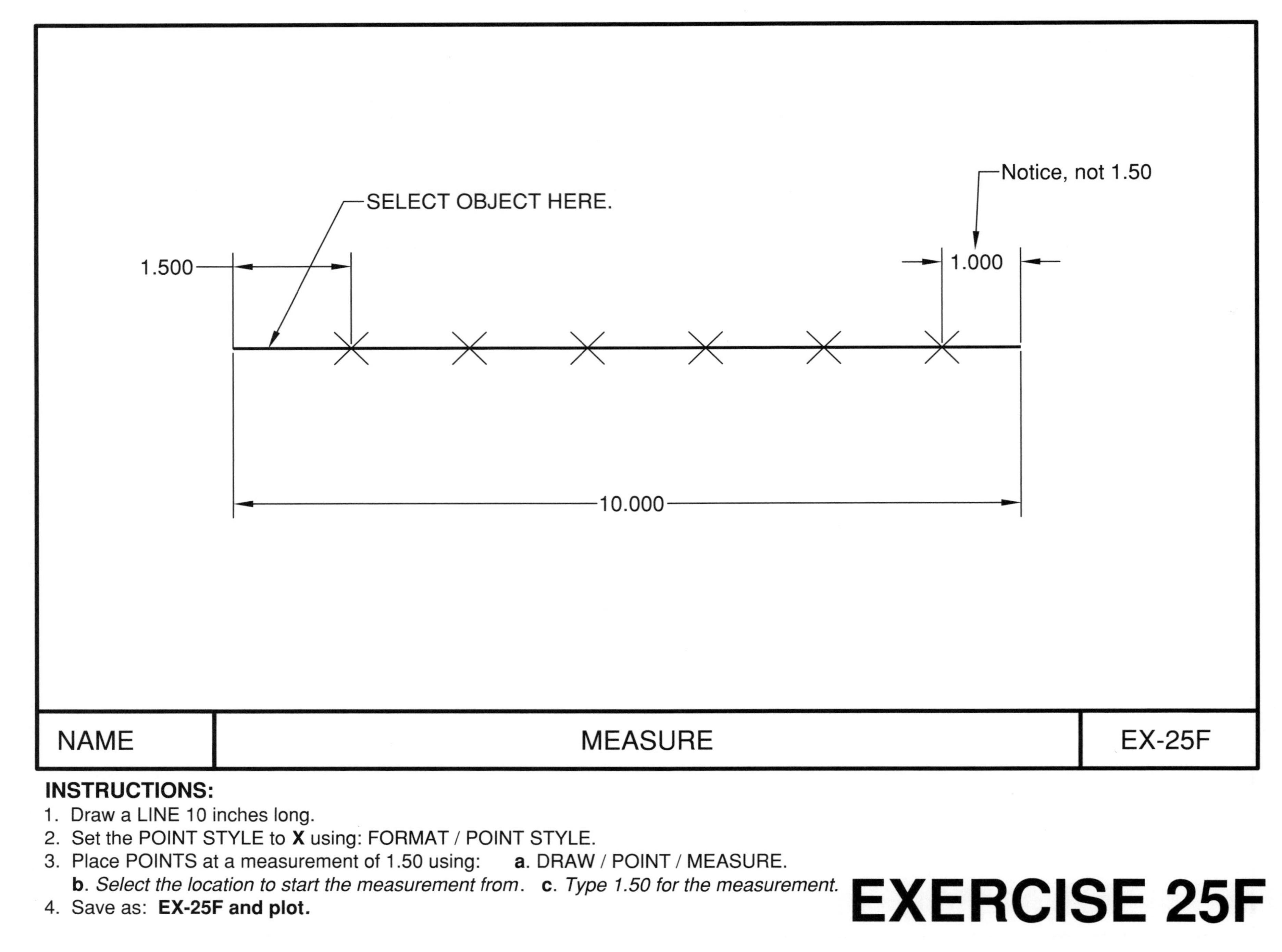

INSTRUCTIONS:

1. Draw a LINE 10 inches long.
2. Set the POINT STYLE to **X** using: FORMAT / POINT STYLE.
3. Place POINTS at a measurement of 1.50 using: **a**. DRAW / POINT / MEASURE.
   b. *Select the location to start the measurement from*. **c**. *Type 1.50 for the measurement.*
4. Save as: **EX-25F and plot.**

EXERCISE 25F

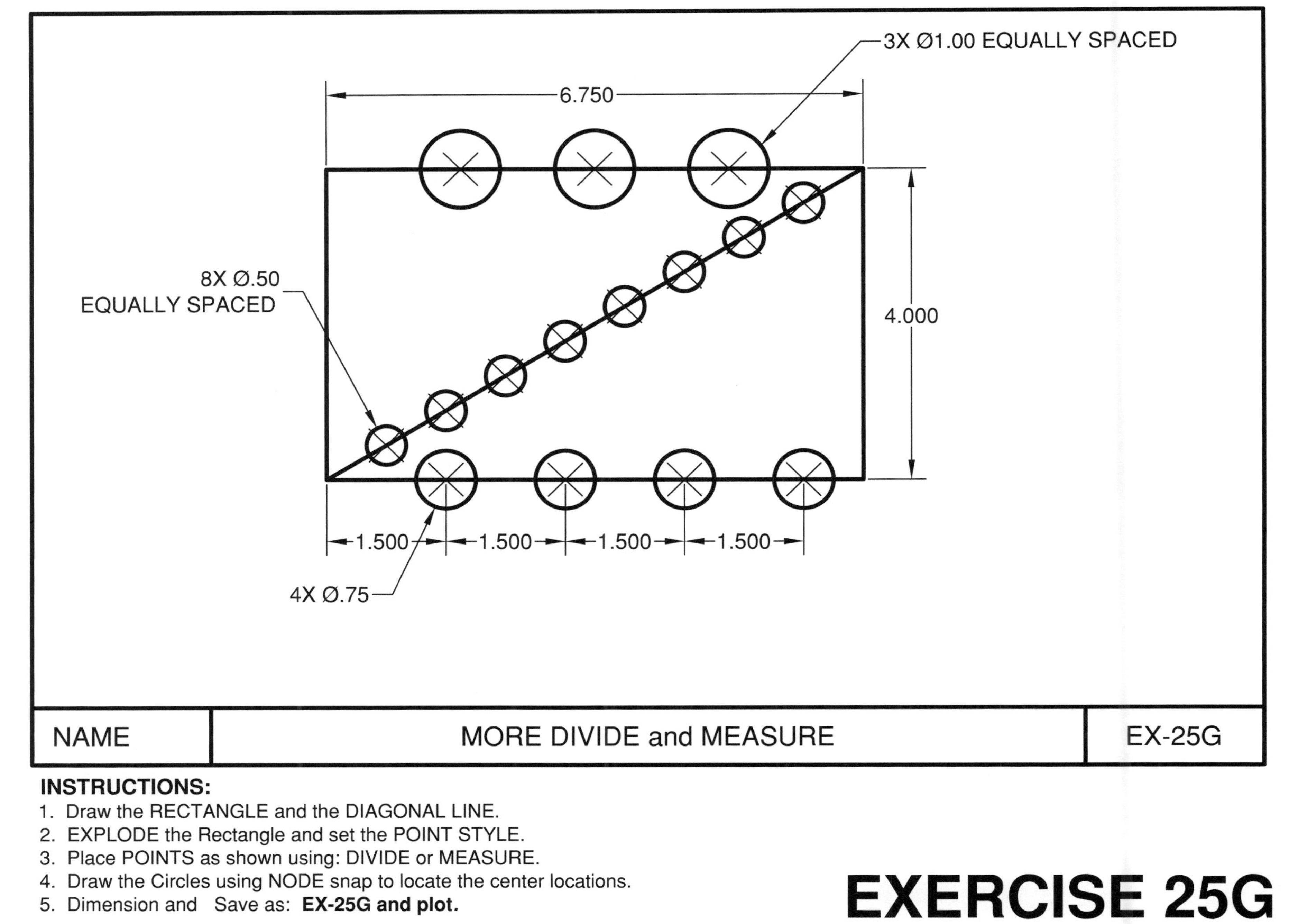

INSTRUCTIONS:

1. Draw the RECTANGLE and the DIAGONAL LINE.
2. EXPLODE the Rectangle and set the POINT STYLE.
3. Place POINTS as shown using: DIVIDE or MEASURE.
4. Draw the Circles using NODE snap to locate the center locations.
5. Dimension and Save as: **EX-25G and plot.**

EXERCISE 25G

NOTES:

LEARNING OBJECTIVES

After completing this lesson, you will be able to:

1. Create your own Layers.
2. Load Linetypes into your drawing.
3. Understand the difference between Model and Layout tabs.
4. Adjust the size of the "Pick" box.
5. Create Floating Viewports
6. Create a Page Setup for plotting your drawings.
7. Create a Decimal "Setup" master file for future use.
8. Create a new Border to use when plotting.

LESSON 26

SERIOUS BUSINESS in this lesson

In the previous lessons you have been having fun learning most of the basic commands that AutoCAD offers and you have been using our drawings that include preset layers, text styles, drawing settings etc. But now it is time to get down to the "serious business" of setting up your own drawing from "scratch".

Starting from scratch means you will need to set or create the following:

Items 1 through 4 you have learned in previous lessons

1. Drawing Units (Lesson 4)
2. Snap and Grid (Lesson 2)
3. Create Text styles (Lesson 25)
4. Create Dimension Styles (Lesson 16)

Items 5 through 9 will be learned in this lesson

5. Create new layers and load linetypes.
6. Create a "Layout" for plotting.
7. Create a "Floating Viewport" in the Layout.
8. Create a "Page Setup" to save plot settings.
9. Create a "Plot Style Table" to control lineweight and color.

*After reading pages 26-3 through 26-16 start Exercise 26A and work your way through to 26D. When you have completed Exercise 26D you will have created a master drawing named, "**My Decimal Setup**". This master drawing will have everything set, created and prepared, ready to use each time you want to create a drawing using decimal units and to be plotted on an 11 x 17 inch sheet.*

This means, for future drawings you merely open "My Decimal Setup" and start drawing. No time consuming setups. It is all ready to go.

In Lesson 27 you will create a master drawing for "feet and inches".

So take it one page at a time and really concentrate on understanding the process.

CREATING NEW LAYERS

1. Select the Layer command using one of the following:

TYPE = LA
PULLDOWN = FORMAT / LAYER
TOOLBAR = OBJECT PROPERTIES

The Layer & Linetype Properties dialog box shown below will appear.

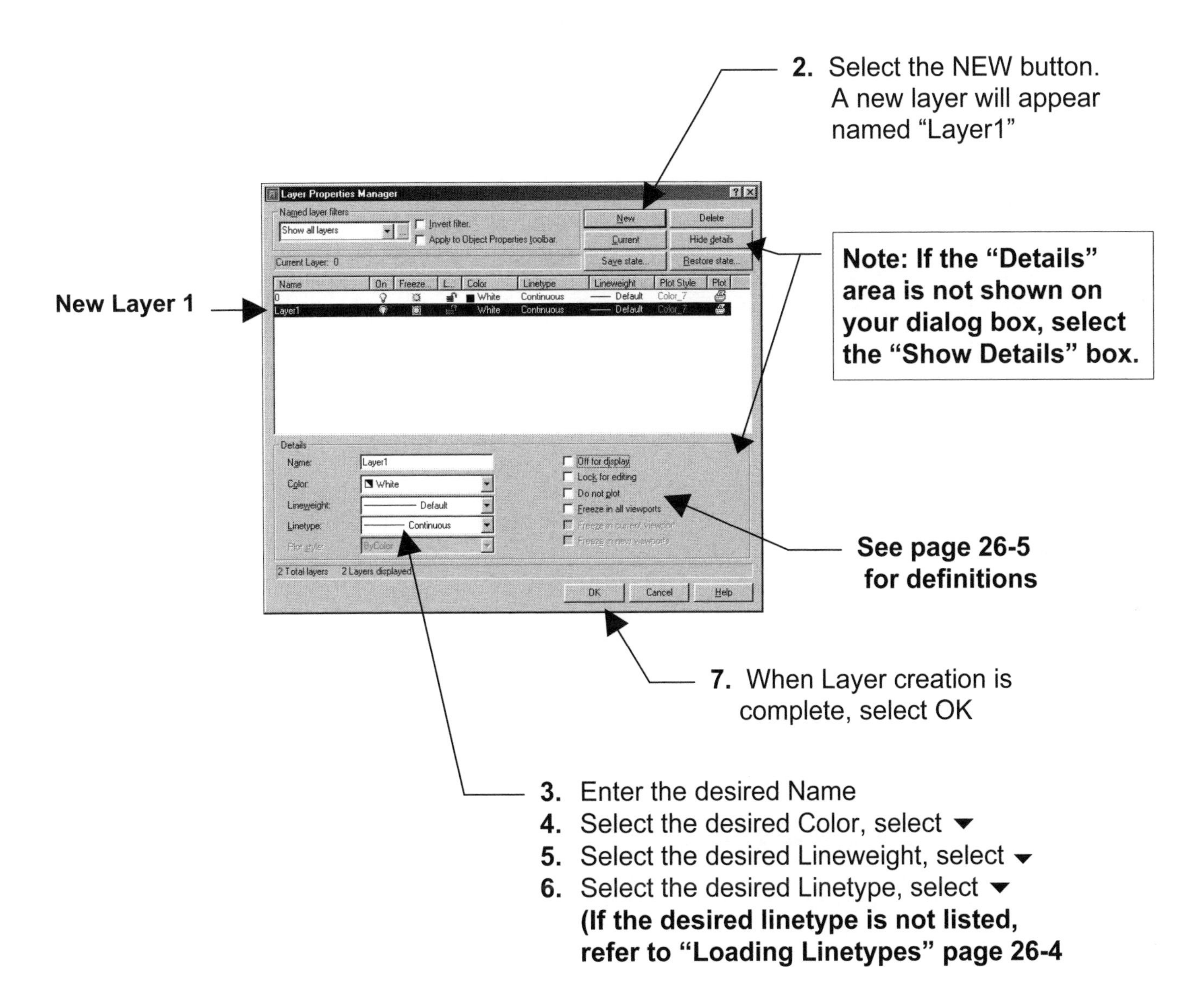

LOADING A LINETYPE

In an effort to conserve data within a drawing file, AutoCAD automatically loads only one linetype called "continuous". If you would like to use other linetypes, such as "dashed", you must "Load" them into the drawing as follows.

1. Select the **LINETYPE** command using one of the following:

   TYPE = LT
   PULLDOWN = FORMAT / LINETYPE
   TOOLBAR = NONE

The Linetype Manager dialog box shown below will appear.

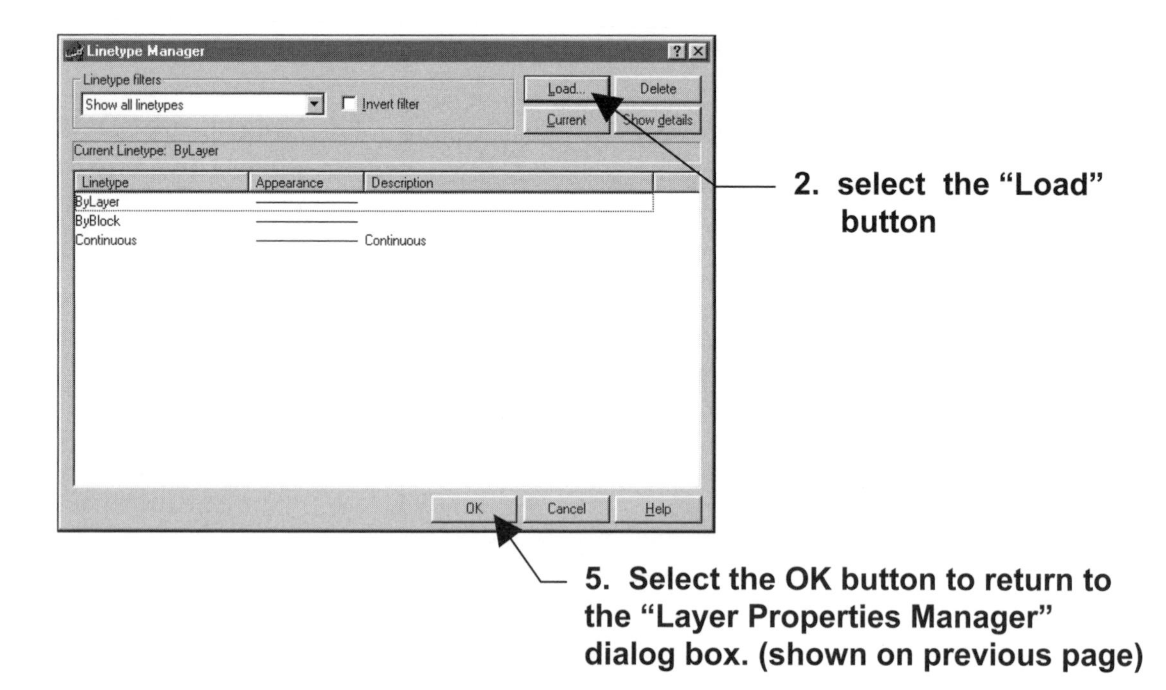

*The **Load or Reload Linetypes** dialogue box shown below will appear.*

3. Scroll thru the list of available Linetypes and select the desired Linetype(s)
<u>Note: To select more than one, hold the Ctrl key down while selecting</u>

4. Select the OK button to return to the "Linetype Manager" dialog box

LAYER CONTROL DEFINITIONS

Off For Display

If a layer is **ON** it is **visible**. If a layer is **OFF** it is **not visible**.
Only layers that are **ON** can be **edited** or **plotted**.
(Warning: Objects on a Layer that is OFF can be accidentally erased even though they are invisible. When you are asked to select objects, in the erase command, if you type **ALL** <enter> all objects will be selected even the invisible ones.)

Lock For Editing

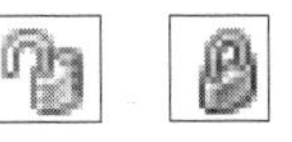

LOCKED layers are visible but cannot be edited. They are visible so they **will** be plotted. (Locked layers cannot be selected by typing ALL.)

Do Not Plot

This option prevents a layer from plotting even though it is visible.

Freeze In All Viewports

Freeze and **Thaw** are very similar to On and Off. A Frozen layer is not visible.
A Thawed layer is visible. Only thawed layers can be edited or plotted.

Additionally:

a. Objects on a Frozen layer **cannot** be accidentally erased by typing All.
b. Freezing saves time, when working with large and complex drawings, because frozen layers are not **regenerated** when you zoom in and out.

Freeze In Current Viewport

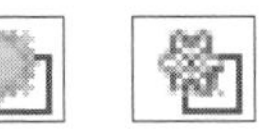

Layers selected will be frozen in the “current” viewport only. Current means the active viewport. Only one viewport can be active at one time.
(Available in Paperspace only)

Freeze In New Viewports

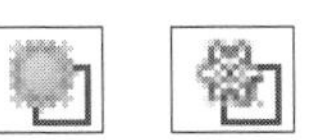

Layers selected will be frozen in the next viewport created. So you are selecting the layers, to be frozen, before you have created the viewport. This one doesn’t get used much.
(Available in Paperspace only)

MODEL and LAYOUT tabs

Read this information carefully. It is very important that you understand this concept.

AutoCAD provides two drawing spaces, **MODEL** and **LAYOUT**. You move into one or the other by selecting either the MODEL or LAYOUT tabs, located at the bottom left of the drawing area.

Model Tab (Also called ***Model Space***)
When you select the Model tab you enter MODEL SPACE.
(This is where you have been drawing for the last 25 lessons)
Model Space is where you **create** and **modify** your drawings.

Layout1 Tab (Also called ***Paper Space***)

When you select a Layout tab you enter PAPER SPACE.
The primary function of Paper Space is to prepare the drawing for plotting.

When you select the Layout tab, for the first time, the "Page Set up" Dialog box will appear. Using this dialog box, you will tell AutoCAD which plotting device and paper size to use for plotting. For now, select the OK button to go on.
(More information on this in "How to create a Page Setup" page 26-11)

Model space disappears and a blank sheet of paper appears on the screen. This sheet of paper is basically lying on top of the Model Space. (See illustration below) To see your drawing (Model Space), while still in paper space, you must cut a hole in this sheet. This hole is called a **"Viewport"** *(Refer to "Viewports" page 26-7.)*

Try to think of this as a picture frame (paper space) lying on top of a photograph. (model space)

Generally, the only objects that should be in paper space are the "Title Block, Border, Dimensions and Notes.

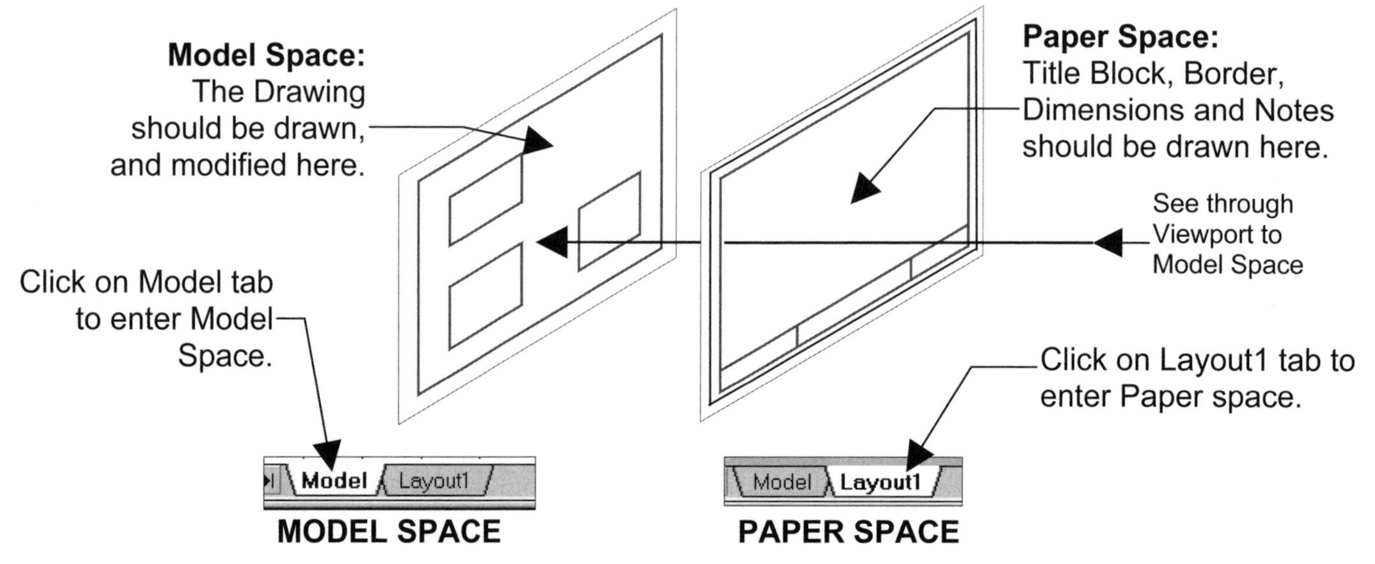

VIEWPORTS

Viewports are only used in Paper Space (Layout tab).
Viewports are holes cut into the sheet of paper displayed on the screen.
Viewports are objects. They can be moved, stretched, scaled, copied and erased.
You can have multiple Viewports and their size and shape can vary.
*You will only be working with **single** viewports in this workbook. Multiple viewports will be explained in the **"Exercise Workbook for Advanced AutoCAD".***

Note: It is considered good drawing management to create a layer for the Viewport "frames" to reside on. This will allow you to control them separately. Such as setting the viewport layer to "No plot" so it will not be visible when plotting.

HOW TO CREATE A VIEWPORT

1. First, create your drawing in Model Space and save it.

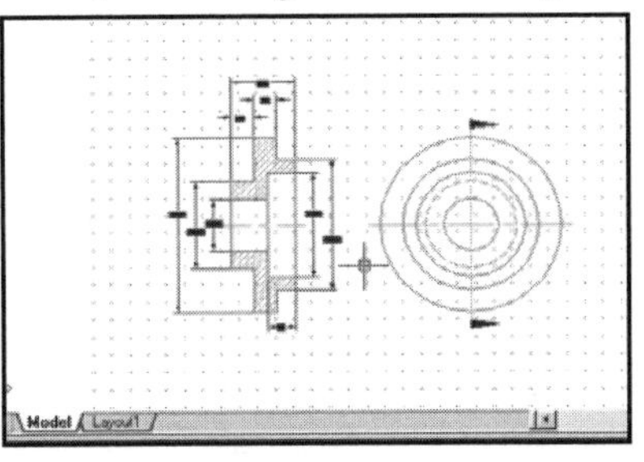

2. Select the "Layout1" tab (If the "Page setup" dialog box appears, select the **OK** button for now or refer to "How to create a Page setup" on page 26-11.)
3. You are now in Paper Space. Model Space appears to have disappeared but actually a blank paper is in front of Model Space, preventing you from seeing your drawing. (Your Border, title block and notes will be drawn on this paper in paper space.)

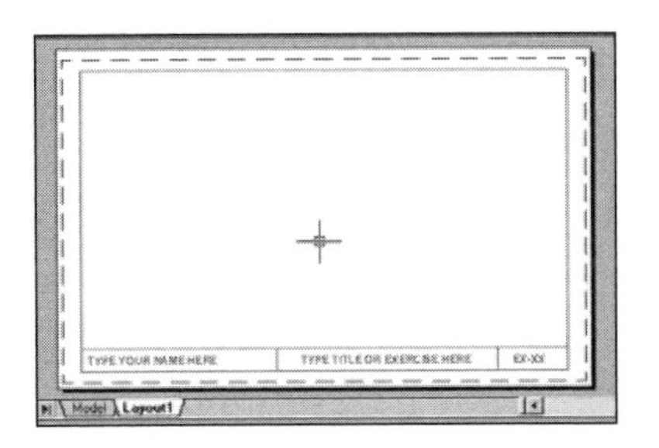

4. Select layer "Viewport" (You may need to create one, refer to 26-3)
5. Open the Viewport toolbar (View / toolbars / viewports)

2004 toolbar

2004 LT toolbar

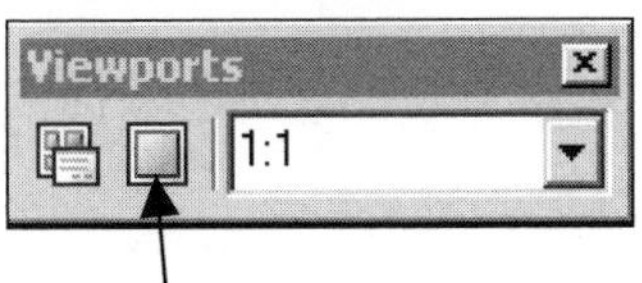

6. Select the "Single Viewport" icon.

7. Draw the Viewport “frame” by specifying the location for the “first corner” and then the opposite corner. (Similar to drawing a Rectangle but **do not** use the Rectangle command)

You should now be able to look through the paper space sheet to Model Space and see your drawing.
(Make sure your grids are ON in Model Space and OFF in paperspace)

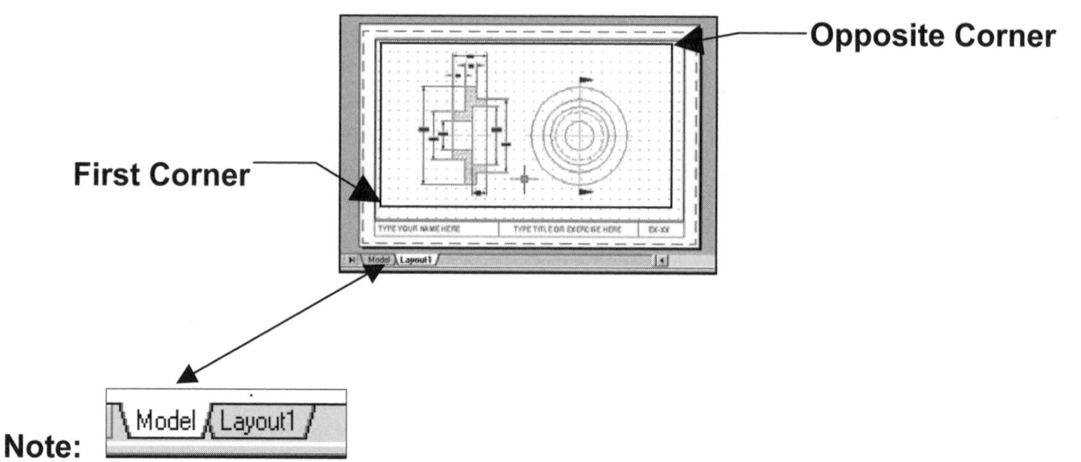

Note:
To go back to Model Space, click on the Model tab.
To return to Paper Space, click on the Layout tab.

EXAMPLE OF WHY WE WANT TO USE PAPER SPACE
I know you are probably wondering why you should bother with paperspace. Paperspace is a great tool to manipulate your drawing for plotting.
Notice the drawing below. The upper viewport displays the entire drawing. The lower right viewport’s scale has been adjusted to get a closer look at that section arrow. The lower left viewport not only has it’s scale adjusted but the dimension layer is frozen in the “current viewport” only. Notice the dimension layer is still thawed in the upper viewport. Can you guess which is the **Active** viewport at the moment?

Experiment and get familiar with Paper space because it is a very useful tool.

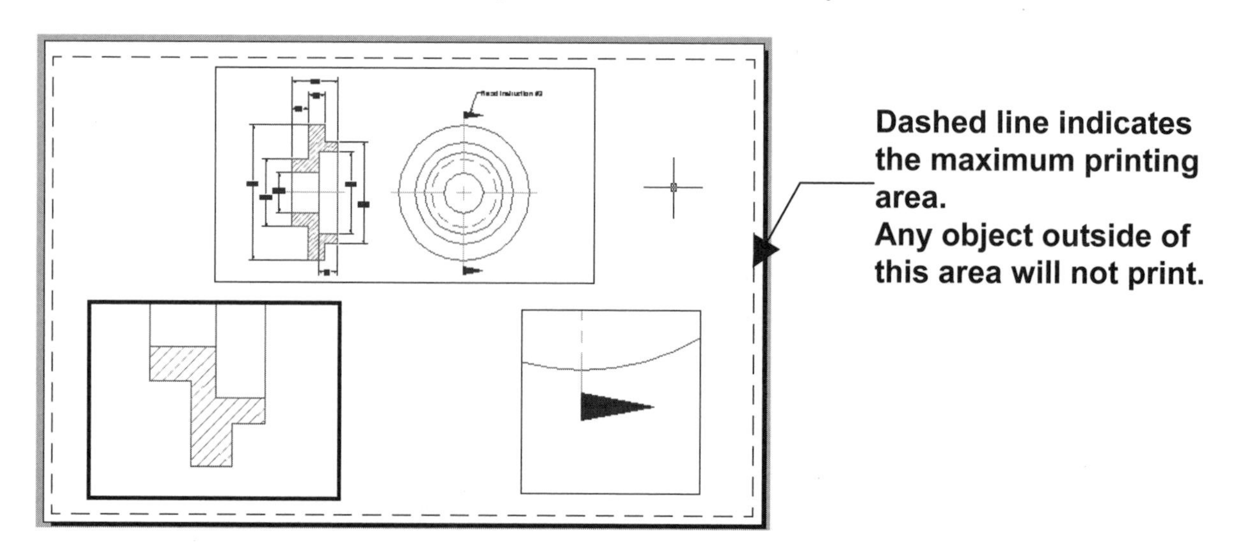

TO REACH THROUGH TO MODEL SPACE WHILE IN PAPER SPACE

Method 1.
Select the **Model / Paper** button on the Status Line.

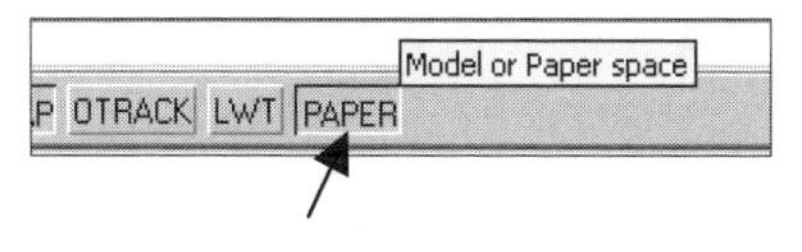

Method 2.
If you double click inside the viewport frame, it will activate the model space and you can reach through the sheet of Paper to Model space. *To return to Paper Space double click in the gray area.*

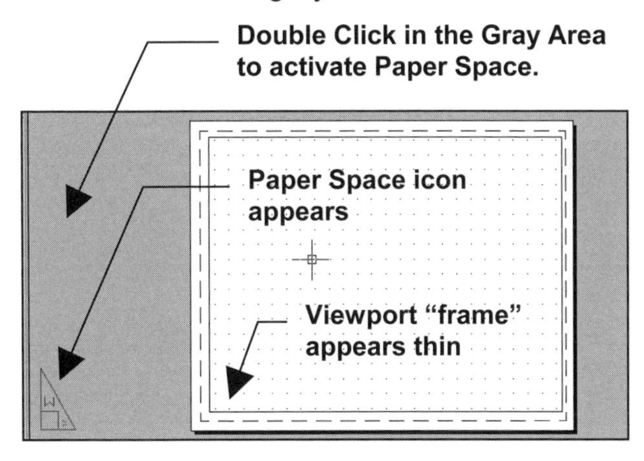

Paper Space Active

Double Click inside the Viewport to activate Model Space

Viewport "frame" appears thick

UCS Icon appears

Model Space Active

You can draw or make changes to the drawing within this Viewport whenever the **UCS icon** is displayed. *(Your UCS icon appearance may be different than the one shown)*

OPTIONS

I prefer to use the Viewport toolbar shown on page 26-10, but you may type "MV <enter>" at the command line, the following options will appear:

ON / OFF Just another option to switch between model and paper space

FIT
a. (default) If you press <enter> this option automatically creates a single rectangular viewport that fills the entire printable area on the sheet
b. Allows you to draw the rectangular viewport by specifying the location of the first corner and then the opposite corner.

HIDEPLOT Prevents hidden lines from plotting in a 3D model.

LOCK Locks the model space and paperspace together. They move and zoom in and out together. The model space scale cannot be changed until you unlock the viewport. (see page 26-10 for other locking methods)

RESTORE Converts saved viewport configurations into individual floating viewports.

OBJECT Converts an object, drawn in paperspace, into a viewport. Objects must be closed, such as: circles, polygons and closed polylines.

POLYGONAL Allows you to draw a viewport outline using polylines. If you do not use CLOSE, AutoCAD will close them automatically.

To "OPEN" the "VIEWPORTS" Toolbar:
1. Select VIEW / TOOLBARS
2. Select the "VIEWPORTS" toolbar from the list.

TOOLBAR FOR 2004 SHOWN BELOW:

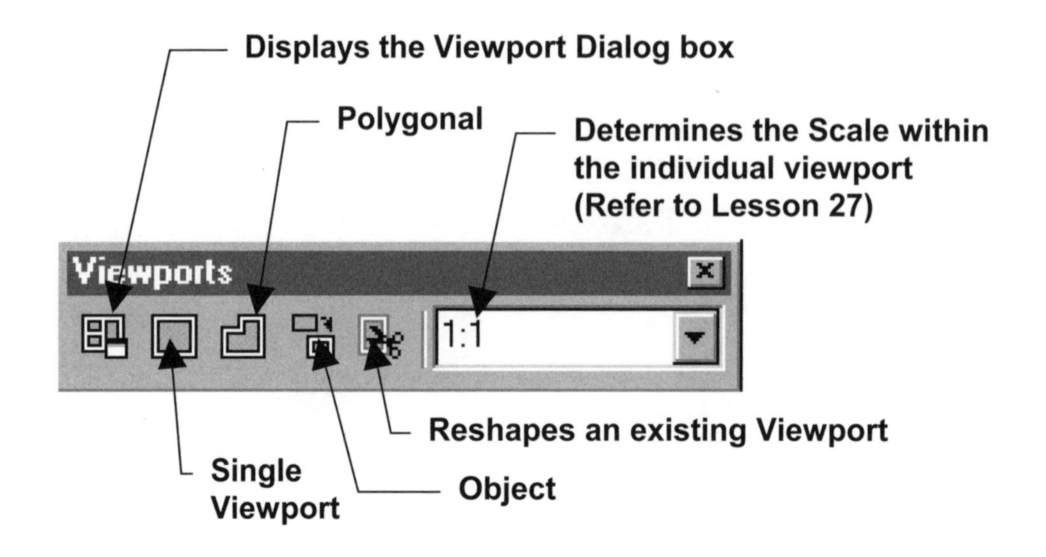

TOOLBAR FOR LT SHOWN BELOW:

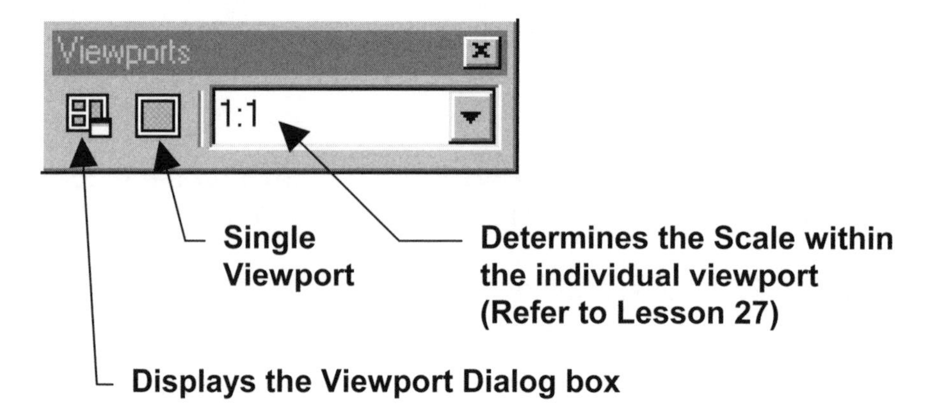

HOW TO LOCK A VIEWPORT

After you have moved and adjusted the scale, within each viewport, you will want to "LOCK" the viewport so it will not change. There are 3 methods.

Method 1. a. Make sure you are in Paper Space.
b. Click once on the Viewport frame
c. Right Click (the short cut menu should appear)
d. Select "Display Locked" - Select Yes or No.

Method 2. Double click on the Viewport frame. The "Properties Palette" will appear. Select Misc / Displayed Locked.

Method 3. Type **MV** at the command line and select "Lock".

HOW TO CREATE A PAGE SET UP

*To plot a drawing you must give the plotter instructions. These instructions are called a "**Page Setup**". You may "**Add**" the page setup to the drawing to save the settings.*

STEP 1. OPEN THE DRAWING YOU WISH TO PLOT.

To plot a drawing it must be displayed on the screen.

STEP 2. SET UP THE PAPERSPACE ENVIRONMENT.

Setting up the Paperspace environment means name the layout tab, select the printer and specify paper size.

IF YOU HAVE PREVIOUSLY COMPLETED THIS STEP, SKIP TO STEP 4.

A. Select a **LAYOUT** tab.

Note: If the Page Setup dialog box shown below does not appear automatically, right click on the Layout tab, then select Page Setup.

1. Type the new name for the Layout tab.
2. Select the "Plot Device" tab.
3. Select the Plotter / Printer.
4. Select the Plot Style table (.ctb) from the list.

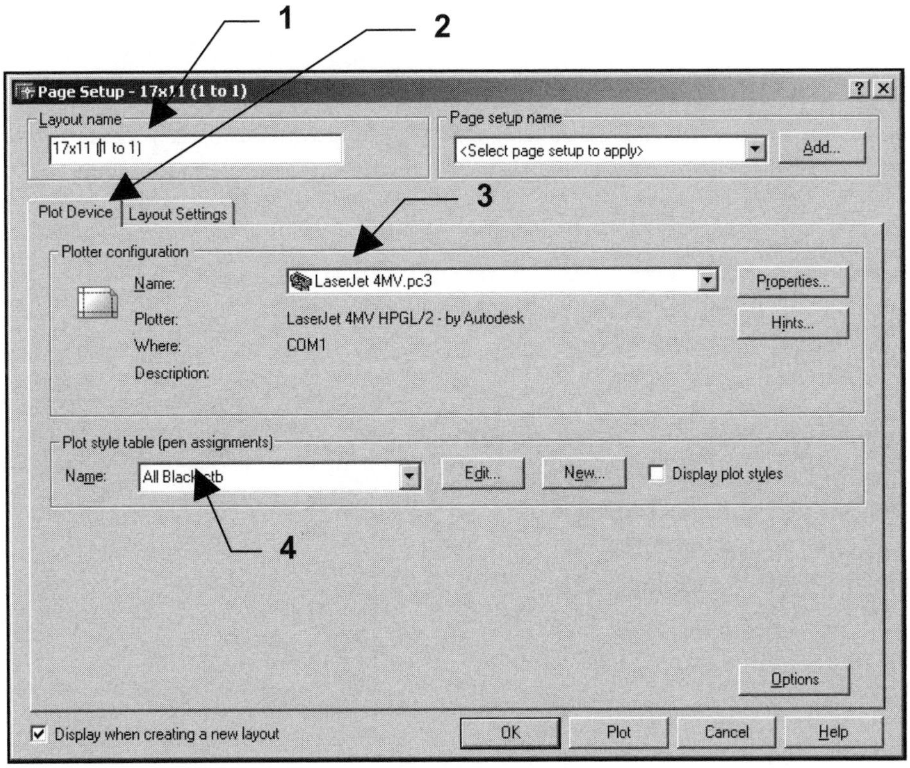

5. Select the “Layout Settings” tab.
6. Select the “Paper Size” from the list.
7. Select the “Scale”.
8. Select “Layout”. Note: You will be instructed later to select “Extents”.
9. Select the “OK” button.

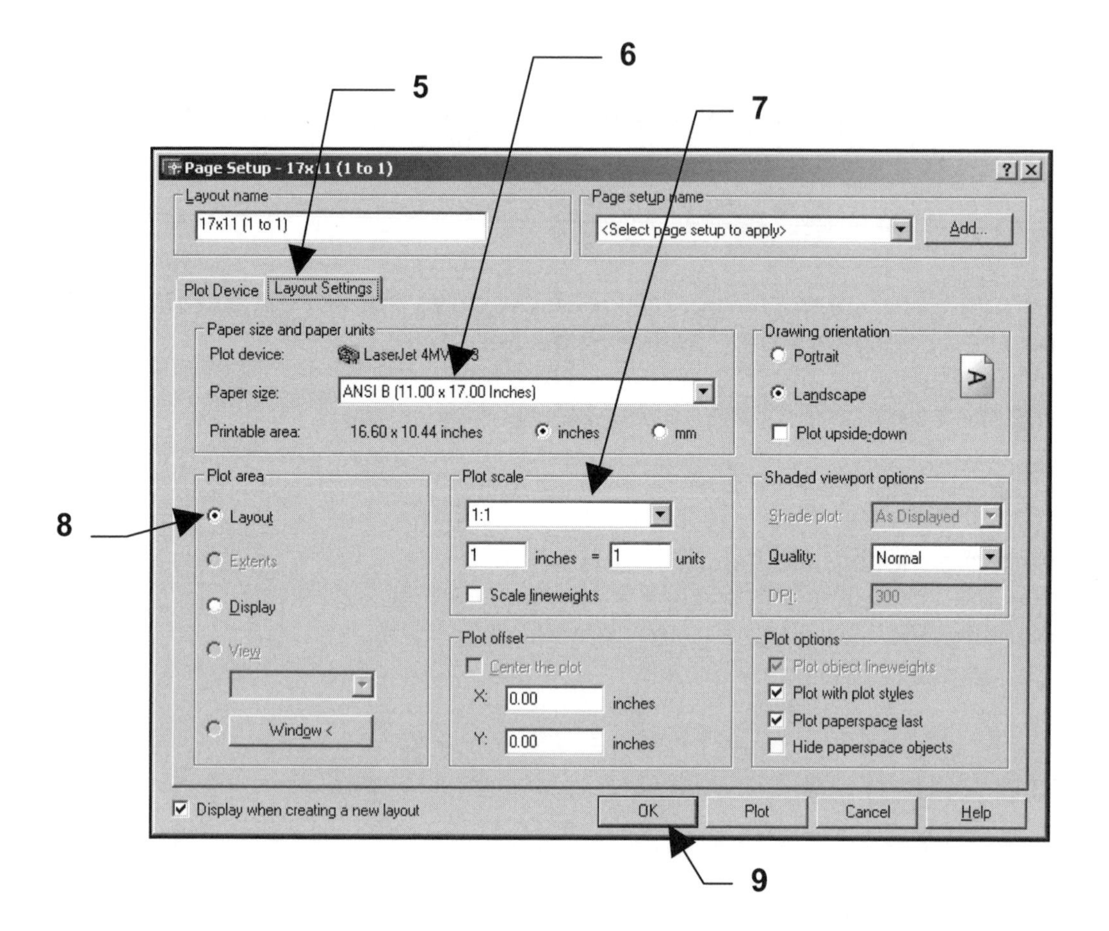

You should now have a sheet of paper displayed on the screen and the Layout tab should now be displayed with it’s new name.

This sheet is in front of “Modelspace”.

STEP 3. CREATE A BORDER, TITLE BLOCK AND CUT A VIEWPORT.

Draw the border, title block and notes in paperspace.
Cut a viewport to see through to modelspace.

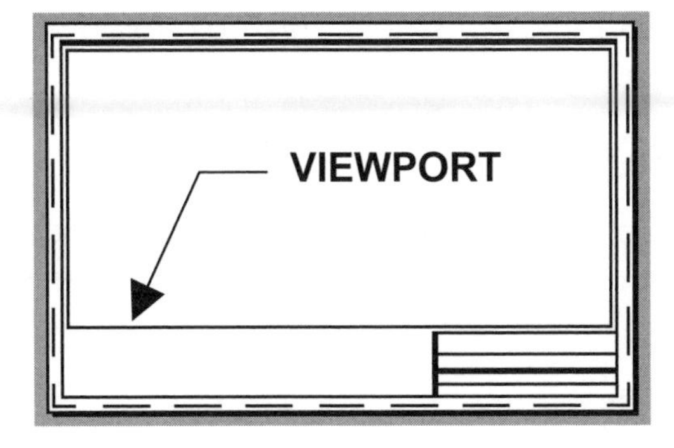

STEP 4. SPECIFY PLOT SETTINGS

Specifying Plot Settings means verify the Layout name, Plot device and paper size. Select what to plot, scale and where to place the drawing on the paper.

A. Select **FILE / PLOT**

(The Plot dialog box shown below should appear)

1
3
2
4

Plot
Layout name
11x17 (1 to 1)
Save changes to layout
Page setup name
<Select page setup to apply>
Add...
Plot Device
Plot Settings
Plotter configuration
Name: LaserJet 4MV.pc3
Properties...
Plotter: LaserJet 4MV HPGL/2 - by Autodesk
Hints...
Where: COM1
Description:
7
Plot style table (pen assignments)
Name: All Black.ctb
Edit...
New...
Plot stamp
On
Settings...
6
What to plot
Current tab
Selected tabs
All layout tabs
Number of copies: 1
5
Plot to file
Plot to file
File name and path:
D:\Architecture PP\Drawing1-11x17 (1 to 1).PLT
Full Preview...
Partial Preview...
OK
Cancel
Help

1. **Layout Name**: Displays the name of the Layout tab you selected. (To change the name of a layout, go back to paperspace, right click on the layout tab and select "Rename" from the short cut menu.)

2. **Save changes to Layout:** Check this box if you want your settings to be saved to the selected Layout tab.

3. Select the **"Plot Device"** tab.

4. Select the **"Plotter Name"** from the list. All previously configured devices are listed. (If your printer / plotter is not listed refer to "Add a Printer / Plotter" Appendix A.

5. **What to Plot:** defines what you want to plot.
   Current tab = plots current Model or Layout tab
   Selected tabs = plots multiple preselected tabs. This option is not available if only one tab is selected.
   All Layouts tabs = plots all layout tabs, selected or not.
   Number of copies = specify number of copies to be plotted.

6. **Plot to File:** Creates a plot file instead of plotting the drawing. If this option is selected, enter the **filename** and specify saving **location**. This is an advanced option, refer to the AutoCAD User's Guide for more information.

7. Select a **"Plot Style Table"** from the list.
   You can also **Create a New** plot style table (Refer to pages 9-7 through 9-13)

B. Select the **"Plot Settings"** tab.

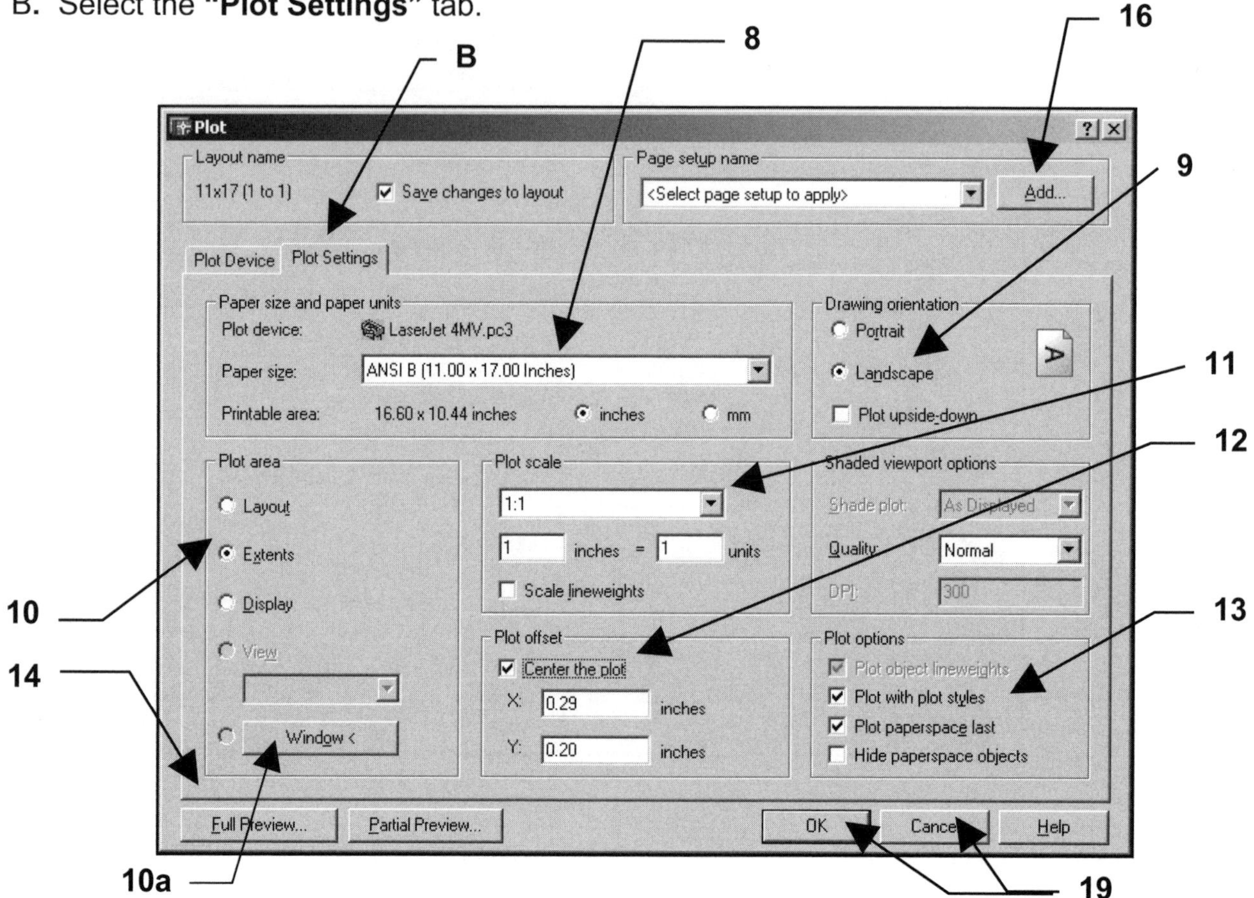

8. Select the desired **Paper Size** from the drop down list. The list should contain all available paper sizes for the plot device you selected.
   (Remember, you must select the plot device first)

9. Select the desired **Drawing Orientation**.
   Landscape = the long edge of the paper represents the top of the page.
   Portrait = the short edge of the paper represents the top of the page.
   (Landscape is the most frequently selected.)

10. Select the **Plot Area**.
    Limits plots the area inside the drawing limits.
    (Select when plotting from Modelspace)
    Layout plots the paper size (Select when plotting a Layout)
    Extents plots all objects in the drawing file even if out of view.
    Display plots the active viewport.
    View plots a previously saved view.
    Window plots objects inside a window. To specify the window, choose the Window button **(10a)** and designate the first and opposite (diagonal) corner of the area you choose to plot. (Similar to the Zoom / Window command)

11. Select a **scale** from the drop down list or enter a custom scale.
    *(If you are plotting from a "**LAYOUT**" tab, you should always use **plot scale 1:1**. Scaling will be discussed in Lesson 27)*

12. **Plot Offset:** Specify where you want the drawing located on the sheet of paper. The **X** and **Y** boxes defines the offset from the lower left corner of the paper. The "**Center the Plot"** box automatically centers the drawing on the paper.

13. **Plot Options:**
    Plot Object Lineweights = plots objects with assigned lineweights.
    Plot with Plot Styles = plots using the selected Plot Style Table.
    Plot paperspace last = plots model space objects before plotting paperspace objects. Not available when plotting from model space.
    Hide Paperspace Objects = used for 3D only. Plots with hidden lines removed.

14. Select **Full Preview** button.
    Full preview displays the drawing as it will plot on the sheet of paper. <u>(Note: If you cannot see through to Model space, you have not cut your viewport yet)</u>

    a. If the drawing is centered on the sheet, press the **Esc** key and go on to step 15.

    b. If the drawing does not look correct, press the **Esc** key and check all your settings, then preview again.

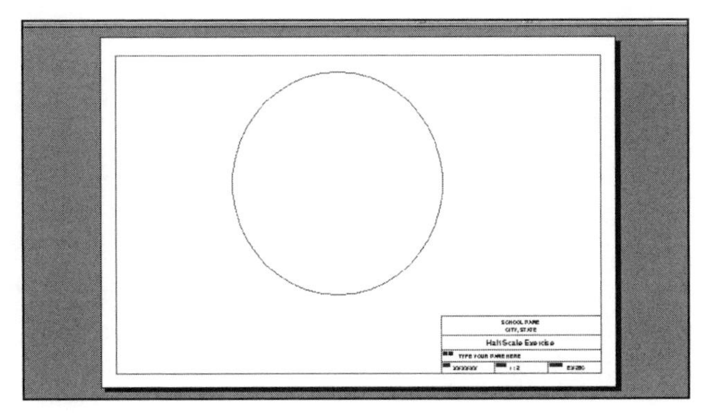

Notice the Viewport "frame" is not visible. The "Viewport" layer is set to "no plot"

STEP 5. NAME THE PAGE SETUP

After you have completed steps 1 through 4 you should save them by specifying a name for the page setup.

16. Select the **"Add"** button (See page 26-14)

17. Type the new page set up name
    Try to give it a name that briefly describes the settings previously selected.

18. Select the **OK** button

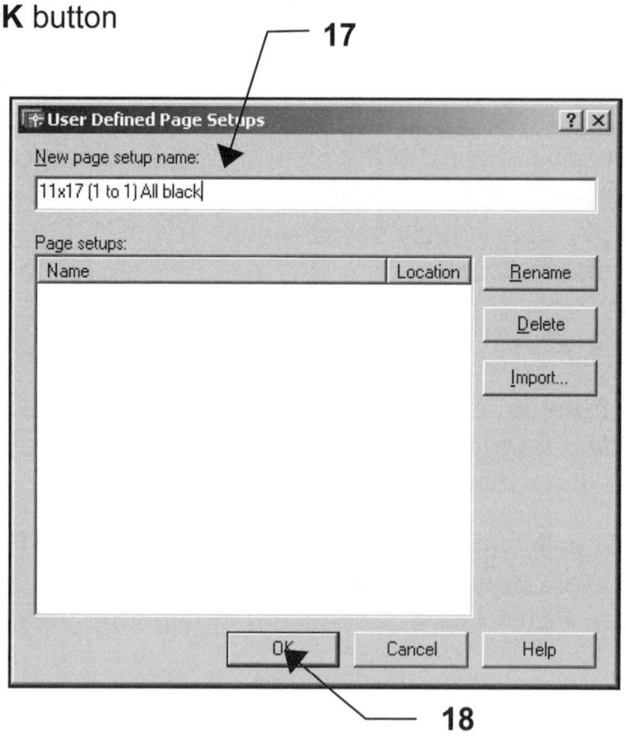

The settings previously selected are now saved as a page setup

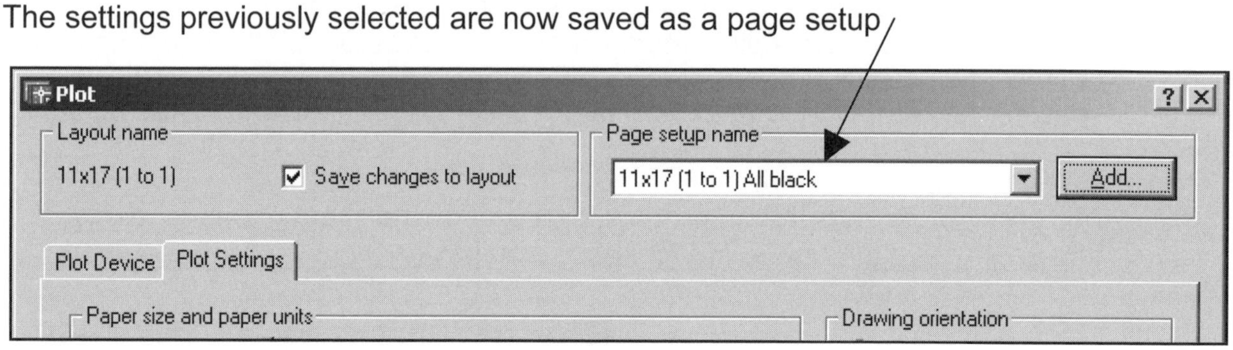

STEP 6. PLOT THE DRAWING or CLOSE THE PLOT DIALOG BOX.

19. ***If your computer is connected to the plotter / printer***, select the **OK** button (19 on page 26-14) to plot, then proceed to step 7.
    If your computer is not connected to the plotter / printer, select the **Cancel** button (19 on page 26-14) to close the Plot dialog box and proceed to step 7.
    Note: Selecting Cancel does not cancel your page setup. It will still be saved to the drawing.

STEP 7. SAVE THE DRAWING AGAIN using FILE / SAVE AS. This will guarantee that the Page Setup you just created will be saved to this drawing for future use.

SETTING THE PICK BOX SIZE

☐ Pick box

When AutoCAD prompts you to ***select objects,*** such as when you are erasing objects, the cursor (crosshairs) turns into a square. This square is called a **Pick box.** The size of the **Pick box** can be changed. Some AutoCAD users prefer large boxes, some like small boxes. The size of the box is your personal preference, however, the smaller the Pick box the more **accurate** you must be when placing the pick box on an object to select it. If the Pick box is too large it could overlap onto other objects that you did not want to select.

HOW TO CHANGE THE "PICK BOX" SIZE.

1. Select **TOOLS / OPTIONS**
2. Select the **SELECTION** tab.

The following dialog box will appear.

3. Adjust the size by sliding the tab (click and drag) to the right (max) or to the Left (Min.). A preview of the new size is displayed.

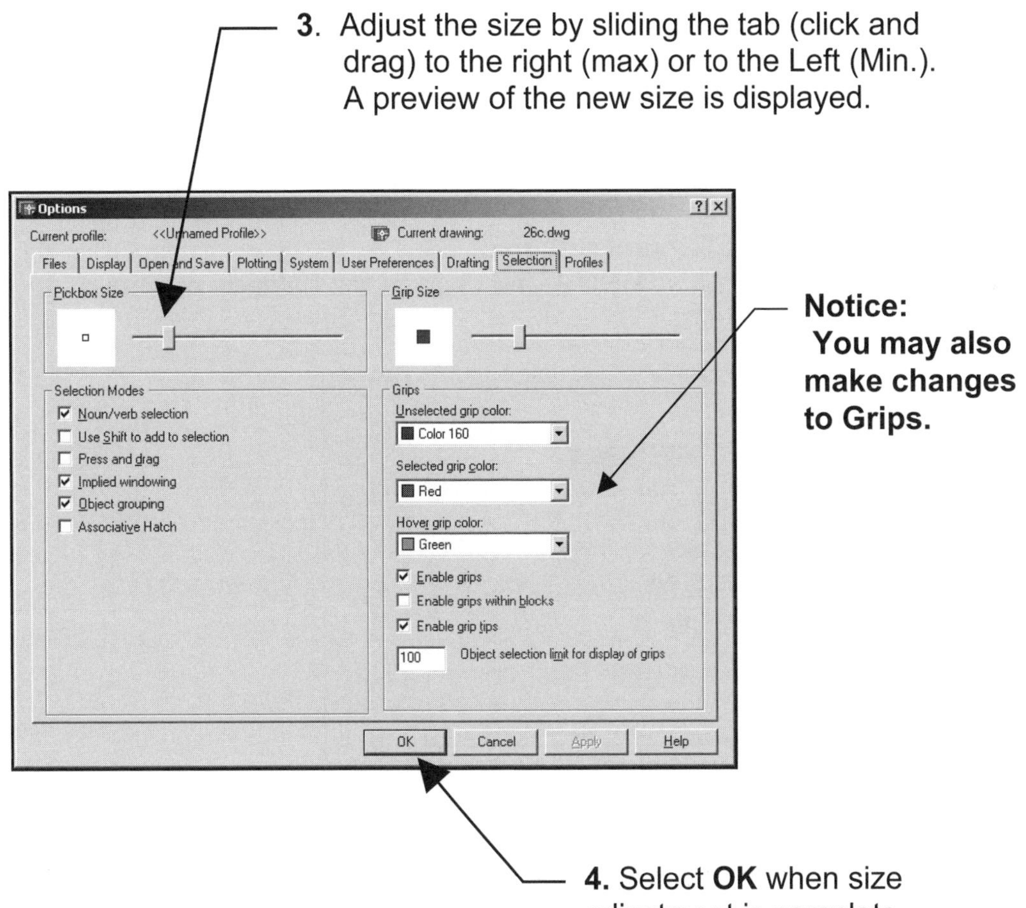

Notice:
You may also make changes to Grips.

4. Select **OK** when size adjustment is complete.

EXERCISE 26A

CREATE A MASTER DECIMAL SETUP DRAWING

The following instructions will guide you through creating a "Master" decimal setup drawing. The "1Workbook Helper" is an example of a Master setup drawing. Even though the screen appears blank, the actual file is full of settings, such as: Units, Drawing Limits, Snap and Grid settings, Layers, Text styles and Dimension Styles. Once you have created this "Master" drawing, you just open it and draw. No more repetitive inputting of settings.

NEW SETTINGS

A. Begin your drawing without a template as follows:

1. Select **"FILE / NEW"**
2. Select **"START FROM SCRATCH"** Box.
3. Select "**OK**".
4. Your screen should be blank, no grids and the current layer is 0

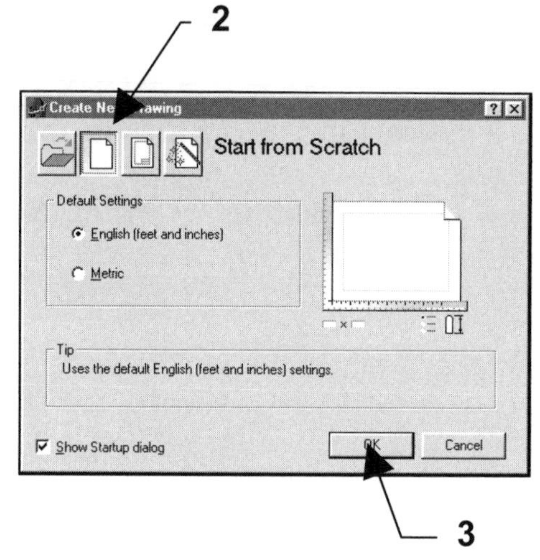

B. Set drawing specifications as follows:

1. Set "**UNITS**" of measurement
   Use "**FORMAT / UNITS** and change the settings as shown then select **OK.**

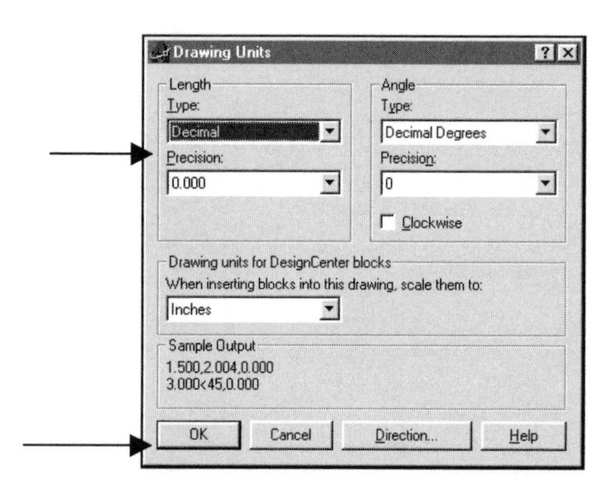

2. Set "**DRAWING LIMITS**" (Size of drawing area)
   USE "**FORMAT / DRAWING LIMITS**
   a. Lower left corner = 0.000,0.000
   b. Upper right corner = 17 , 11
   c. Use **"VIEW / ZOOM / ALL"** to generate the new limits
   d. Set your **Grids** to **ON** to display the paper size.

3. Set **"SNAP AND GRID"**
   Use "**TOOLS / DRAFTING SETTINGS**

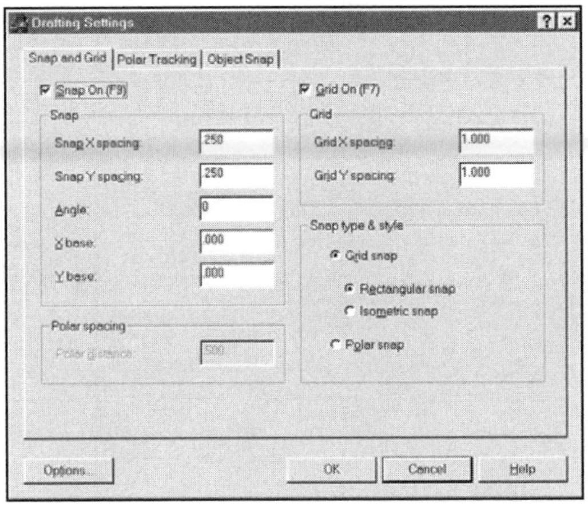

4. Set **"PICK BOX"** size (your preference)
   See page 26-17 for instructions.

NEW LAYERS

C. Create new layers

1. First the **Load** linetypes listed below. (See page 26-4 for instructions)
   CENTER2
   HIDDEN
   PHANTOM2

2. Assign names, colors, linetypes and plotability. (See page 26-3 for instructions)

NAME	**COLOR**	**LINETYPE**	**LWT**	**PLOT**
BORDER	RED	CONTINUOUS	.039	YES
CENTER	CYAN	CENTER2	Default	YES
CONSTRUCTION	WHITE	CONTINUOUS	Default	NO
DIMENSION	BLUE	CONTINUOUS	Default	YES
HATCH	GREEN	CONTINUOUS	Default	YES
HIDDEN	MAGENTA	HIDDEN	Default	YES
OBJECT	RED	CONTINUOUS	.024	YES
PHANTOM	MAGENTA	PHANTOM2	Default	YES
SECTION	WHITE	PHANTOM2	.031	YES
TEXT HEAVY	WHITE	CONTINUOUS	Default	YES
TEXT LIGHT	BLUE	CONTINUOUS	Default	YES
THREADS	GREEN	CONTINUOUS	Default	YES
VIEWPORT	GREEN	CONTINUOUS	Default	NO

NEW TEXT STYLE

D. Create a text style

1. Select **"FORMAT / TEXT STYLE"**
2. Make the changes shown in the dialog box.

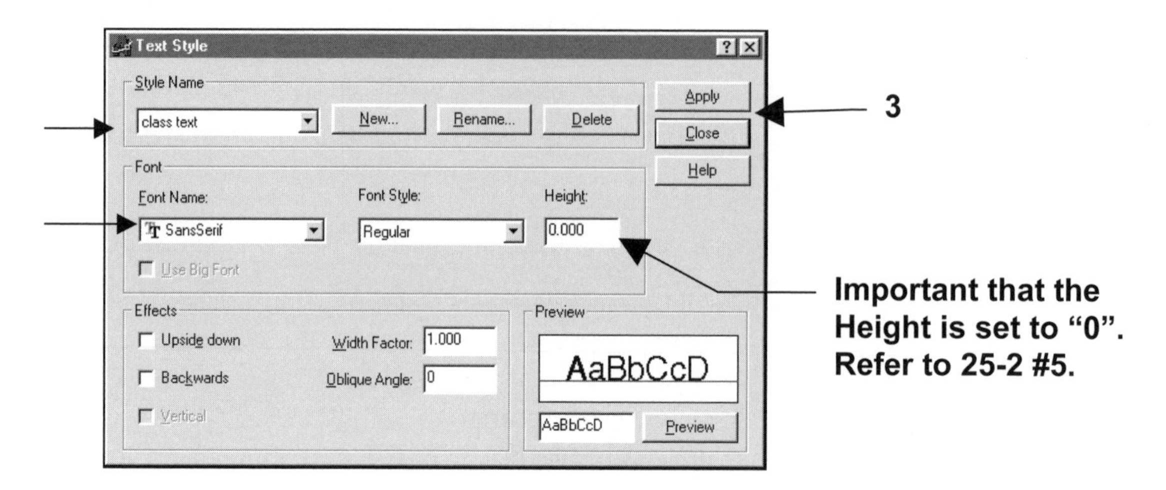

3. When complete, select **APPLY** then **CLOSE**.

NEW DIMENSION STYLE

E. Create a new Dimension Style named ***Class Style***
(Refer to Page 16-7 for step by step instructions)

NEW DIMENSION SUB-STYLE

F. Create a new Dimension Sub-Style for ***Radius***
(Refer to Page 18-6 for step by step instructions)

THIS NEXT STEP IS VERY IMPORTANT..

G. **SAVE ALL THE SETTINGS YOU JUST CREATED**
1. Select File / Save as
2. Save as: **My Decimal Setup**

H. Now continue on to Exercise 26B.

EXERCISE 26B

CREATE A BORDER FOR PLOTTING

The following instructions will guide you through creating a Border drawing that will be used in combination with "My Decimal Setup" when plotting. You will create a Layout and draw a border with a title block. All of this information will be saved and you will not have to do this again.

A. Open **My Decimal Setup**

B. Select a **LAYOUT** tab. Model Layout1

Note: If the Page Setup dialog box shown below does not appear automatically, right click on the Layout tab, then select Page Setup.

1. Type the new name:
11x17 (1 to 1)

2. Select the "*Plot Device*" tab.

3. Select the Plotter.
(Use this plotter for this exercise. Refer to Appendix A for instruction if it does not appear in the list.)

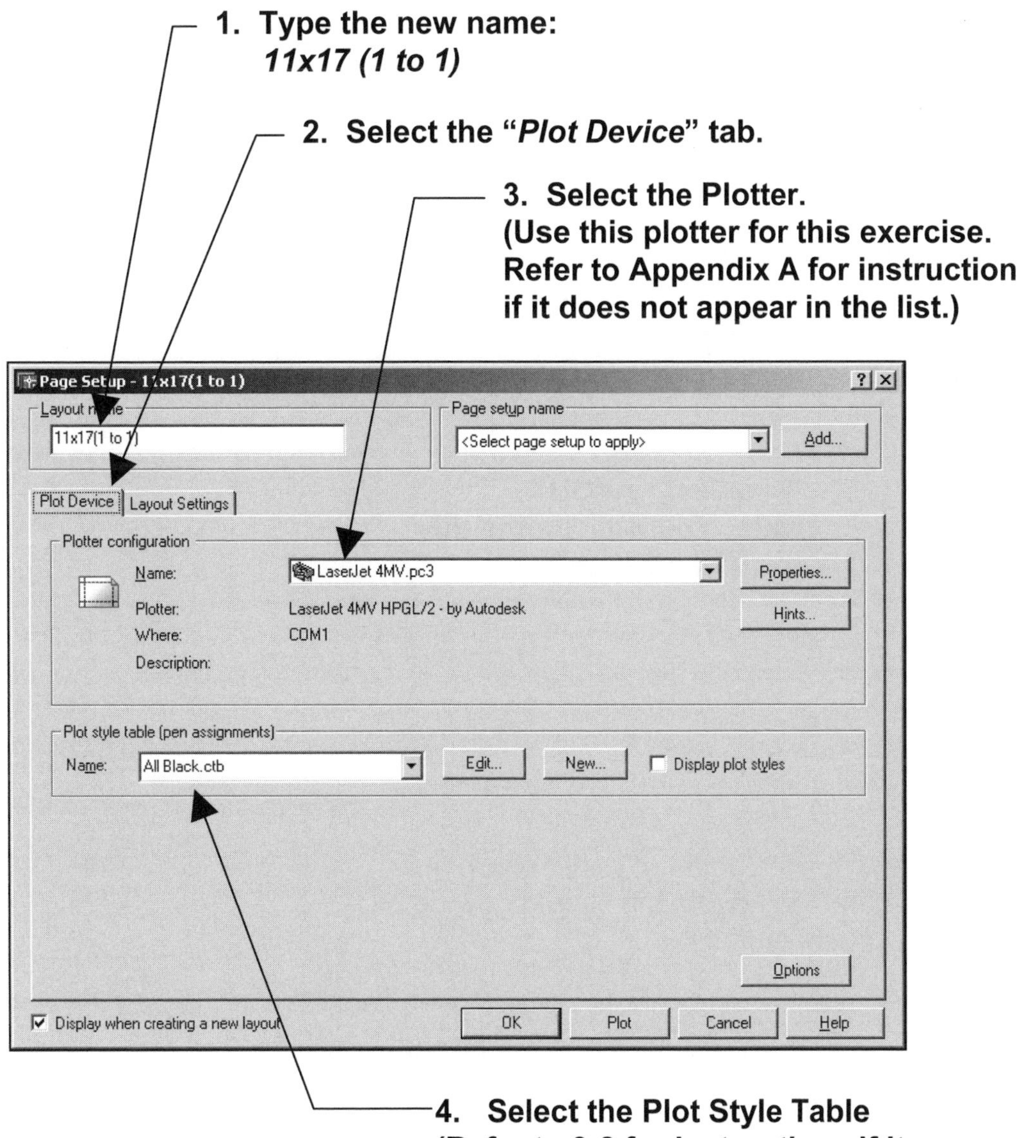

4. Select the Plot Style Table
(Refer to 9-8 for instructions if it does not appear in the list.)

5. Select the "*Layout Settings*" tab.

6. Select the "*Paper Size*".

7. Select scale "1:1"

Page Setup - 11x17(1 to 1)

Layout name: 11x17(1 to 1)

Page setup name: <Select page setup to apply> — Add...

Plot Device | Layout Settings

Paper size and paper units
Plot device: LaserJet 4M.pc3
Paper size: ANSI B (11.00 x 17.00 Inches)
Printable area: 16.60 x 10.44 inches — inches — mm

Drawing orientation
Portrait
Landscape
Plot upside-down

Plot area
Layout
Extents
Display
View
Window <

Plot scale
1:1
1 inches = 1 units
Scale lineweights

Shaded viewport options
Shade plot: As Displayed
Quality: Normal
DPI: 300

Plot offset
Center the plot
X: 0.00 inches
Y: 0.00 inches

Plot options
Plot object lineweights
Plot with plot styles
Plot paperspace last
Hide paperspace objects

Display when creating a new layout

OK | Plot | Cancel | Help

8. Select LAYOUT
Note: You will change this to "Extents" later.

9. Select the *OK* button

You should now have a sheet of paper displayed on the screen and the Layout tab should now be displayed as "11x17 (1 to 1)".

This sheet is in front of "Model". In Exercise 26C, you will cut a hole (viewport) in this sheet so you can see through to Model.

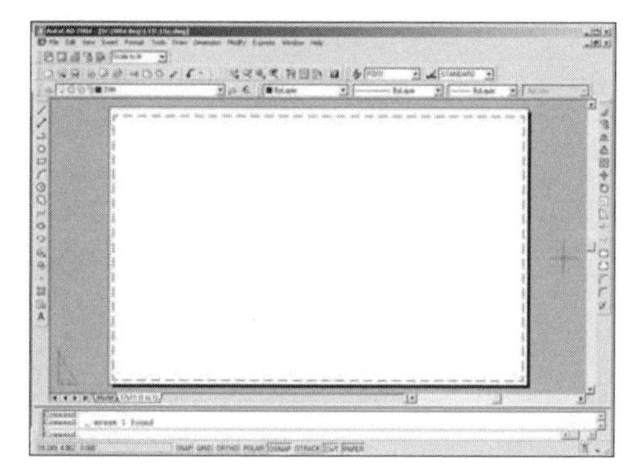

C. Draw the Border with title block, shown below, on the sheet of paper shown on the screen.

D. When you have completed the Border, shown below:

1. Select File / Save as
2. Save as: **My Decimal Setup** (Again)

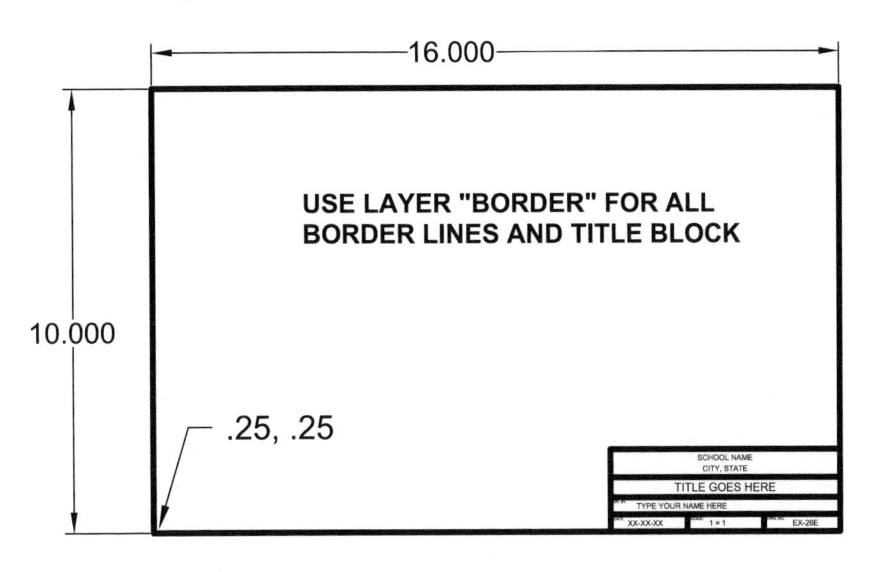

Important: Use "Single Line Text" to place the text in the Title Block below.

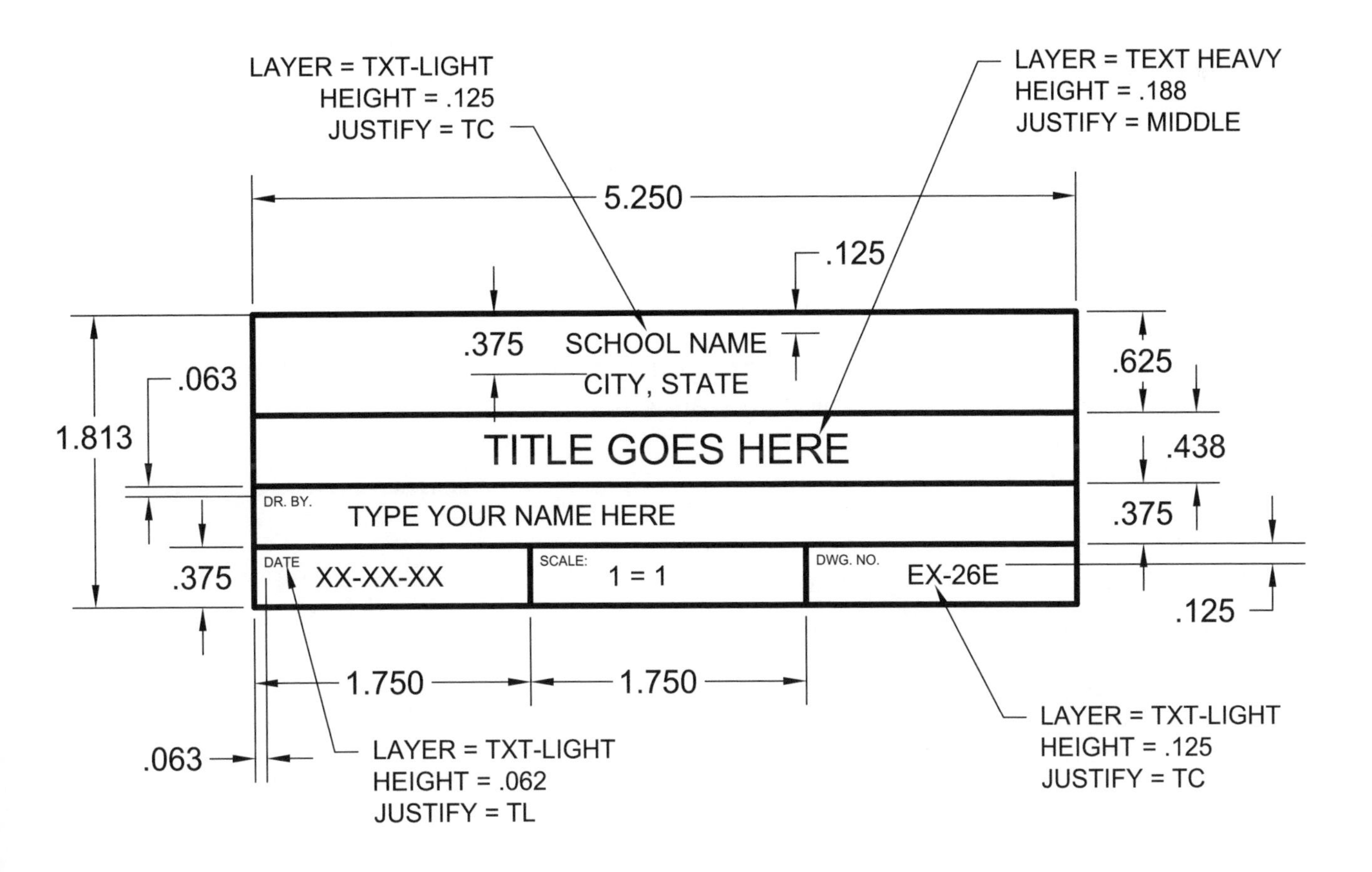

EXERCISE 26C

CREATE A VIEWPORT

The following instructions will guide you through creating a VIEWPORT in the Border Layout sheet. Creating a viewport has the same effect as cutting a hole in the sheet of paper. You will be able to see through the viewport frame (hole) to Model.

A. Open **My Decimal Setup**
B. Select the **11-17 (1 to 1)** tab.
C. Select layer "Viewport"

Single Viewport icon (LT's toolbar looks a little different but it works the same)

D. Select the Single Viewport icon from the Viewports Toolbar or Type MV <enter>.

E. Draw a Single viewport approximately as shown.

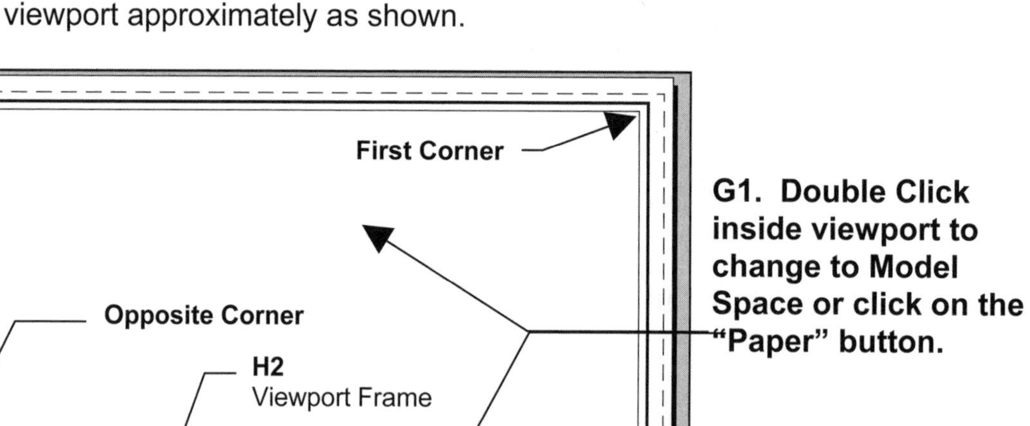

F. After successfully creating the Viewport, you should now be able to see through to Model. (Your grids should appear if they are ON.)

G. Adjust the Model space scale.
   1. Select the **Paper** button so it changes to "Model" or double click inside the Viewport Frame.
   2. Do **Zoom / All** before adjusting the scale.
   3. Select 1:1 in the **VIEWPORT** toolbar.

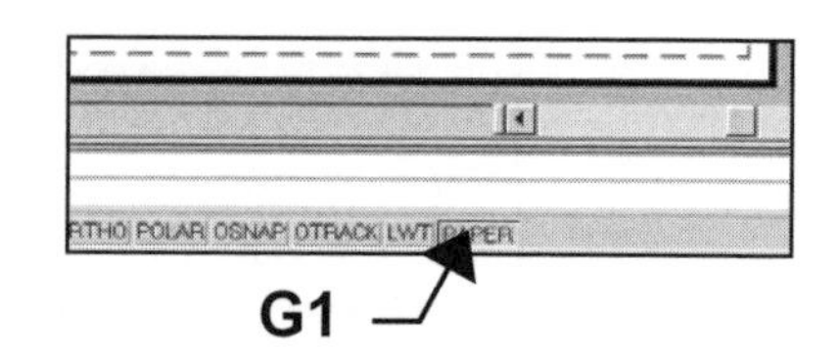

H. Lock the Viewport
   Refer to 26-10 for Locking instructions.

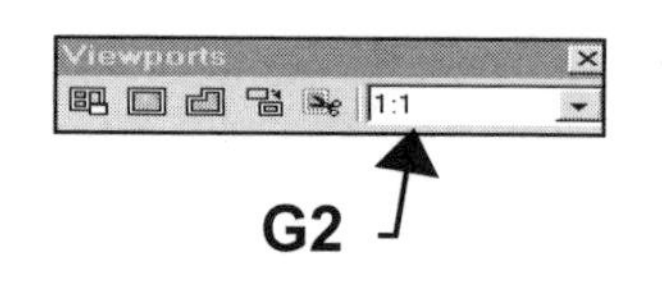

J. Save as: **My Decimal Setup**

K. Continue on to 26-D.

EXERCISE 26D

COMPLETING THE PAGE SETUP

The following instructions will guide you through the final steps to completing the **Page Setup**. This Page Setup will stay with **My Decimal Setup** and you will be able to use it over and over again. (Refer to Page 26-13 Step 4 for more detailed explanations of each area.)

A. Open **My Decimal Setup** if it isn't already open.

B. Select the **11x17 (1 to 1)** layout tab.

You should be looking at your Border and Title Block now.

C. Select **File / Plot** or place the cursor on the **11x17 (1 to 1)** tab and press the right mouse button. Select **Plot** from the short cut menu.

The Plot dialog box should appear

D. Select the **Plot Device** tab and select the options 1 thru 3 below:

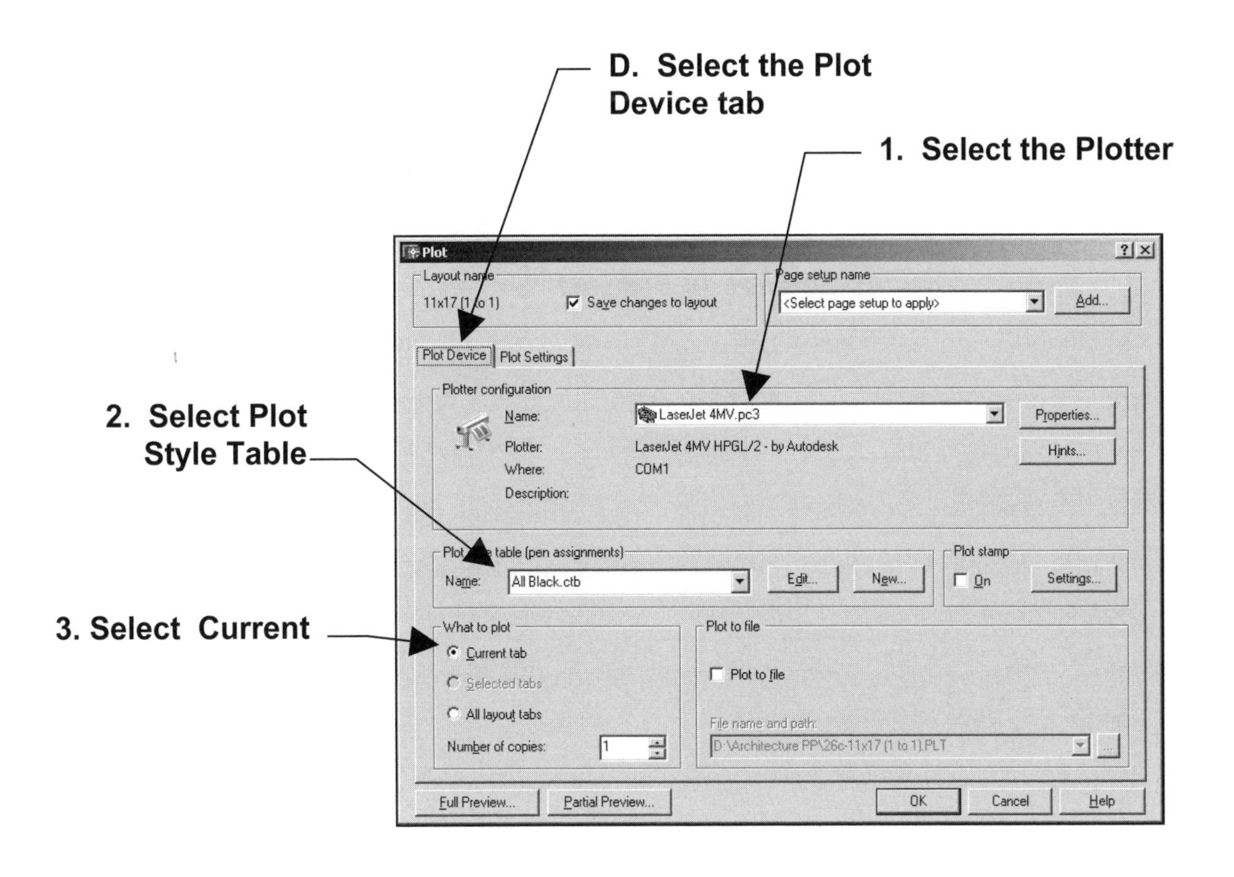

E. Select the **Plot Settings** tab then select options 1 thru 7 below:

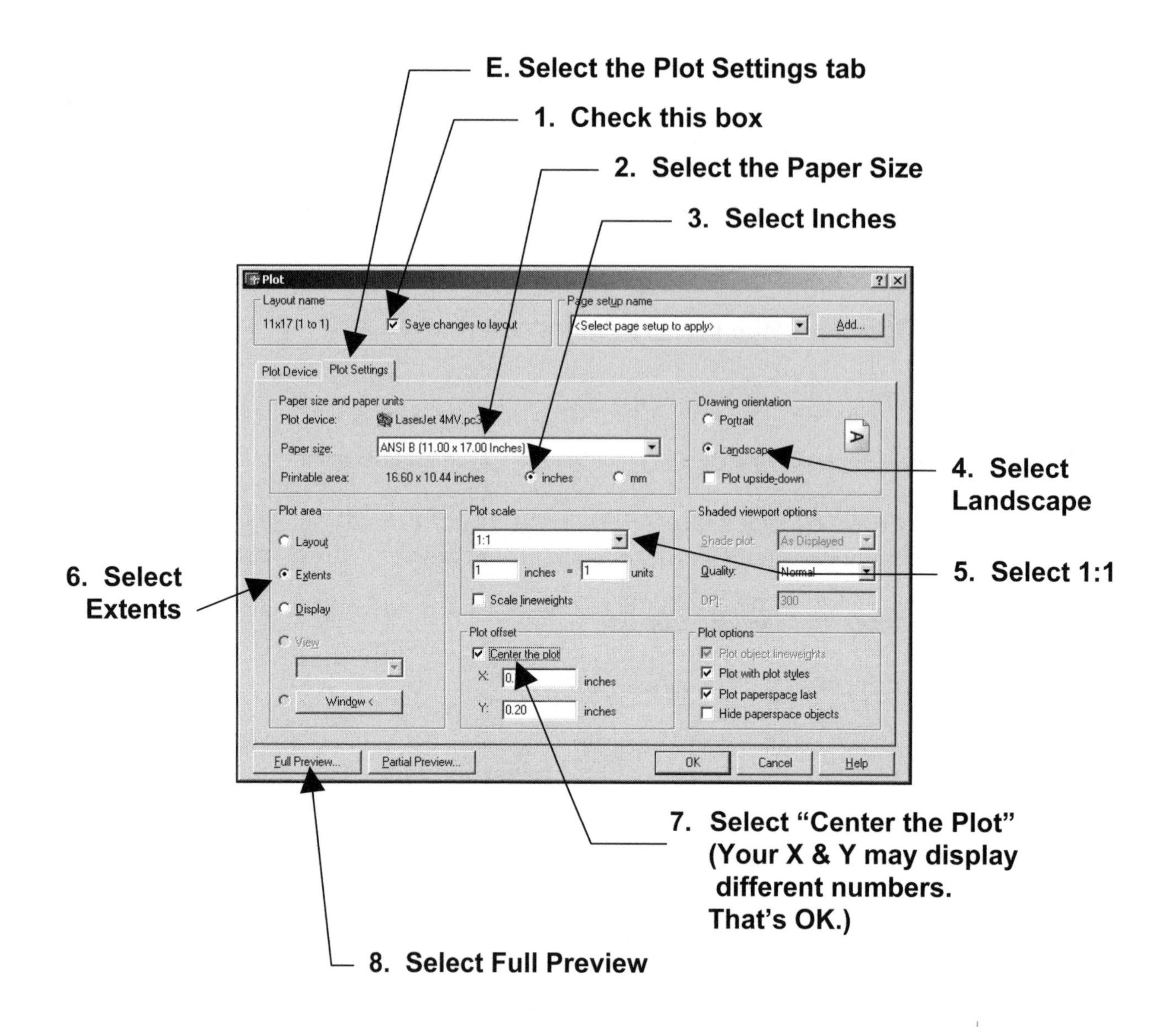

F. Preview the Layout

1. If the drawing is centered on the sheet, press the Esc key and continue on to **G**.

2. If the drawing does not look correct, press the Esc key and check all your settings, then preview again.

G. Select the **ADD** button.

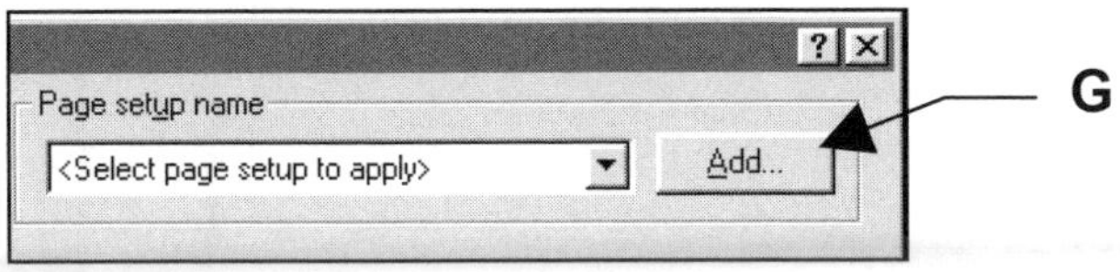

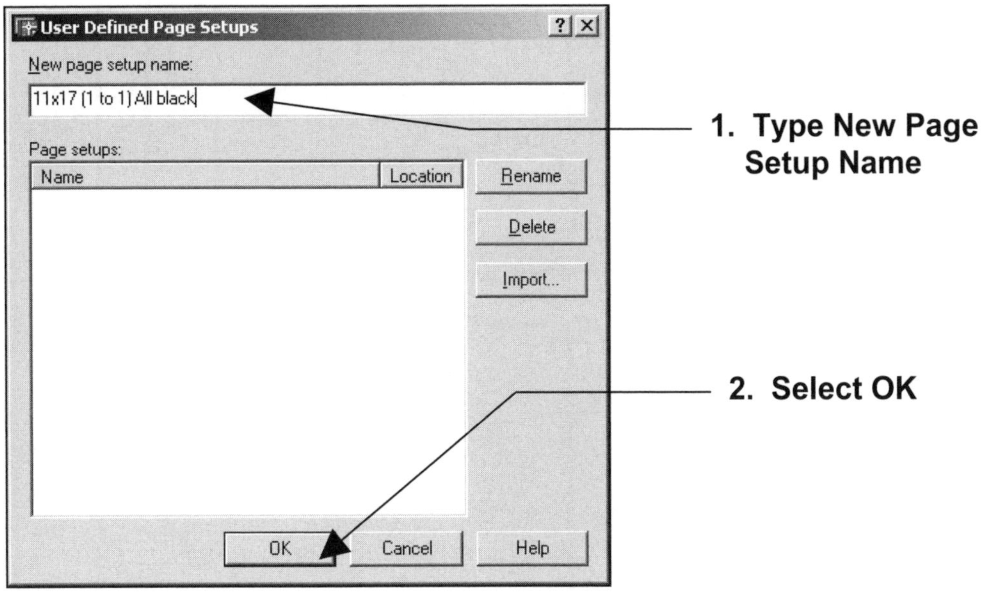

G

1. **Type New Page Setup Name**

2. **Select OK**

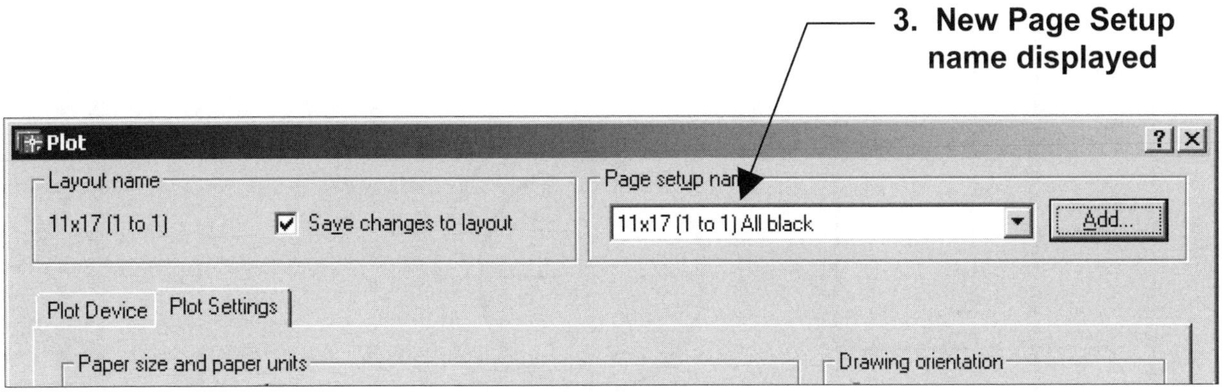

3. **New Page Setup name displayed**

H. If your computer **is** connected to the Plotter / Printer, select the **OK** button to plot.

I. If your computer **is not** connected to the Plotter / Printer, select the **CANCEL** button. (Your settings **will not** be lost)

J. **Save** this file one more time:
   1. Select **File / Save** as
   2. Save as: **My Decimal Setup**

*You have now completed the **Page Setup** for the **My Decimal Setup**. Now you are ready to use this master file to create many drawings in the future. In fact, you have one on the very next page.*

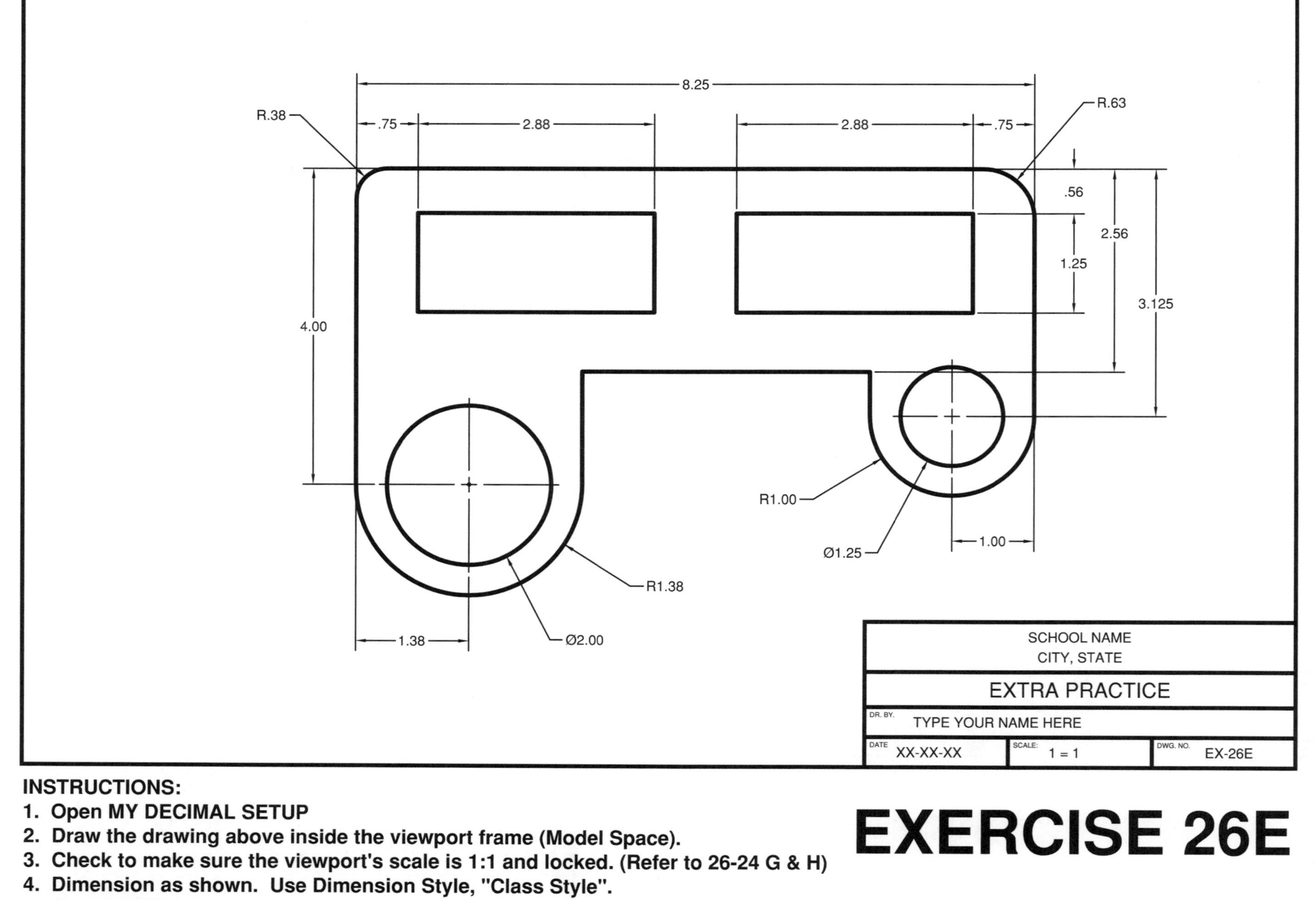

INSTRUCTIONS:

1. **Open MY DECIMAL SETUP**
2. **Draw the drawing above inside the viewport frame (Model Space).**
3. **Check to make sure the viewport's scale is 1:1 and locked. (Refer to 26-24 G & H)**
4. **Dimension as shown. Use Dimension Style, "Class Style".**
5. **Save as: EX-26E**
6. **Plot using "11x17 (1 to 1) All Black" Page Setup.**

EXERCISE 26E

LEARNING OBJECTIVES

After completing this lesson, you will be able to:

1. Understand scaled drawings.
2. Adjust the scale within a viewport.
3. Calculate the Drawing Scale Factor.
4. Understand how scale affects Text, Hatch and Dimensions.
5. Dimension a scaled drawing.
6. Create an Feet-Inches "Setup" file for future use.
7. Create a new Architectural Border to use when plotting.

LESSON 27

CREATING and PLOTTING SCALED DRAWINGS

In the lessons, previous to lesson 26, you worked only in Model space. Then in lesson 26 you learned that AutoCAD actually has another environment called Paper space or Layout. In this lesson we need to learn more about why we need 2 environments and how they make plotting drawings easier.

A very important rule in CAD you must understand is; *"all objects are drawn full size."* In other words, if you want to draw a line 20 feet long, you actually draw it 20 feet long. If the line is 1/8" long, you actually draw it 1/8" long.

Drawing and Plotting objects that are very large or very small.

In the previous lessons you created basically medium sized drawings. Not too big, not too small. But what if you wanted to draw a house? Could you print it, to scale, on a 17 X 11 piece of paper? How about a small paper clip. Could you make it big enough to dimension? Let's start with the house.

Drawing something large such as a house.

1. Start a new drawing from scratch. (26-18)
2. Set the units for the drawing to Architectural.
3. Set the grids to: 12 Set the snap to: 3
4. Set the drawing limits, in model space, to:
   Lower left corner: 0, 0 Upper Right Corner: 45', 35' (Zoom / All)
   (Big enough for the entire house to be drawn full size.)
5. Draw a rectangle: 30' L X 20' W (representing the house) and a pretend roof.

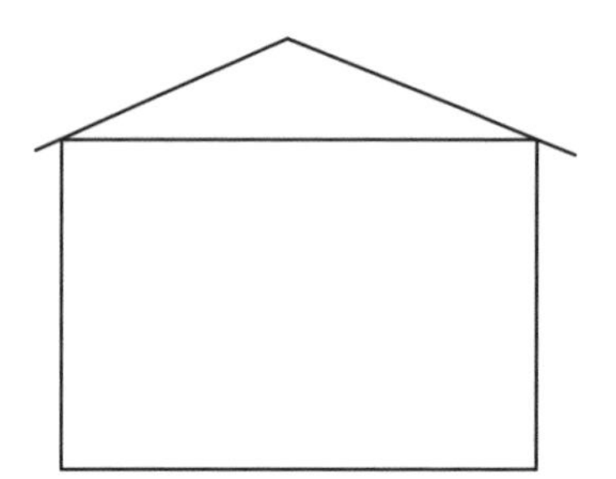

Plotting the house drawing

Now you want to plot this house drawing on a 17 x 11 piece of paper. This is where paper space makes it easy.

6. Select the Layout tab (paper space).
7. Specify the plotter LaserJet 4MV and the paper size 11 x 17.
8. Open the "Viewport" tool bar and cut a viewport so you can see through Viewport to the house drawing that is in model space.
9. Double click inside the Viewport to get into model space.
10. Select View / Zoom / Extents so the entire house is visible within the viewport frame.

Now here is where the magic happens.

11. Adjust the scale of model space. (See 27-6)
    Scroll down the scales and select ¼" = 1'.

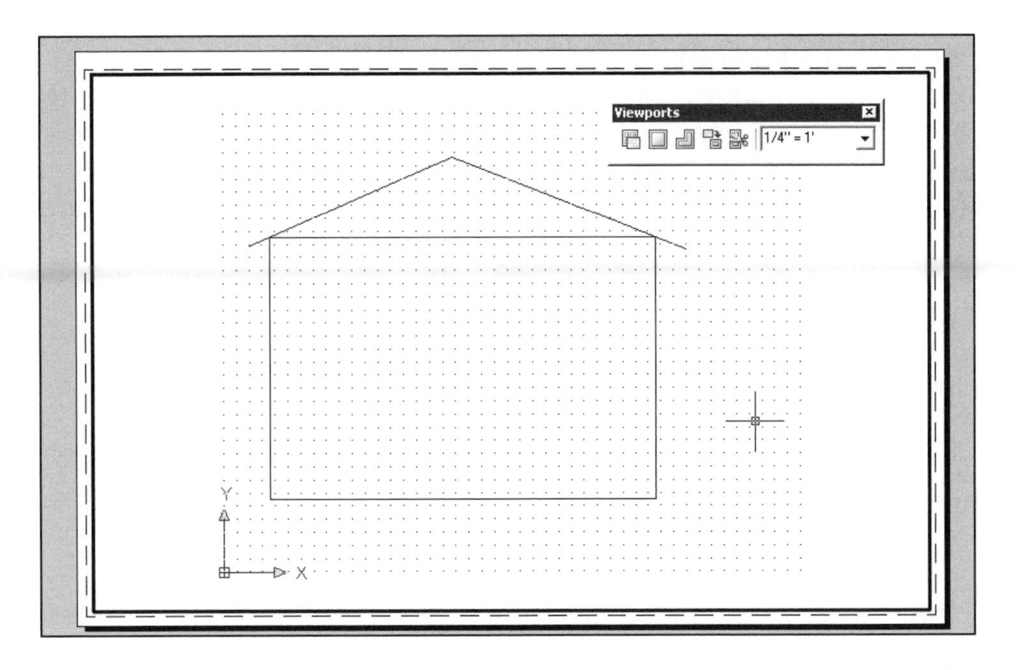

Wow, a 30ft x 20ft house fit on to an 11" x 17" sheet of paper.

Let's talk about what really happened?

Remember the photo and picture frame example I suggested in lesson 26? (Refer to 26-6) This time try to picture you standing in front of your house with an empty picture frame in your hands. Look at your house through the picture frame. The house is way too big to fit in the frame. So you walk across the street and look through the picture frame again. Does the house appear smaller? If you could walk far enough away from the house, it would eventually appear small enough to fit in the picture frame. But....the house did not actually change size did it? It only appears smaller because you are farther away from it.

This is the concept that you need to understand.

When you adjust the scale of model space, it does not actually change size, it just gets farther away and appears smaller.

When you dimension the house, the dimension values will be the actual size of the House. In other words, the 30 ft. width will have a dimension of 30'-0".

12. Lock the viewport (Refer to 26-10)
    When you have adjusted the model space scale to your satisfaction you should "lock" the viewport so the scale will not change when you zoom.

13. Save this drawing as "HOUSE". *You will need it for the "Drawing Scale Factor" information.*

14. Plot the paper space / model space combination using File / Plot.
    The Plot scale should be 1 = 1 (Remember you have already adjusted the model space scale, you do not need to scale the combination, it is just fine.

15. Don't forget to Preview before plotting.

Drawing something small such as a paperclip.

When plotting something smaller you have to move the picture frame closer to model space rather than farther away. So in that case, you would adjust the model space scale to a scale larger than 1 : 1 until the object in model space appeared large enough to see easily and dimension. Remember, even though the object appears larger, when you dimension it the dimension values will be the correct information.

1. Open **My Decimal Setup**
2. Select the “Model” tab.
3. Draw a Rectangle 2” L X 1” W (We won’t actually draw the paperclip)

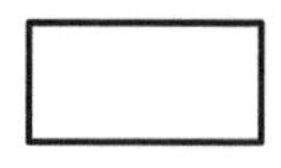

4. Select the “Layout2” tab.
5. Specify the plotter LaserJet 4MV and the paper size 11 x 17.
6. Open the “Viewport” tool bar and cut a viewport so you can see through Viewport to the house drawing that is in model space.

The Rectangle will fill the viewport.

7. Double click inside the Viewport to get into model space.
8. Adjust the scale, of model space first to 1 : 1 .

The Rectangle appears too small and would be difficult to dimension.

9. Adjust the scale, of model space, this time to 4 : 1

Now the Rectangle is large enough to see clearly and dimension.

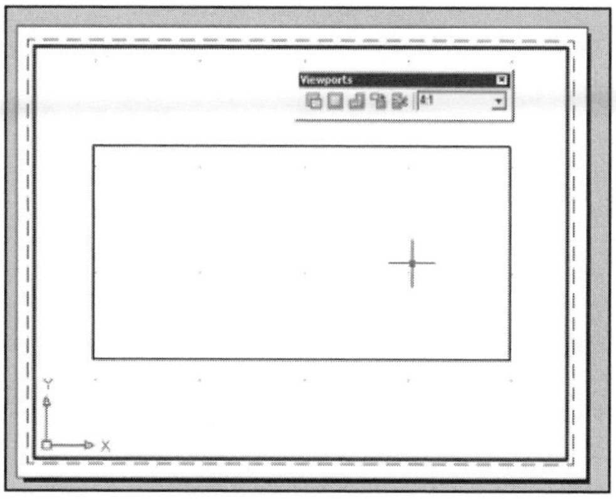

You will understand this concept better after you have completed the exercises in this lesson.

ADJUSTING THE SCALE INSIDE A VIEWPORT

The following will walk you through the process of adjusting the scale within a viewport.

NOTE: DO NOT ADJUST THE SCALE OF "PAPER SPACE". PAPER SPACE SCALE SHOULD REMAIN 1 : 1.

1. **Open** a drawing.
2. Select a **Layout** tab. (paper space)
3. Specify a plotter and paper size. (If you haven't already)
4. Create a Viewport.
5. Open the Viewport toolbar
6. Double click inside of the Viewport to change to Model Space.
   (It is very important that you be inside the Viewport.
   You scale the Model geometry not the Layout)
7. Select **View / Zoom / All** before adjusting the scale of the viewport.
8. ***Adjust the scale*** of the Model geometry by selecting a scale from the **"Viewports"** toolbar.

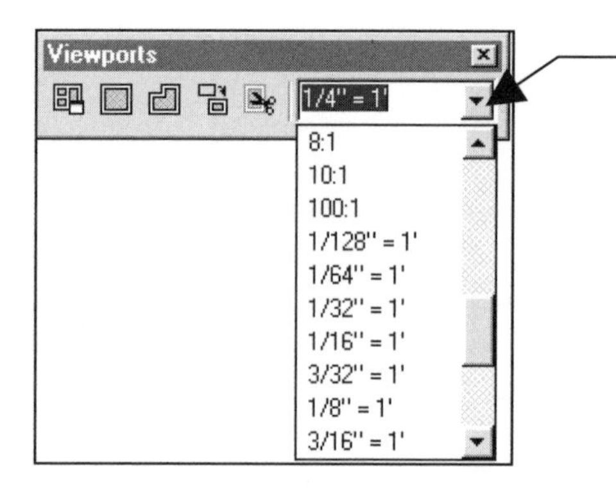

Select the scale from the drop down menu.

(There are 31 different decimal and architectural scales.)

If you would like a scale other than one listed in the drop down list you may use the Zoom XP command as follows:

a. Select View / Zoom / Scale
b. Enter a scale factor (nX or nXP): ***enter the "scale" XP***
   (XP means "times Paperspace"

The Scale for 2 : 1 is 2
The Scale for 1/4"= 1' is 48.
(Refer to page 27-7 for "How to calculate the drawing scale factor")

CALCULATING THE DRAWING SCALE FACTOR

Think about this....
When you adjust the scale within a Viewport all objects within the Viewport appear either larger or smaller. ALL OBJECTS, including Text, Hatch and Dimension features such as arrowheads. It does not affect the dimension "value". Even the spacing for non-continuous linetypes such as "dashed" appear larger or smaller.

If the objects appear smaller, you must increase the size of the Text height, Hatch scale or spacing and Dimension entities.

If the objects appear larger, you must decrease the size of the Text height, Hatch scale or spacing and Dimension entities.

How to determine how much to increase or decrease.
To determine how much to increase or decrease you need to calculate the **drawing scale factor**. The drawing scale factor (DSF) means "How many times smaller or larger did the drawing get when you adjusted the scale".

Here are a few of the most commonly used scales and their drawing scale factors:

Scale	DSF	Scale	DSF
1" = 1'	12	1 : 2	2
1/4" = 1'	48	2 : 1	1/2
1/8" = 1'	96	4 : 1	1/4

HOW TO CALCULATE DRAWING SCALE FACTOR (DSF):

The drawing scale factor is the reciprocal of the adjusted scale.

For example, if the scale has been adjusted to: 1/4" = 1'
calculate the scale factor as follows:
1. Adjusted scale: ¼" = 1'
2. Convert to decimals: .25 = 12
3. Divide 12 by .25 : 12 ÷ .25 (The Drawing scale factor is 48)

This means that text height, hatch scale/spacing and dimension scale must be adjusted by 48. (Refer to more detailed instructions on the following pages)

Another example if the scale has been adjusted to 4 : 1
1. Adjusted scale: 4 : 1
2. Convert to decimals: 4 = 1
(if necessary)
3. Divide 1 by 4 1 ÷ 4 = 1/4 (The Drawing scale factor is 1/4)

This means that text height, hatch scale/spacing and dimension scale must be adjusted by 1/4. (Refer to more detailed instructions on the following pages)

HOW THE DSF EFFECTS TEXTS

When you adjust the scale within a Viewport any text within the Viewport will appear either larger or smaller depending on the scale.

To calculate the correct text height, do the following:
1. Decide what height you would like the text to be when plotted on the paper.
2. Calculate the Drawing Scale Factor.
3. Multiply the (Drawing Scale Factor) X (Plotted Text height)

For example:
1. The text height after plotting should be 1/8".
2. The adjusted scale within the viewport is 4 : 1 (DSF calculated = 1/4)
3. Multiply (1/8) X (1/4) = 1/32"

So the text height you use is 1/32". Now I know that seems like it will be very small, but remember, the viewport scale has been adjusted to 4 : 1 (4X it's original appearance. So if you set a text height of 1/32" it will be enlarged 4X and have an appearance of 1/8". (The height you originally wanted)

The following is an example of how the text will appear when plotted.

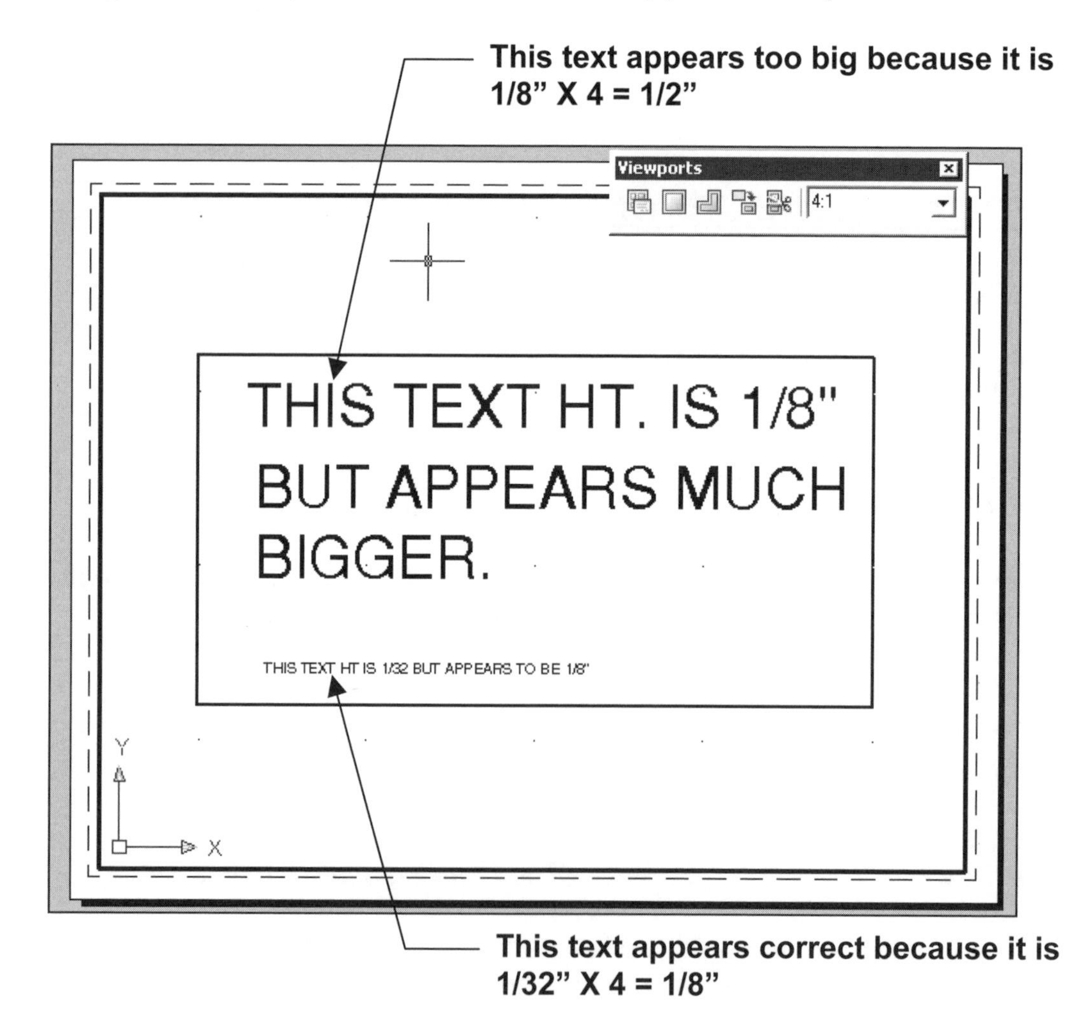

HOW THE DSF EFFECTS HATCH

When you adjust the scale within a Viewport the spacing between the Hatch lines, within the Viewport, will appear either larger or smaller depending on the scale.

PREDEFINED

Calculate the Drawing Scale Factor and enter the DSF in the Scale box.

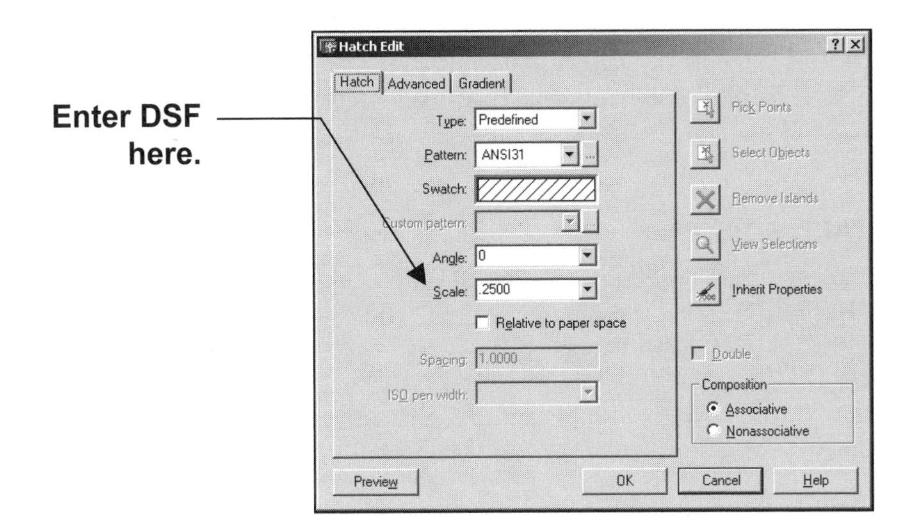

USER DEFINED

Calculate the Drawing Scale Factor and enter the DSF in the SPACING box.

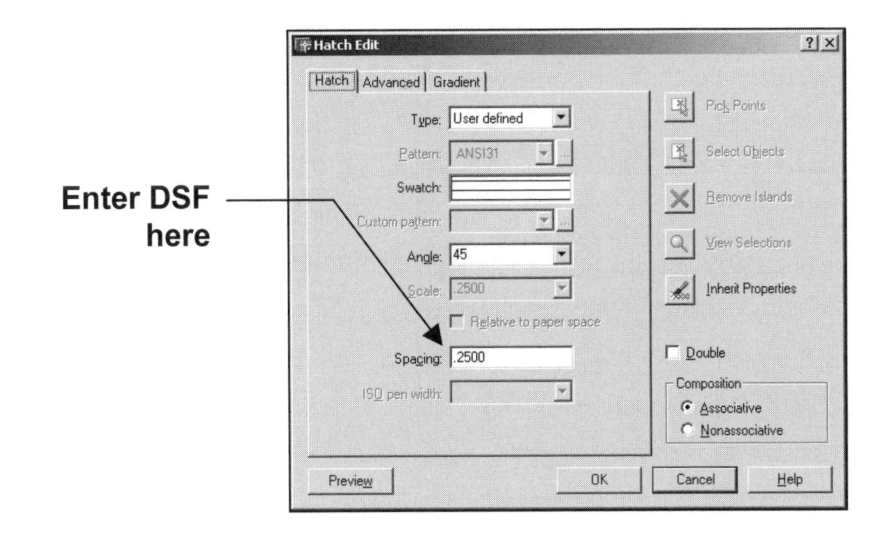

HOW THE DSF EFFECTS DIMENSIONING

Note: the following only affects dimensions in model space.
Refer to 26-11 for dimensions in paper space.

When you adjust the scale within a Viewport, dimension features will increase or decrease in appearance also.

Do not misunderstand, it will not affect the dimension "value", only the size of the dimension features, such as arrowheads.

All you need to do is set the Dimension "overall scale" box to the Drawing Scale Factor. Dimension features sizes should remain as you normally have them set. Then AutoCAD will automatically adjust the size of all the dimension features as you draw the dimensions in the viewport.

HOW TO SET THE "OVERALL SCALE".

1. Select "Format / Dimension Style"
2. Select the dimension style that you want to change.
3. Select the "Modify" button.
4. Select the "Fit" tab.
5. Select "Use overall scale of:"
6. Enter the DSF in the box.
7. Select OK and close the Dimension Style dialog box.

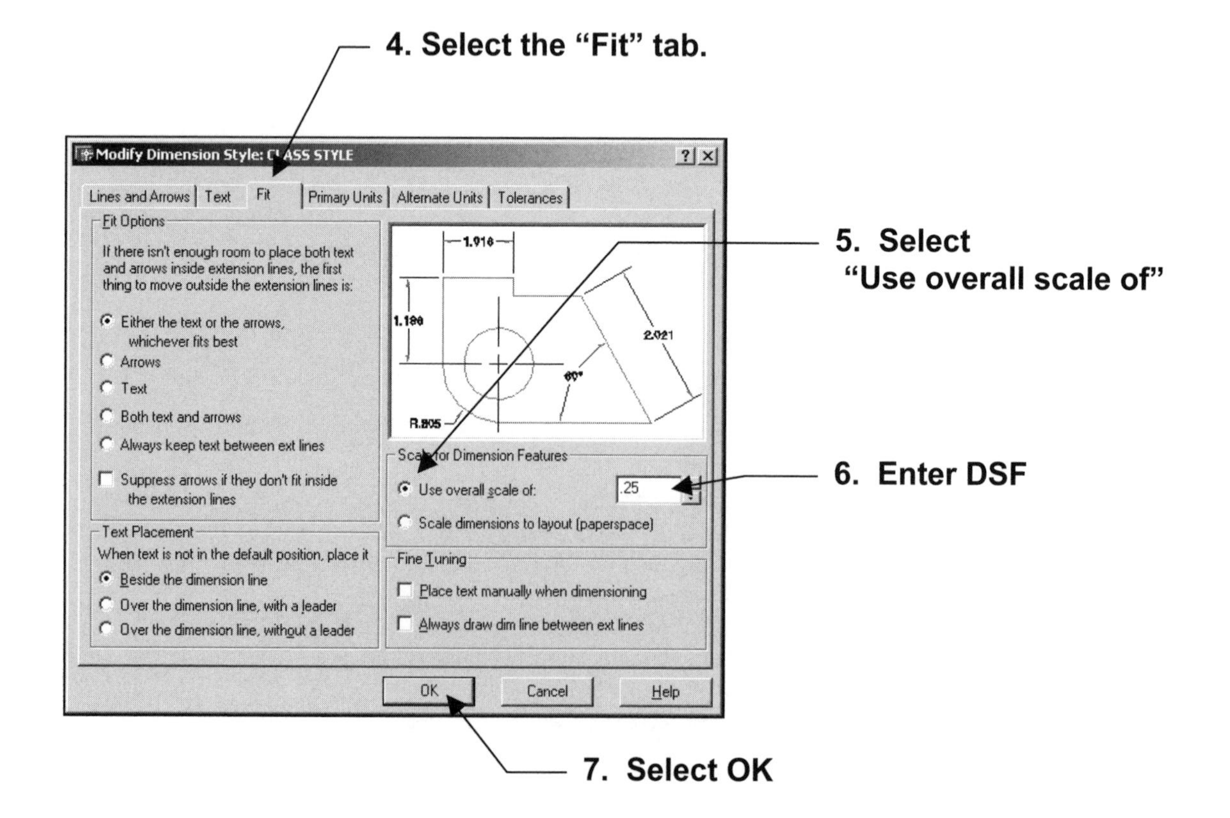

TRANS-SPATIAL DIMENSIONING

In Lesson 16 you learned about True Associative dimensioning. (Refer to 16-2) Remember, when using True Associative dimensioning, the dimension is actually attached to the object. If the object changes, the dimension changes. True Associative dimensioning is a very important and powerful tool within AutoCAD.

True Associative dimensioning gets even better because it can also be trans-spatial. Trans-spatial means that you have the ability to place a dimension in paper space while the object you are dimensioning is in model space. Even though the dimension is in paper space, it is actually attached to the object, in model space.

For example,

1. You draw a house in model space.
2. Now select the layout tab.
3. Cut a viewport so you can see the drawing of the house.
4. Go to model space, adjust the scale of the viewport (model space) and lock it.
5. Now go to paper space and dimension the house.

Why is Trans-spatial dimensioning so great?

If you dimension in Paper Space you do not have to be concerned with the Drawing Scale Factor. The overall scale described on page 27-10 is set to 1.

SHOULD YOU DIMENSION IN PAPERSPACE OR MODELSPACE?

It is possible to dimension in either Paperspace or Modelspace. Personally I dimension in either space depending on the situation. Here are some things to consider.

Paperspace dimensioning

Pro's

1. You never have to worry about the drawing scale factor. [Overall scale (page 27-10) will remain set to 1.]
2. Dimensions do not change appearance if you adjust the scale of the viewport.
3. All dimensions will have the same appearance in all viewports.

Con's

1. Qdim is not trans-spatial. (You can only use it in modelspace.)
2. Dimensions sometimes temporarily float away from objects if you move the objects. (Use "Dimregen" to put them back in place)

Modelspace dimensioning

Pro's

1. Dimensions never float away from objects.
2. Qdim works in modelspace.

Con's

1. You must change the overall scale (page 27-10) to match the drawing scale factor.
2. Dimensions will appear different in each viewport if the viewports have different adjusted scales.

LINETYPE SCALE

AutoCAD has many Linetypes, as you learned in Lesson 26. (26-4) A Linetype is a series of dashes, lines and spaces. Each linetype has specific dimensions for the lines, dashes and spaces. When you are in paper space and you adjust the scale of the model space, AutoCAD automatically adjusts the scale of the linetype dashes, lines and spaces. So you do not have to do any modifications unless you desire.

Changing individual objects
If you want to change the linetype scale for an individual object use the Properties Palette.

Changing the entire drawing
If you want to change the linetype scale for the entire drawing:

1. On the command line type: ***LTS <enter>***
2. LTSCALE Enter new linetype scale factor <1.0000>: ***enter the new value***

NOTES:

If you are in Paperspace you will only need to change the value to 2 or .5 etc. AutoCAD attempts to adjust the linetype scale automatically to the drawing scale factor. But sometimes you want to tweek it a little.

Linetypes may not appear the same if you are in the model tab rather than the Layout tab. Linetype scale is handled differently in Model space.

Linetype scaling is a visual preference. There is not a rule for spacing in cad.

PSLTSCALE
This variable controls paper space linetype scaling. This means that all linetype dash lenghs and spaces will be scaled to the paper space scale.

1 = the viewports can have different adjusted scales and the linetype scale will be the same in all viewports. (this is the default setting)

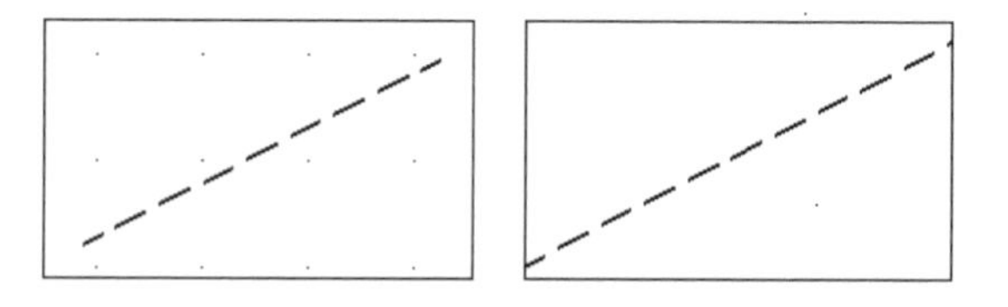

0 = viewports with different adjusted scales will also appear to have differing linetype scales.

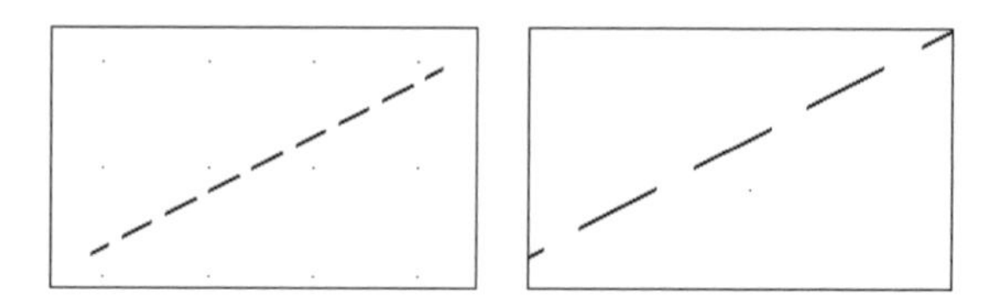

EXERCISE 27A

CREATE AN MASTER FEET-INCHES SETUP DRAWING

The following instructions will guide you through creating a "Master" feet-inches setup drawing. The "1Workbook Helper" is an example of a Master setup drawing. Even though the screen appears blank, the actual file is full of settings, such as; Units, Drawing Limits, Snap and Grid settings, Layers, Text styles and Dimension Styles. Once you have created this "Master" drawing, you just open it and draw. No more repetitive inputting of settings.

NEW SETTINGS

A. Begin your drawing without a template as follows:

1. Select **"FILE / NEW"**
2. Select **"START FROM SCRATCH"** Box.
3. Select "**OK**".
4. Your screen should be blank, no grids and the current layer is 0

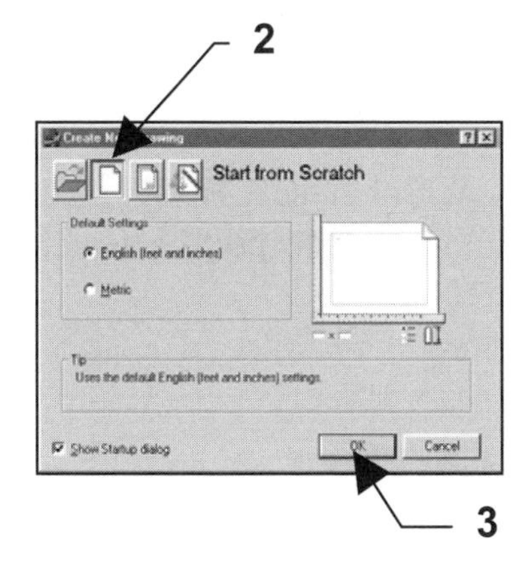

B. Set drawing specifications as follows:

1. Set **"UNITS"** of measurement
   Use "**FORMAT / UNITS** and change the settings as shown below, then select **OK.**

Drawing Units
Length
Type:
Architectural
Precision:
0'-0 1/16"
Angle
Type:
Decimal Degrees
Precision:
0
Clockwise
Drawing units for DesignCenter blocks
When inserting blocks into this drawing, scale them to:
Inches
Sample Output
1 1/2",2",0"
3"<45,0"
OK | Cancel | Direction... | Help

2. Set **"DRAWING LIMITS"** (Size of drawing area)
   USE "**FORMAT / DRAWING LIMITS**
   a. Lower left corner = 0'-0", 0'-0"
   b. Upper right corner = 68', 44' (Notice this is feet not inches)
   c. Use **"VIEW / ZOOM / ALL"** to generate the new limits
   d. Set your **grids** to **ON** to display the paper size.

3. Set **"SNAP AND GRID"**
   Use "**TOOLS / DRAFTING SETTINGS**

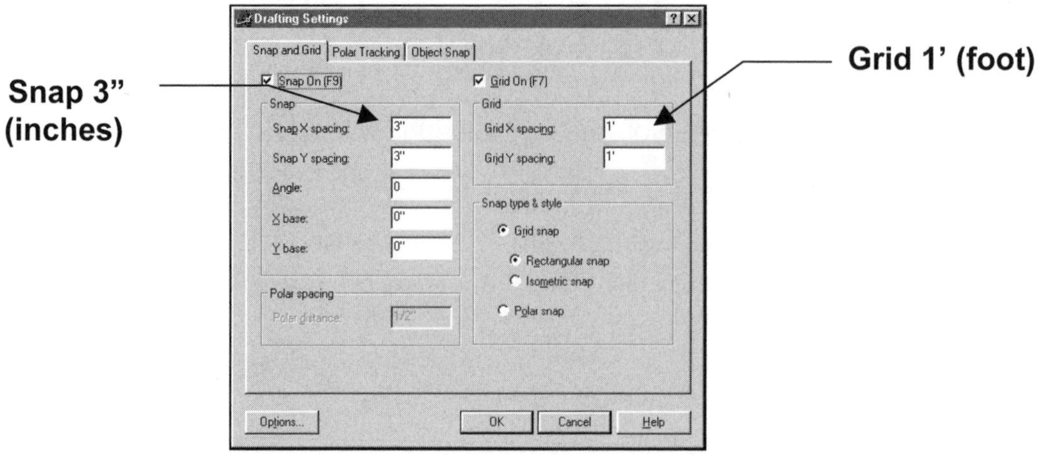

Snap 3" (inches)

Grid 1' (foot)

4. Set **"PICK BOX"** size (your preference)
   See page 26-17 for instructions.

NEW LAYERS

C. Create new layers

1. **Load** linetype (See page 26-4 for instructions)
   DASHED

2. Assign names, colors, linetypes and plotability. (See page 26-3 for instructions)

NAME	COLOR	LINETYPE	LWT	PLOT
BORDER	RED	CONTINUOUS	.039	YES
CABINETS	CYAN	CONTINUOUS	.016	YES
CONSTRUCTION	WHITE	CONTINUOUS	Default	NO
DIMENSION	BLUE	CONTINUOUS	Default	YES
DOORS	GREEN	CONTINUOUS	.016	YES
ELECTRICAL	CYAN	CONTINUOUS	.010	YES
FURNITURE	MAGENTA	CONTINUOUS	.016	YES
PLUMBING	9	CONTINUOUS	.010	YES
TEXT HEAVY	WHITE	CONTINUOUS	Default	YES
TEXT LIGHT	BLUE	CONTINUOUS	Default	YES
VIEWPORT	GREEN	CONTINUOUS	Default	NO
WALLS	RED	CONTINUOUS	.024	YES
WINDOWS	GREEN	CONTINUOUS	.016	YES
WIRING	CYAN	DASHED	.010	YES

NEW TEXT STYLE

D. Create a 2 text styles named ***ARCH TEXT*** and ***CLASS TEXT***

1. Select **"FORMAT / TEXT STYLE"**
2. Make the changes shown in the dialog boxes below.

ARCH TEXT **CLASS TEXT**

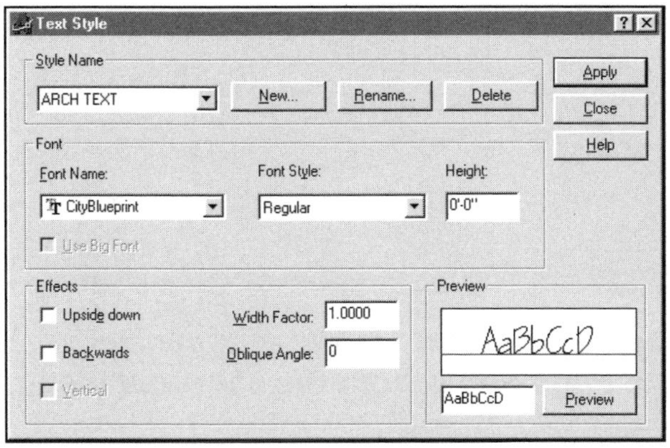

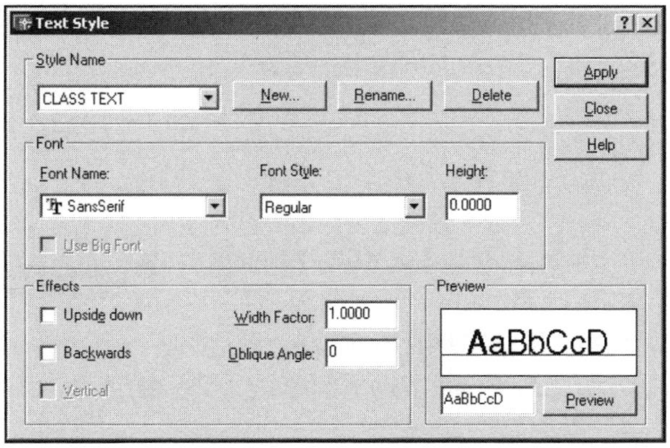

3. When complete, select **APPLY** then **CLOSE**.

NEW DIMENSION STYLE

E. A new dimension style should also be created but that will be discussed separately in Exercise 27C, page 27-19.

THIS NEXT STEP IS VERY IMPORTANT.

F. SAVE ALL THE SETTINGS YOU JUST CREATED

1. Select File / Save as
2. Save as: **My Feet-Inches Setup**

EXERCISE 27B

CREATE AN ARCHITECTURAL BORDER FOR PLOTTING

The following instructions will guide you through creating a Border drawing that will be used in combination with "My Feet-Inches Setup" when plotting. You will create a Layout and draw a border with a title block. All of this information will be saved and you will not have to do this again.

A. Open **My Feet-Inches Setup**

B. Select the **LAYOUT1** tab.

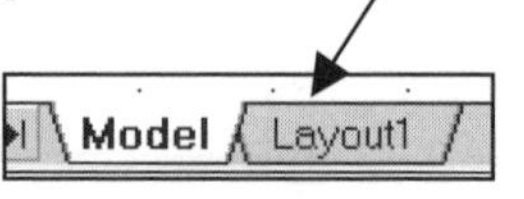

Note: If the Page Setup dialog box shown below does not appear automatically, right click on the Layout tab, and then select Page Setup from the short cut menu.

1. Type the new name:
Qtr equals foot

2. Select the "*Plot Device*" tab.

3. Select the Plotter.

Select "Plot style table"

5. Select the "*Layout Settings*" tab.

6. Select the "*Paper Size*".

7. Select scale "1:1"

Page Setup - Qtr equals foot

Layout name: Qtr equals foot

Page setup name: <Select page setup to apply> Add...

Plot Device | Layout Settings

Paper size and paper units
Plot device: LaserJet .pc3
Paper size: ANSI B (11.00 x 17.00 Inches)
Printable area: 16.60 x 10.44 inches | inches | mm

Drawing orientation: Portrait | Landscape | Plot upside-down

Plot area: Layout | Extents | Display | View | Window <

Plot scale: 1:1 | 1 inches = 1 units | Scale lineweights

Shaded viewport options: Shade plot: As Displayed | Quality: Normal | DPI: 300

Plot offset: Center the plot | X: 0.00 inches | Y: 0.00 inches

Plot options: Plot object lineweights | Plot with plot styles | Plot paperspace last | Hide paperspace objects

Display when creating a new layout | OK | Plot | Cancel | Help

8. Select the *OK* button

You should now have a sheet of paper displayed on the screen and the Layout tab should now be displayed as **"Qtr equals foot".**

This sheet is in front of "Model". In Exercise 27D, you will cut a hole in this sheet so you can see through to Model.

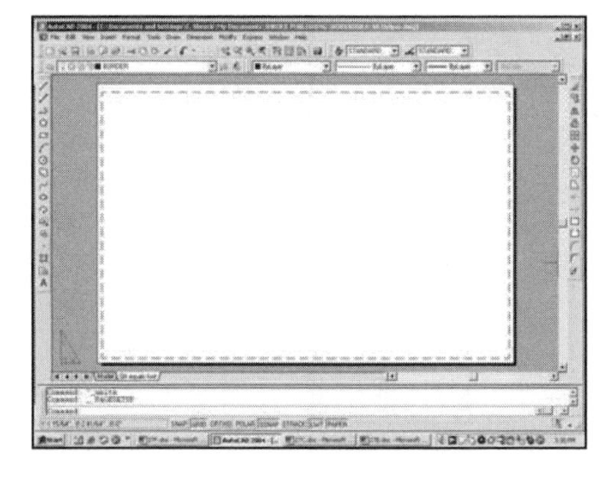

C. Draw the Border with title block, shown below, on the sheet of paper shown on the screen.

D. When you have completed the Border, shown below:

1. Select File / Save as

2. Save as: **My Feet-Inches Setup** (Again)

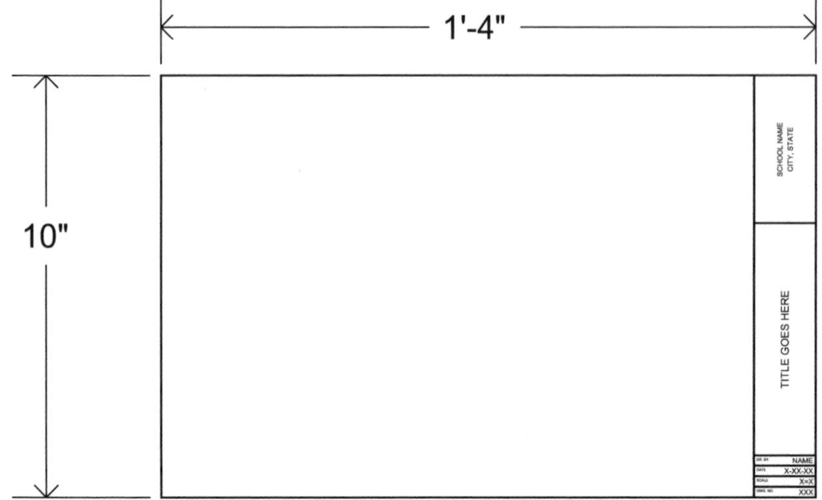

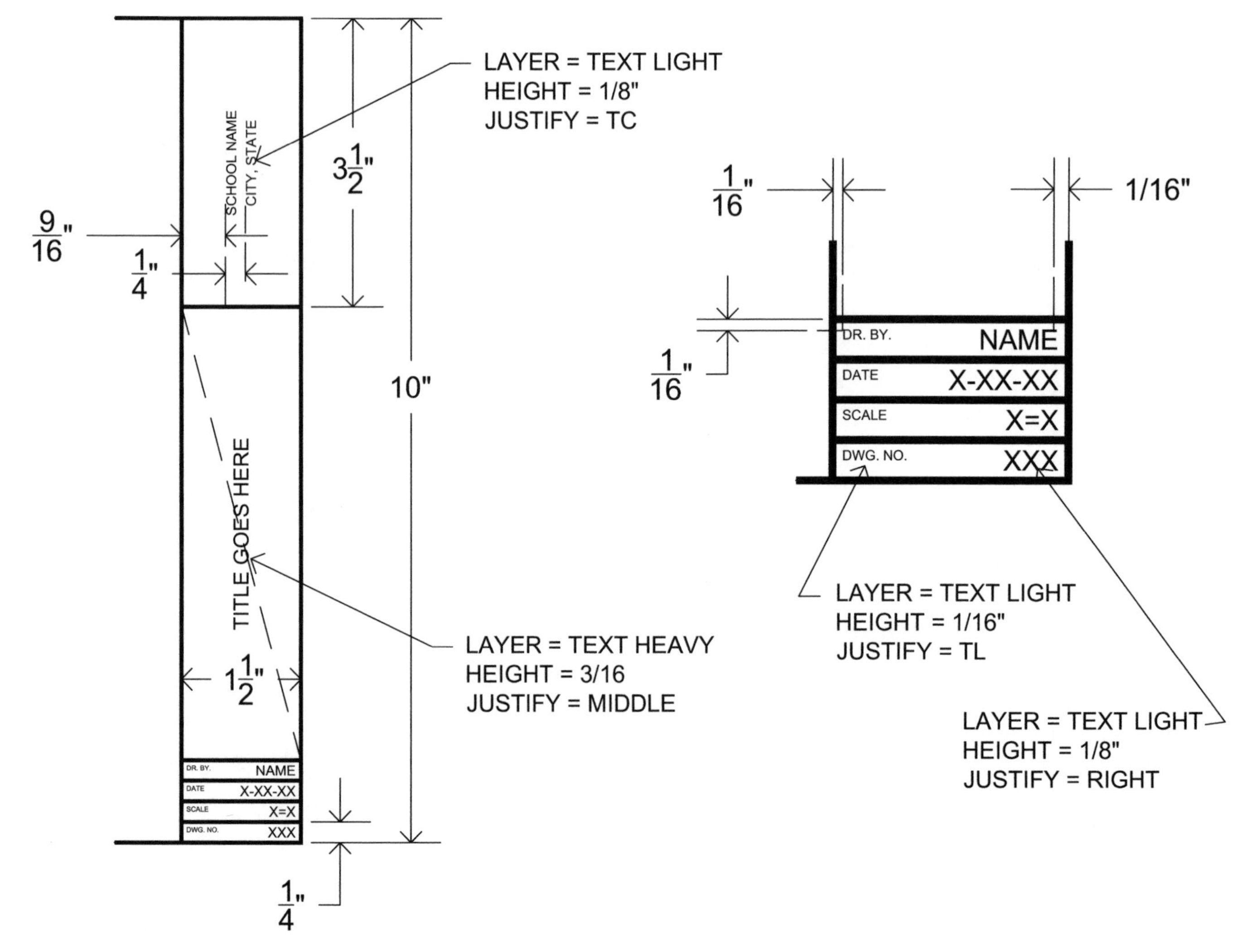

EXERCISE 27C

CREATE A NEW ARCHITECTURAL DIMENSION STYLE

1. Open **My Feet-Inches Setup.**
2. Set **DIMASSOC** to **2**
3. Select **FORMAT / DIMENSION STYLE**
4. Select the **NEW** button.
5. New Style Name: **ARCH DIM**
6. Start With: **Standard**
7. Use For: **All dimensions**
8. Select the **CONTINUE** box.

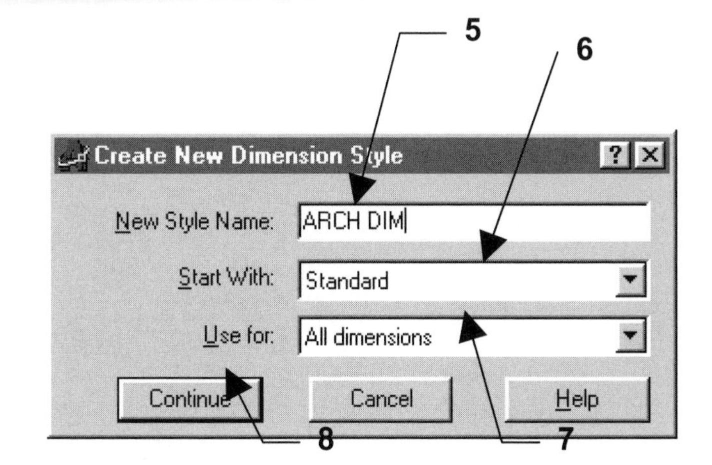

9. Select the **"Primary Units"** tab and make the following changes.

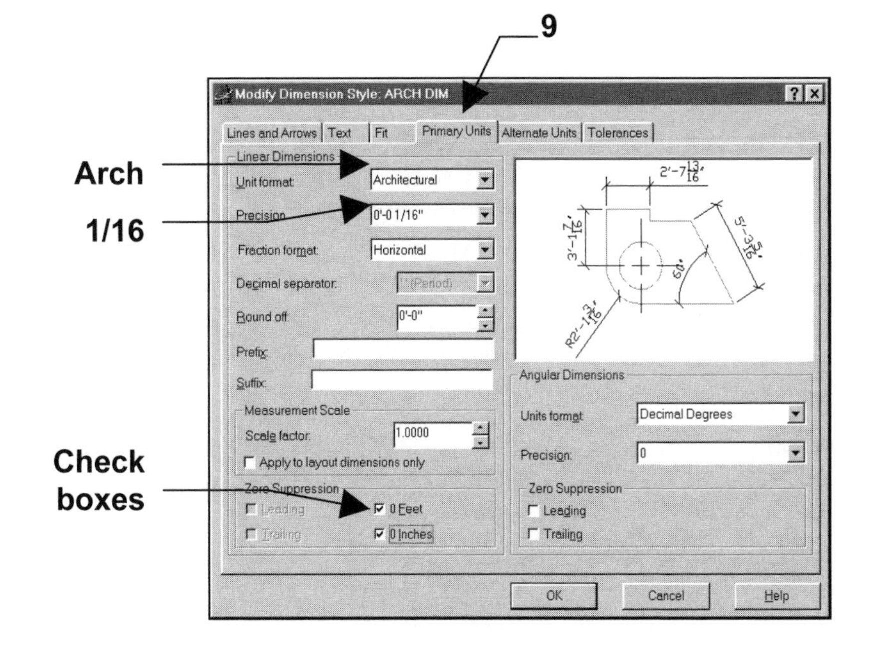

DO NOT SELECT THE OK BUTTON YET

10. Select the ***"Lines and Arrows"*** tab and make the following changes.

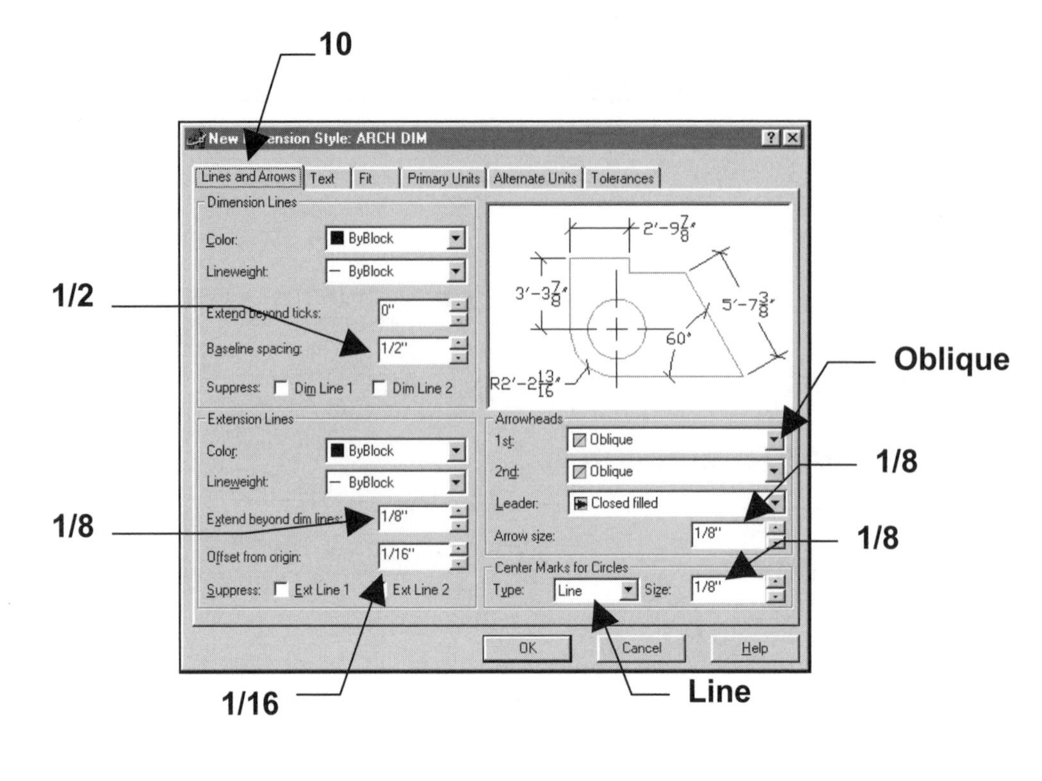

DO NOT SELECT THE OK BUTTON.

11. Select the **"Text"** tab and make the following changes .

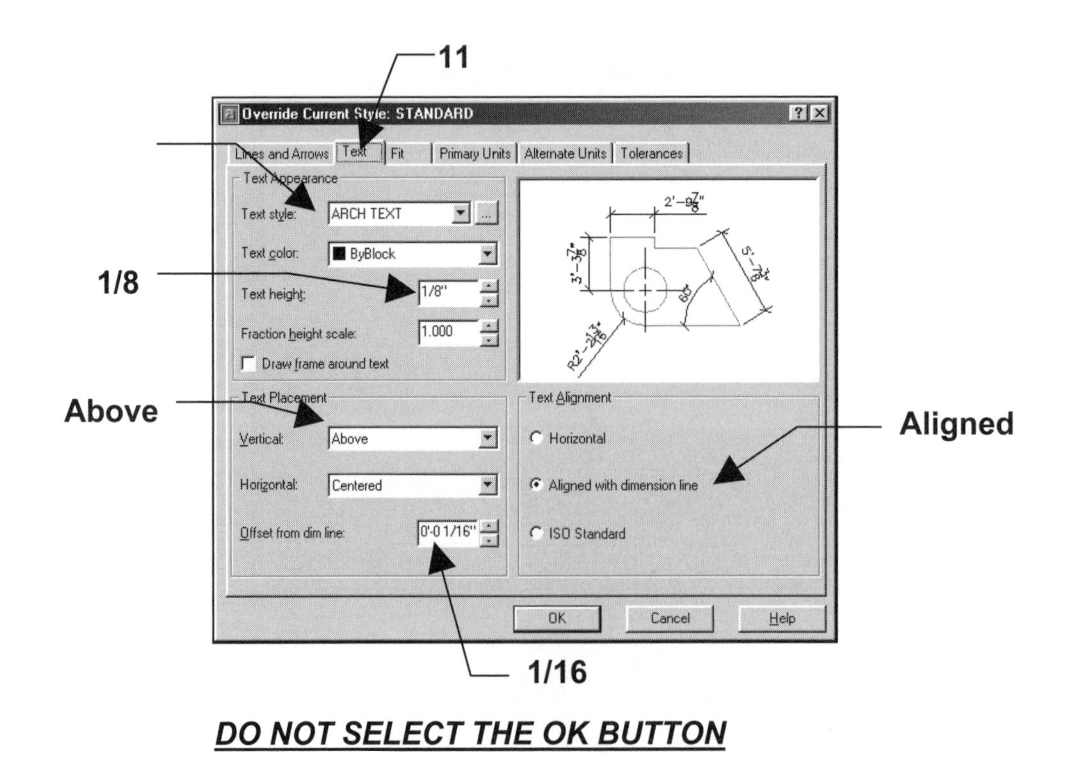

DO NOT SELECT THE OK BUTTON

12. Select the **"Fit"** tab and make the following change.

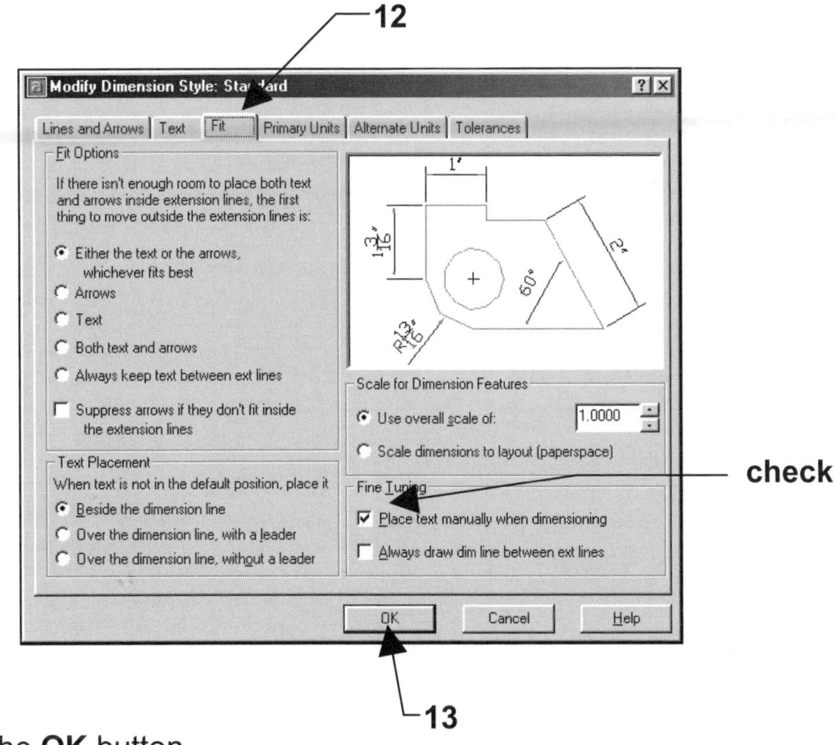

13. **NOW** select the **OK** button.

14. *Your new style "**ARCH DIM**" should be listed.*

15. Select the **"Set Current"** button to make your new style "**ARCH DIM**" the style that you will use.

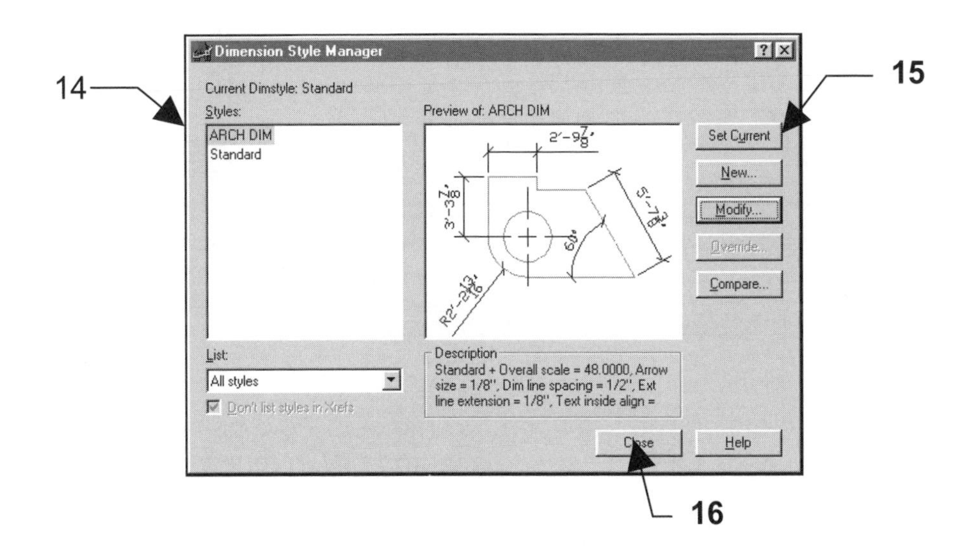

16. Select the **Close** button to **exit**.
17. Continue to save your drawing as **My Feet-Inches Setup**

EXERCISE 27D

CREATE A VIEWPORT

The following instructions will guide you through creating a VIEWPORT in the Border Layout sheet. Creating a viewport has the same effect as cutting a hole in the sheet of paper. You will be able to see through the viewport frame (hole) to Model.

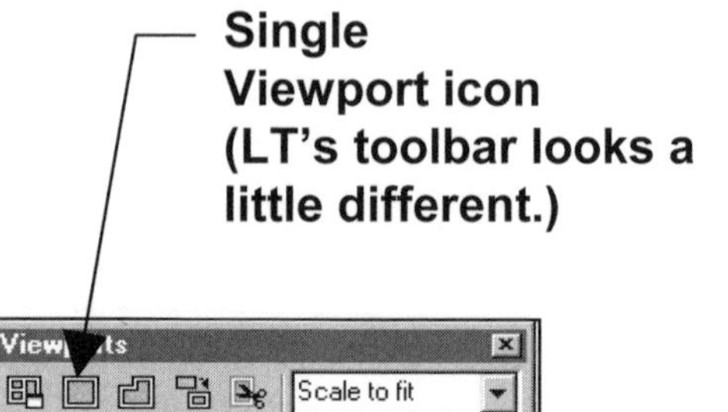

A. Open **My Feet-Inches Setup**
B. Select the **Qtr equals foot** tab.
C. Select layer "Viewport"

D. Select the Single Viewport icon from the Viewports Toolbar or Type MV <enter>

E. Draw a Single viewport approximately as shown.

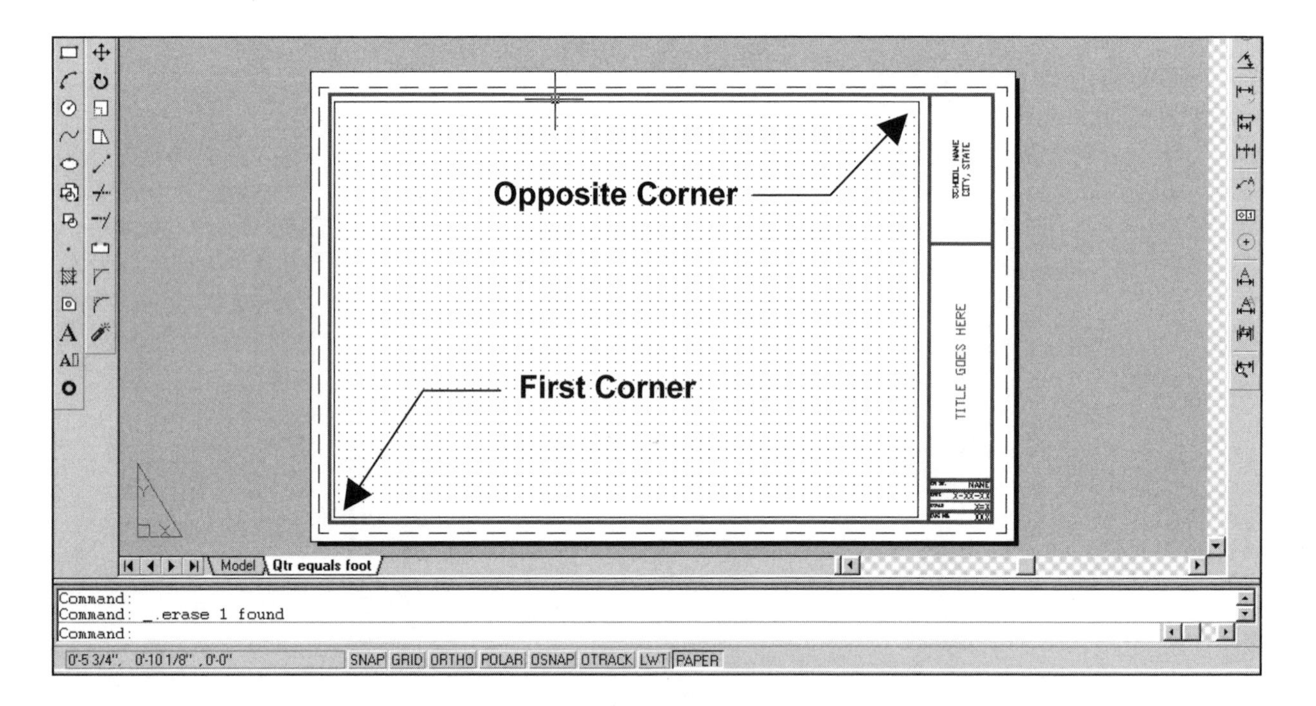

F. After successfully creating the Viewport, you should now be able to see through to Model. (Note: Turn grids OFF in paperspace and ON in modelspace.)

G. File / Save as: **My Feet-Inches Setup**

EXERCISE 27E

ADJUSTING THE SCALE INSIDE THE VIEWPORT

1. Open **My Feet-Inches Setup**.

2. Move to the **Qtr equals foot** tab.

3. Double click inside of the Viewport ***(It is very important that you be inside the Viewport. You want to scale the Model geometry not the Layout)***

4. Select View / Zoom / All (This puts your drawing limits in the correct location)

5. ***Adjust the scale*** of the Model geometry as follows:

   a. Select 1/4" = 1' from the **"Viewports"** toolbar.

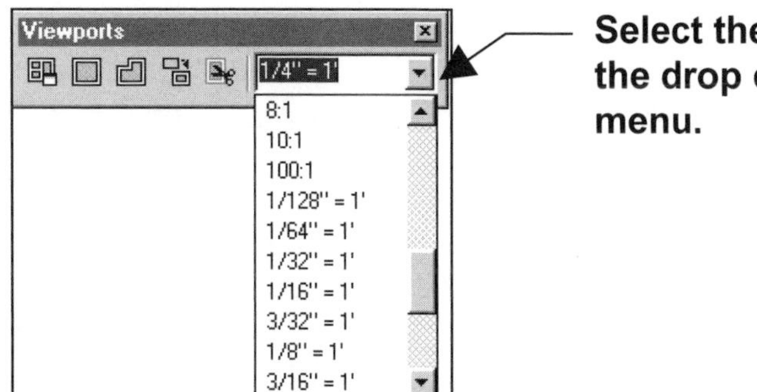

Select the scale from the drop down menu.

The Model geometry should have changed. You will not notice too much of a change other than your grid spacing should appear smaller. Now your "Qtr equals foot" layout is ready to accept a very large drawing. You will understand better after you complete Exercise 27G.

6. **Lock the Viewport scale**
   Refer to page 26-10

EXERCISE 27F

COMPLETING THE PAGE SETUP

The following instructions will guide you through the final steps to completing the **Page Setup**. This Page Setup will stay with **My Feet-Inches Setup** and you will be able to use it over and over again. (Refer to Page 26-13 Step 4 for more detailed explanations of each area.)

A. Open **My Feet-Inches Setup** if it isn't already open.

B. Select the **Qtr equals Foot** layout tab.

You should be looking at your Border and Title Block now.

C. Select **File / Plot** or place the cursor on the **Qtr equals Foot** tab and press the right mouse button and then select **Plot** from the short cut menu.

The Plot dialog box should appear

D. Select the **Plot Device** tab and select the options 1 thru 3 below:

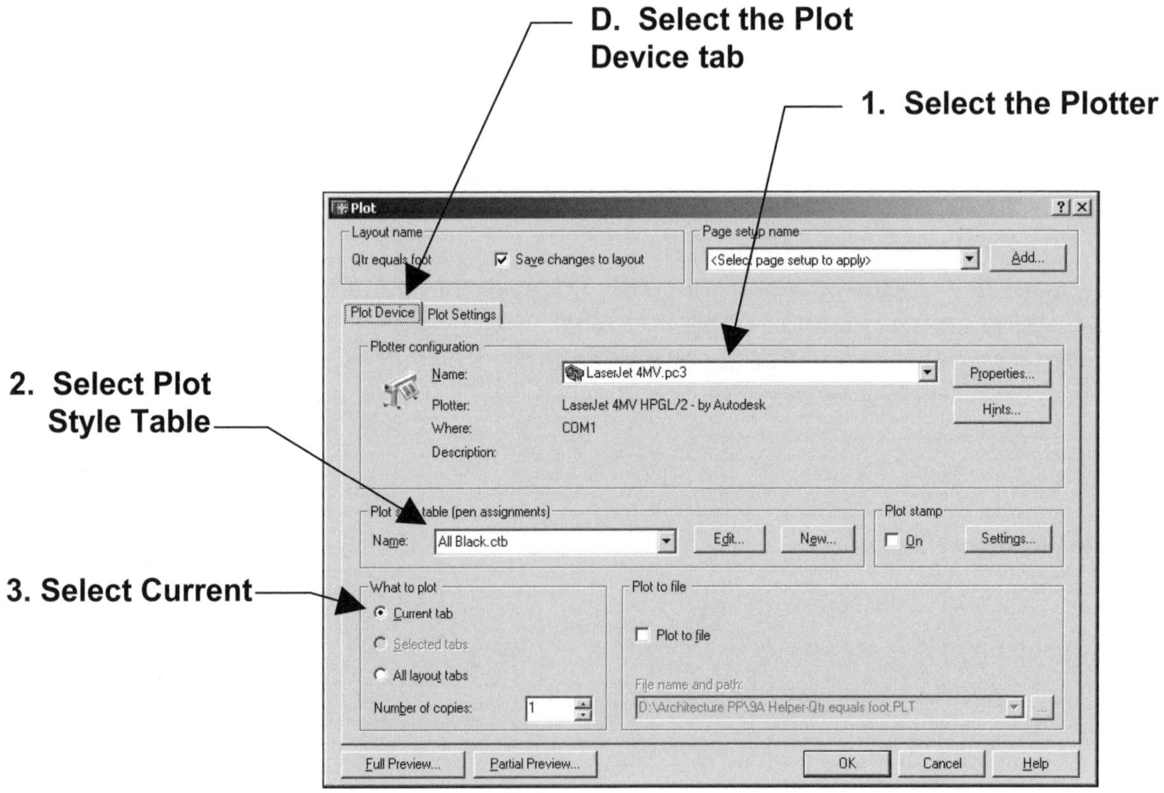

D. Select the Plot Device tab

1. Select the Plotter

2. Select Plot Style Table

3. Select Current

E. Select the **Plot Settings** tab then select options 1 thru 7 below:

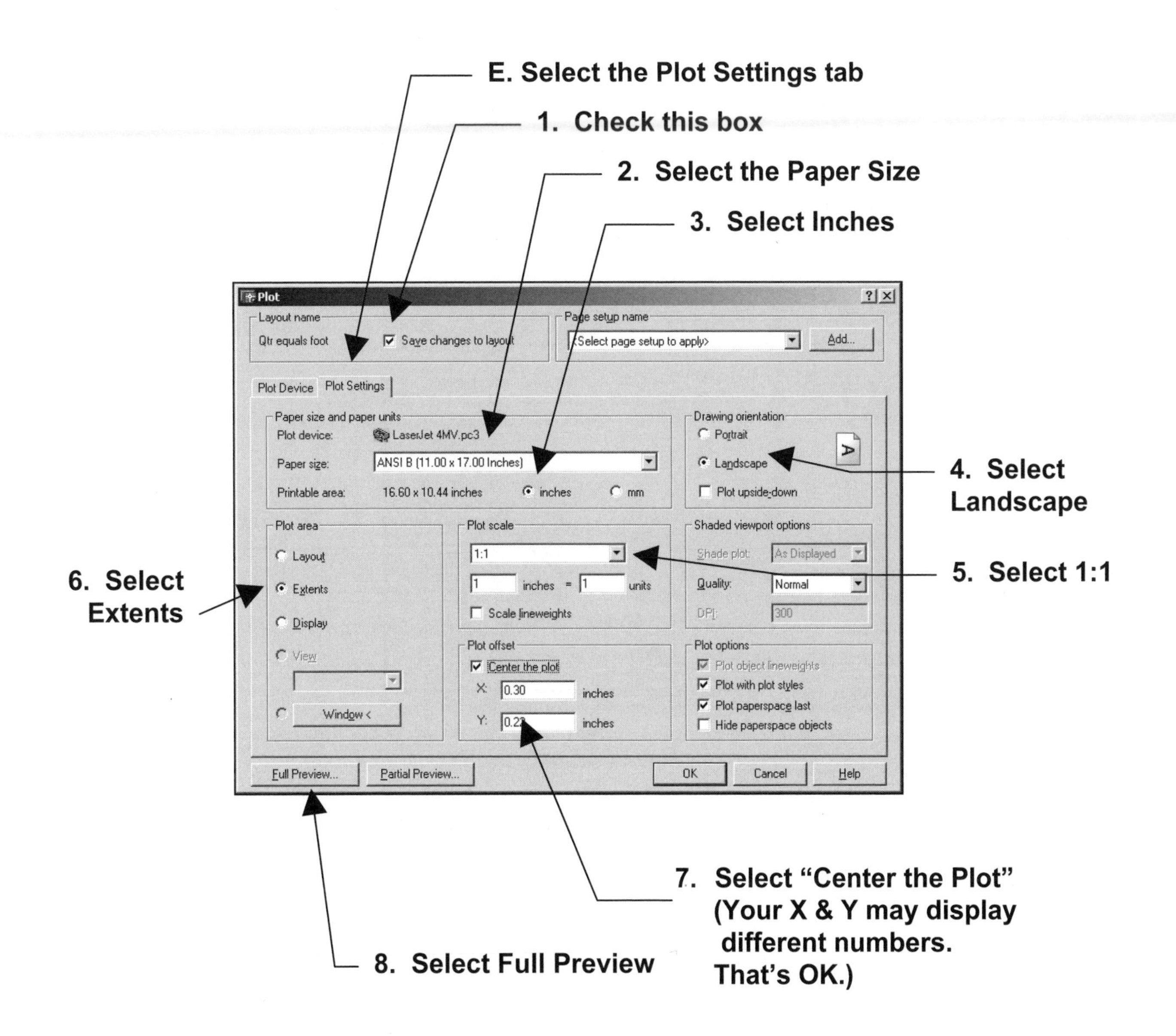

F. Preview the Layout

1. If the drawing is centered on the sheet, press the Esc key and continue on to **G**.

2. If the drawing does not look correct, press the Esc key and check all your settings, then preview again.

G. Select the **ADD** button.

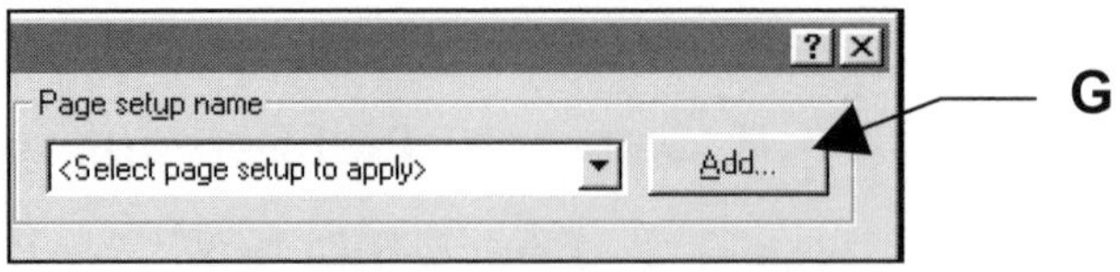

Page setup name
<Select page setup to apply>
Add...

G

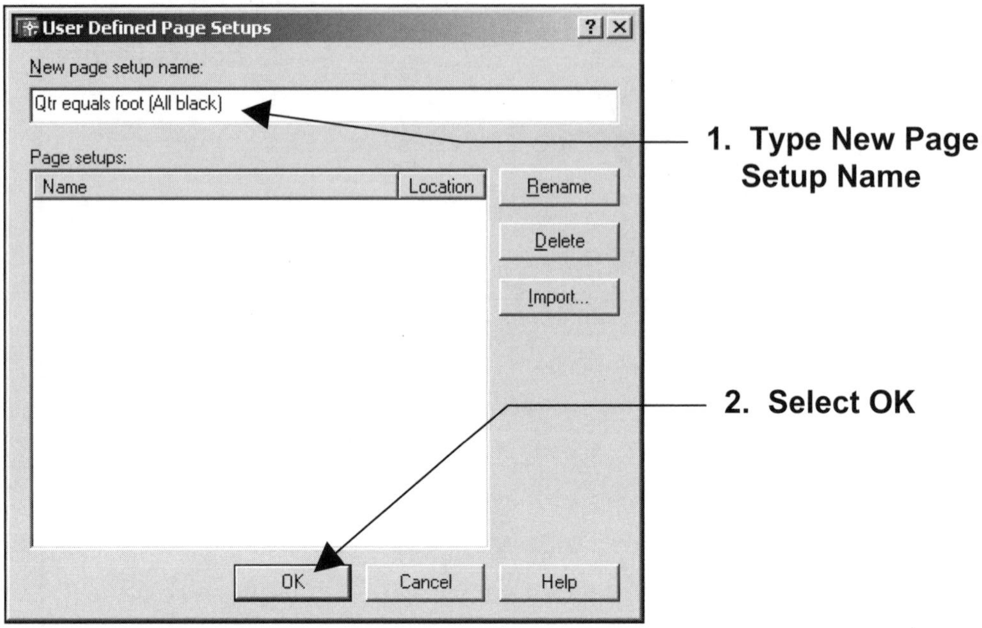

User Defined Page Setups

New page setup name:

Qtr equals foot (All black)

Page setups:

Name | Location

Rename

Delete

Import...

OK Cancel Help

1. Type New Page Setup Name

2. Select OK

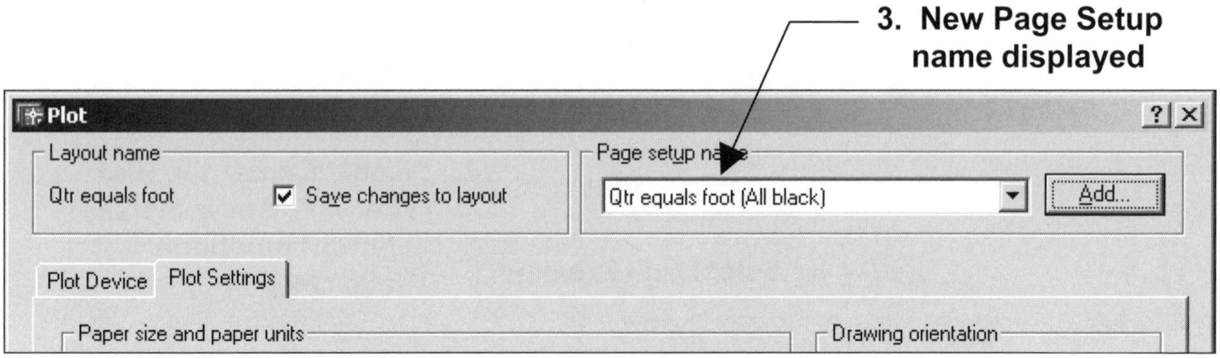

3. New Page Setup name displayed

Plot

Layout name

Qtr equals foot ☑ Save changes to layout

Page setup name

Qtr equals foot (All black)

Add...

Plot Device | Plot Settings

Paper size and paper units

Drawing orientation

H. If your computer **is** connected to the Plotter / Printer, select the **OK** button to plot.

I. If your computer **is not** connected to the Plotter / Printer, select the **CANCEL** button. (Your settings **<u>will not</u>** be lost)

J. **Save** this file <u>one more time</u>:
   1. Select **File / Save** as
   2. Save as: **My Feet-Inches Setup**

*You have now completed the **Page Setup** for the **My Feet-Inches Setup**. Now you are ready to use this master file to create many drawings in the future.*
<u>In fact, you have one on the very next page.</u>

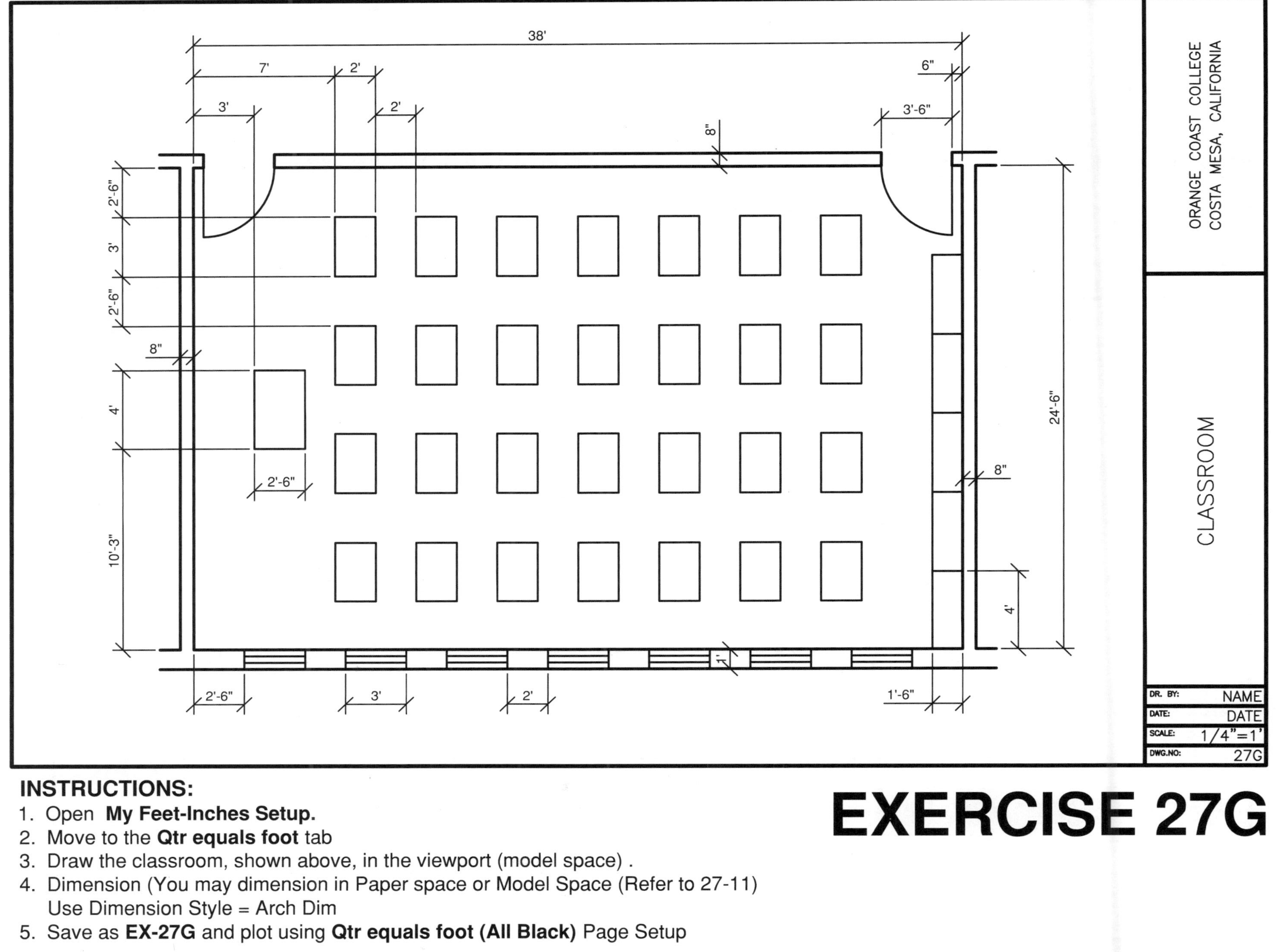

EXERCISE 27G

INSTRUCTIONS:

1. Open **My Feet-Inches Setup.**
2. Move to the **Qtr equals foot** tab
3. Draw the classroom, shown above, in the viewport (model space) .
4. Dimension (You may dimension in Paper space or Model Space (Refer to 27-11)
   Use Dimension Style = Arch Dim
5. Save as **EX-27G** and plot using **Qtr equals foot (All Black)** Page Setup

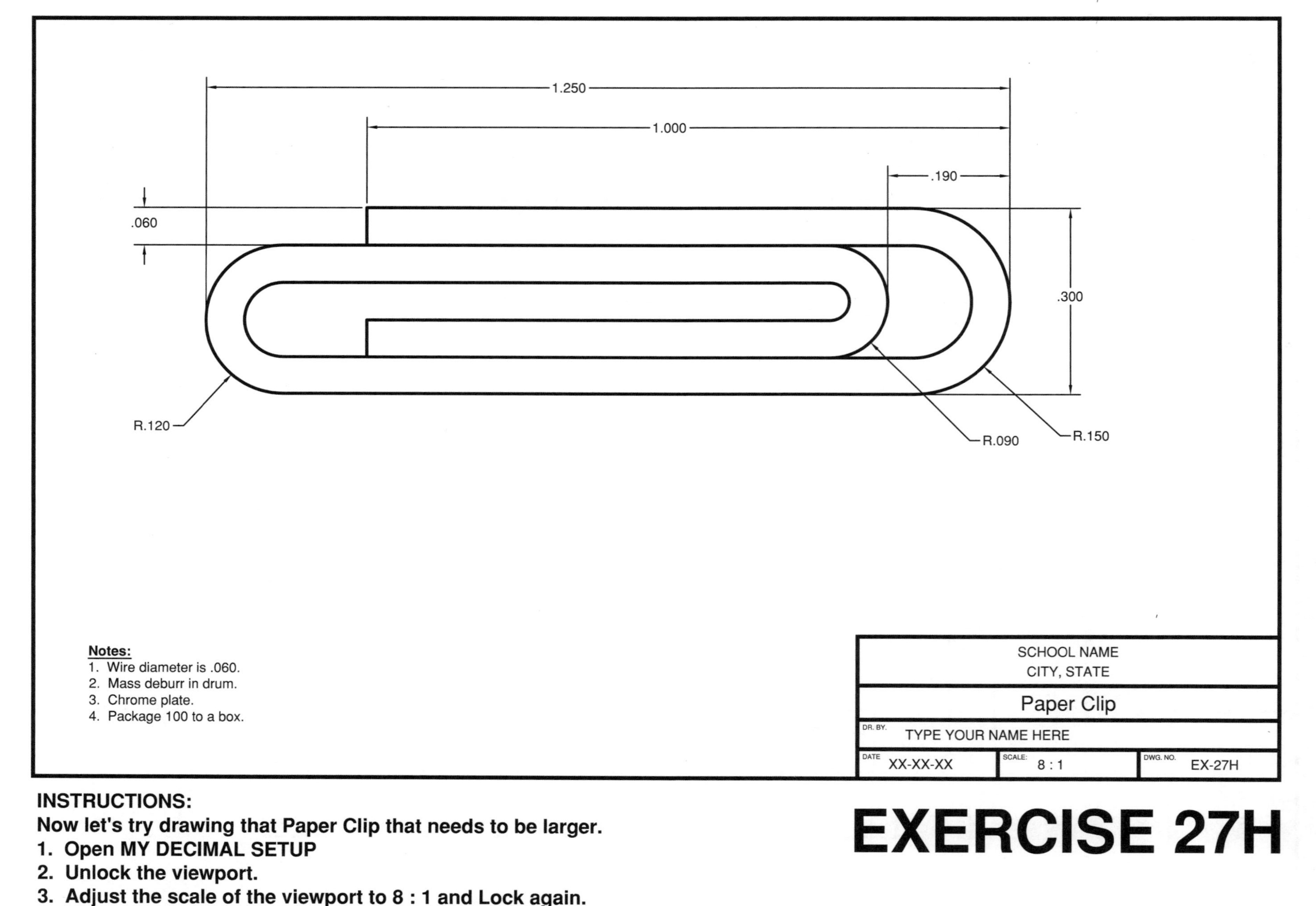

EXERCISE 27H

INSTRUCTIONS:

Now let's try drawing that Paper Clip that needs to be larger.

1. **Open MY DECIMAL SETUP**
2. **Unlock the viewport.**
3. **Adjust the scale of the viewport to 8 : 1 and Lock again.**
4. **Draw the drawing above in the viewport (Model Space).**
5. **Dimension and Notes (1/8 ht.) in Paper space Use Dim. style: Class style)**
6. **Save as: EX-27H**
7. **Plot using "11x17 (1 to 1) All Black" Page Setup.**

LEARNING OBJECTIVES

After completing this lesson, you will be able to:

1. Understand what are Blocks.
2. Create a Block.
3. Insert a Block into your drawing.
4. Understand the rules governing color and linetype.

LESSON 28

BLOCKS

A BLOCK is a group of objects that have been converted into ONE object. A Symbol is a typical application for the block command. First a BLOCK must be created. Then it can be INSERTED into the drawing. An inserted Block uses less file space than a set of objects copied.

CREATING A BLOCK

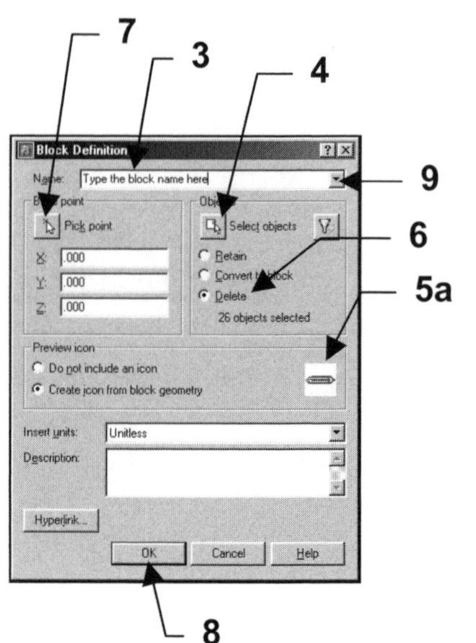

1. First draw the objects that will be converted into a Block. (Refer to page 28-3 **Color and Linetype.**)

2. Select the **BMAKE** command using one of the following:

   TYPE = B
   PULLDOWN = DRAW / BLOCK / MAKE
   TOOLBAR =DRAW

   (the dialog box, on the right, will appear)

3. Enter the New Block name in the **Name** box.

4. Select the **SELECT OBJECTS** button.

   The Block Definition box will disappear and you will return temporarily to the drawing.

5. Select the objects you want in the block, then press <enter>
   a. The Block Definition box will reappear and the objects you selected should be illustrated in the Preview Icon area.

6. Select **Delete** (Refer to 28-3 for definitions)

7. Select the **Pick Point** button. (Or you may type the X and Y coordinates)
   The Block Definition box will disappear again and you will return temporarily to the drawing.

   Select the location where you would like the insertion point for the Block. Later when you insert this block, it will appear on the screen attached to the cursor at this insertion point. Usually this point is the CENTER, MIDPOINT or ENDPOINT of an object.

8. Select the **OK** button.
   The objects will disappear but the new block is now stored in the drawing's block definition table.

9. To verify the creation of this Block, select the Block command, select the Name (▾).
   A list of all the blocks, in this drawing, will appear.

ADDITIONAL DEFINITIONS OF OPTIONS.

Retain
If this option is selected, the original objects will stay visible on the screen after the block has been created.

Convert to block
If this option is selected, the original objects will disappear after the block has been created, but will immediately reappear as a block. It happens so fast, you won't even notice the original objects disappeared.

Delete
If this option is selected, the original objects will disappear from the screen, after the block has been created.

Do not include an icon or Create icon from block geometry
These options determine whether a " preview thumbnail" sketch is created. This option is used with the "Design Center" to drag and drop the blocks into a drawing. The Design Center is an advanced option and is not discussed in this book.

Insert Units
You may define the units of measurement for the block. This option is used with the "Design Center" to drag and drop with Autoscaling. The Design Center is an advanced option and is not discussed in this book.

COLOR and LINETYPE

If a block is created on Layer 0:

When the block is inserted, it will assume the Properties, color, linetype etc. of the layer that is current at the time of insertion.
The block will also reside on the layer that was current at the time of insertion.
If the Block is then **Exploded**, the objects included in the block, will go back to their original color, linetype and layer.

If a block is created on Specific layers:

When the block is inserted, it will retain its own Properties, color, linetype etc. It **will not assume** the color and linetype of the layer that is current.
But the block **will reside** on the current layer at the time of insertion.
If the Block is then **Exploded**, the objects included in the block, will go back to their original layer. The color and linetype remain the same.

INSERTING BLOCKS

A **BLOCK** can be inserted at any location on the drawing. When inserting a Block you can **SCALE** or **ROTATE** it.

1. Select the INSERT command using one of the following:
   TYPE = DDINSERT
   PULLDOWN = INSERT / BLOCK
   TOOLBAR =DRAW

The INSERT dialog box will appears.

2. Select the BLOCK name then select the OK button.
   a. If the block is in the drawing that is open on the screen:
      select the block from the drop down list shown above
   b. The Browse button can be used to find an entire drawing or a WBlock.
      Wblocks will be discussed in the Advanced workbook.

   This returns you to the drawing and the selected block should be attached to the cursor.

3. Select the location for the block by pressing the left mouse button or typing coordinates.

NOTE: If you want to scale or rotate the block before you actually place the block, press the right hand mouse button, and you may select an option from the short cut menu or select an option from the command line menu shown below.

Command: _insert
Specify insertion point or **[Scale/X/Y/Z/Rotate/PScale/PX/PY/PZ/PRotate]:**

You may also "preset" the insertion point, scale or rotation. This is discussed on Page 28-5.

PRESETTING THE "INSERTION POINT", "SCALE" OR "ROTATION"

You may want to specify the **Insertion point, Scale or Rotation** in the **"INSERT"** box instead of at the command line.

1. Remove the check mark from the **"SPECIFY ON-SCREEN"** box.
2. Fill in the appropriate information describe below:

Insertion point
Type the X and Y coordinates from the Origin. The Z is for 3D only.
The example below indicates the blocks insertion point will be 5 inches in the X direction and 3 inches in the Y direction, from the Origin.

Scale
You may scale the block proportionately by typing the scale factor in the X box and then check the Uniform Scale box.
If the block will be scaled non-proportionately, type the different scale factors in both X and Y boxes.
The example below indicates that the block will be scale proportionate at a factor of 2.

Rotation
Type the desired rotation angle relative to its current rotation angle.
The example below indicates the block will be rotated 45 degrees from it's originally created angle orientation.

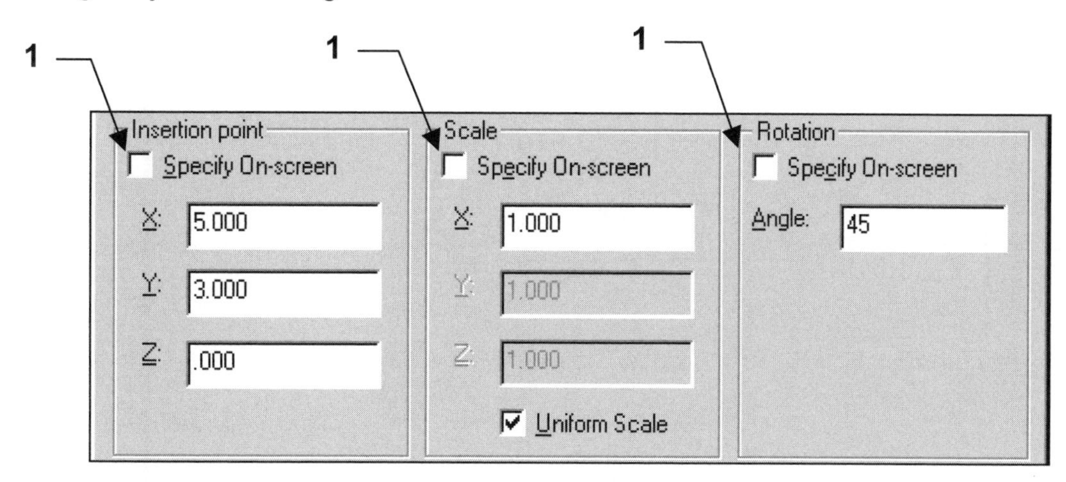

Blocks have many uses and will be discussed further in the "Exercise Workbook for Advanced AutoCAD 2004" along with "DesignCenter" and "Tool Palettes".

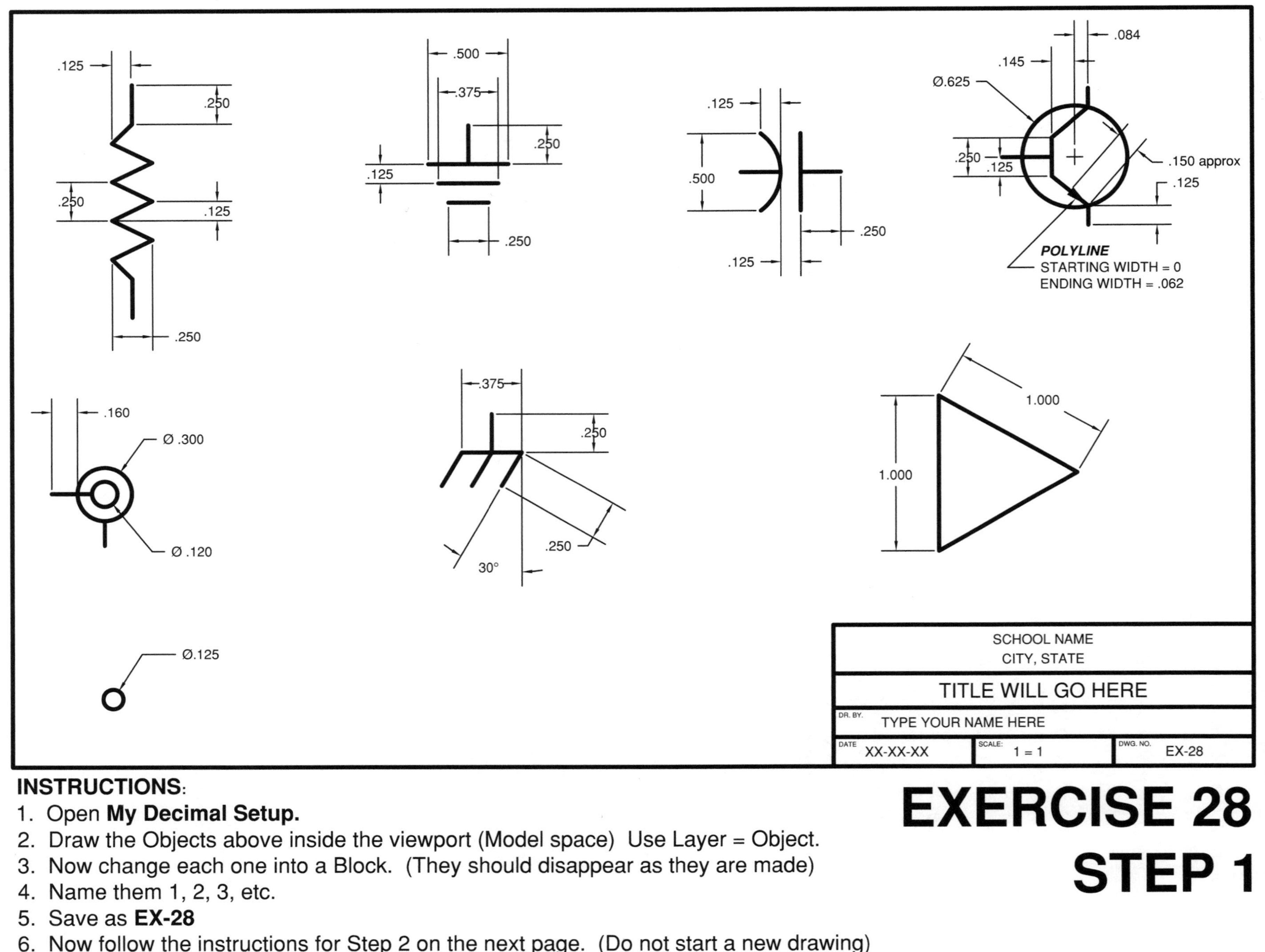

INSTRUCTIONS:

1. Open **My Decimal Setup.**
2. Draw the Objects above inside the viewport (Model space) Use Layer = Object.
3. Now change each one into a Block. (They should disappear as they are made)
4. Name them 1, 2, 3, etc.
5. Save as **EX-28**
6. Now follow the instructions for Step 2 on the next page. (Do not start a new drawing)

EXERCISE 28
STEP 1

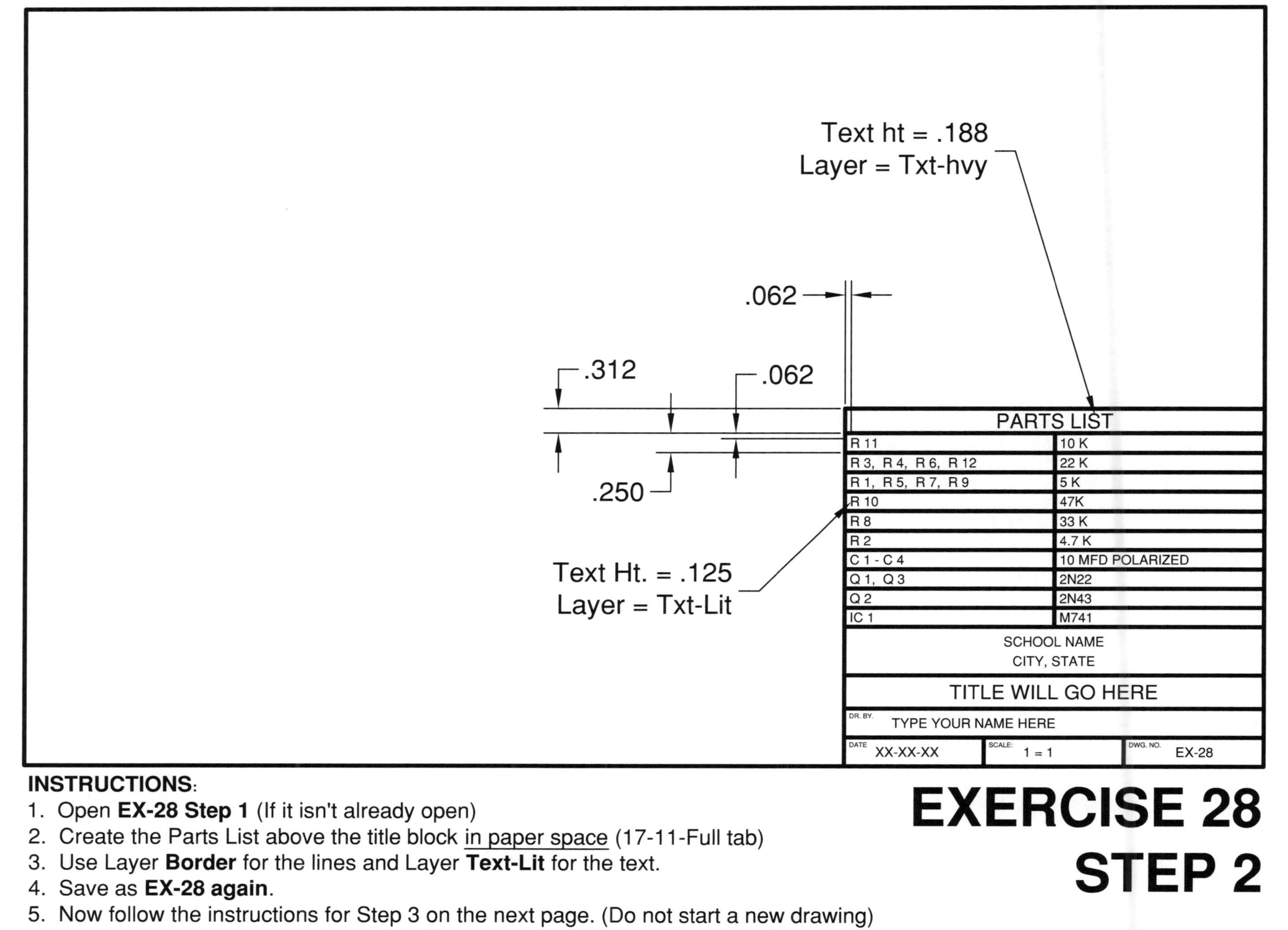

EXERCISE 28
STEP 2

INSTRUCTIONS:
1. Open **EX-28 Step 1** (If it isn't already open)
2. Create the Parts List above the title block in paper space (17-11-Full tab)
3. Use Layer **Border** for the lines and Layer **Text-Lit** for the text.
4. Save as **EX-28 again**.
5. Now follow the instructions for Step 3 on the next page. (Do not start a new drawing)

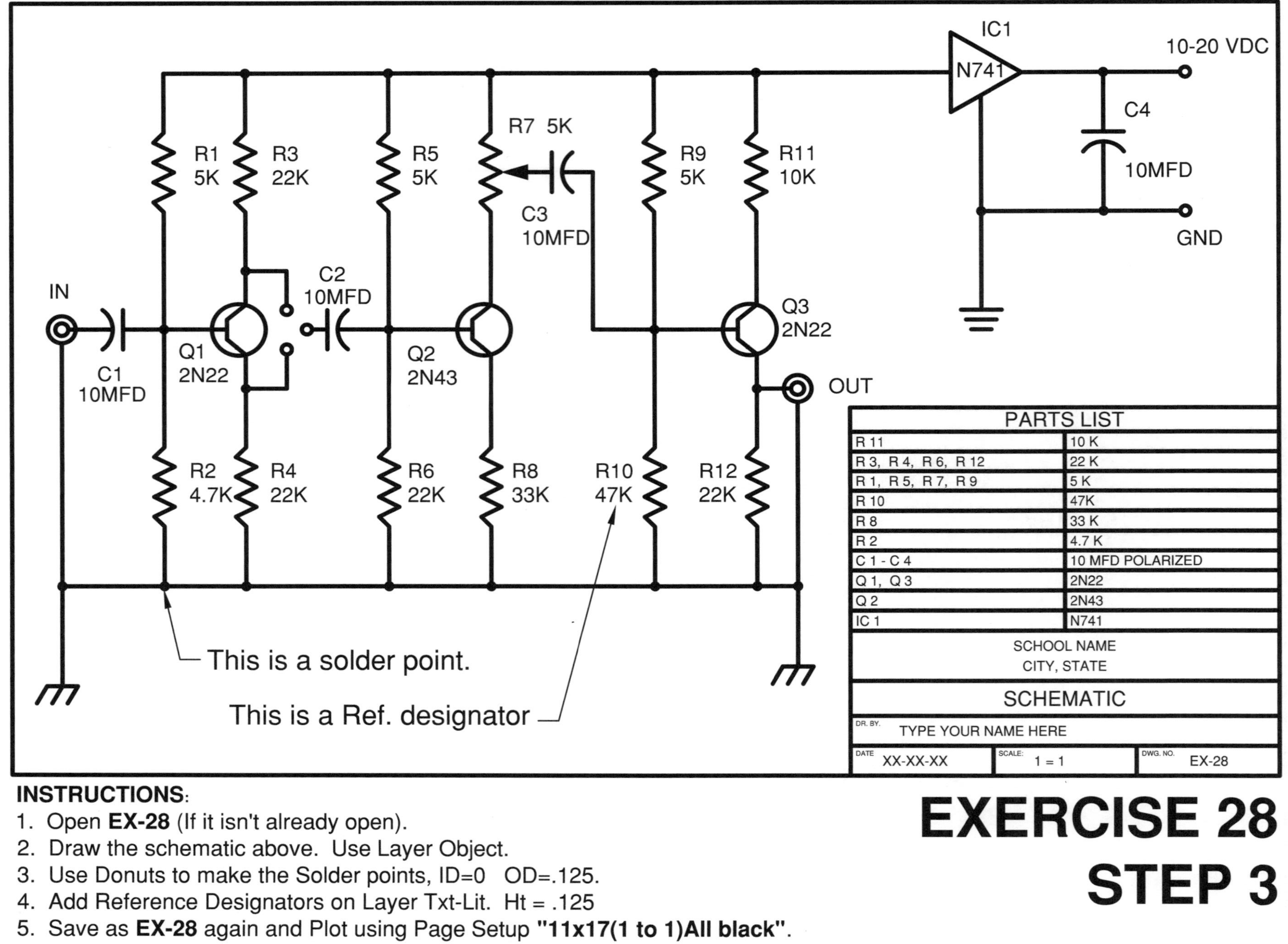

PARTS LIST	
R 11	10 K
R 3, R 4, R 6, R 12	22 K
R 1, R 5, R 7, R 9	5 K
R 10	47K
R 8	33 K
R 2	4.7 K
C 1 - C 4	10 MFD POLARIZED
Q 1, Q 3	2N22
Q 2	2N43
IC 1	N741

SCHOOL NAME
CITY, STATE

SCHEMATIC

DR. BY. TYPE YOUR NAME HERE

DATE XX-XX-XX | SCALE: 1 = 1 | DWG. NO. EX-28

INSTRUCTIONS:
1. Open **EX-28** (If it isn't already open).
2. Draw the schematic above. Use Layer Object.
3. Use Donuts to make the Solder points, ID=0 OD=.125.
4. Add Reference Designators on Layer Txt-Lit. Ht = .125
5. Save as **EX-28** again and Plot using Page Setup **"11x17(1 to 1)All black"**.

EXERCISE 28
STEP 3

LEARNING OBJECTIVES

After completing this lesson, you will be able to:

1. Use the Spacetrans command to adjust text to the scale of Paper space.

Before starting this lesson, review the following pages:

Tools / Inquiry commands. (page 9-6)
Object snap settings. (page 4-2)
Linetype scale command. (page 27-12)

LESSON 29

'SPACETRANS

In the previous lessons you learned the difference between model space and paper space. You learned that paper space is always 1 : 1 but model space can be adjusted to any scale.

Now we need to discuss what do you do if you need to add text to a model space that has had its scale adjusted. How do you determine what text height to specify?

For example, in Exercise 27G, the model space has been adjusted to 1/4 " = 1'.
That means that model space is 48 times smaller than actual size.
Remember, model space is far away from the viewport frame, so it appears smaller.
Now, what if you wanted to put text in the drawing to name the objects, such as "desks".
What text height would you use?

Try this:

I would like the text height to be 1/4" on the paper after the drawing is plotted.

1. Open EX-27G
2. Double click inside the Viewport to get into model space.
3. Select **Draw / Text / Single line Text**
4. Select a location beside one of the desks for the text start point.
5. Enter text ht. 1/4 and rotation 0
6. Type: **DESK** <enter>

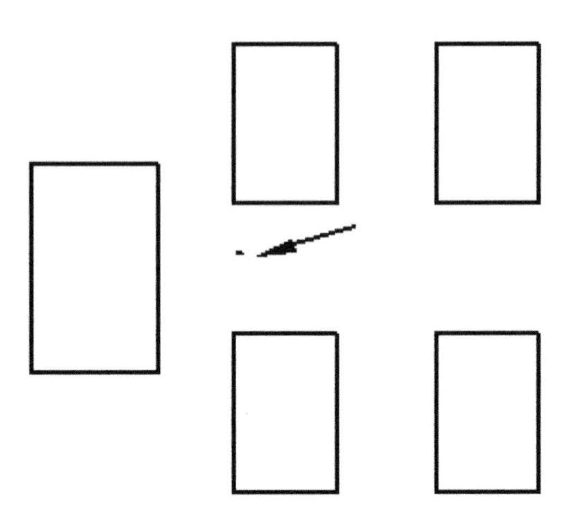

Can you see the text??

Pretty teeny huh??

That's because the text is very far away from you in model space.

Remember model space scale, in this drawing, has been adjusted to 1/4" = 1'.

The information on the next page will tell you how to solve this problem.

The following will explain the 3 methods you can use to get the correct height.

Method 1. (*Most difficult*)

Before entering the text height, calculate the text height by multiplying the desired text height times the drawing scale factor.
If model space was 1/4" = 1' the drawing scale factor would be 48.
So the Text height calculation should be: (1/4) X (48) = 12" (Refer to page 27-7)
So the Text height entered should be 12.

Method 2. (*Probably the easiest but with some caution*)

Place the text in Paper space not model space. Just like dimensions.
If the desired text height is 1/4, the text height entered would be 1/4 because paper space scale has not been adjusted, it is 1:1.
(But caution, if you move the objects in model space the text in paper space will not move automatically.)

Method 3. (The new command for this lesson: **'Spacetrans**)

Use the command '**Spacetrans**.
'Spacetrans will automatically calculate what height the text should be in model space.

SINGLE LINE TEXT

1. When prompted for the text ht., type **'spacetrans <enter>**
   ** **(Note: the apostrophe is necessary)**
2. Type the desired text height, such as 1/4"
3. Enter rotation angle <enter>
4. Type the text.

MULTILINE TEXT

1. Place the first corner of the text area.
2. Select the Height option by typing "H" <enter>
3. Type **'spacetrans <enter>**
4. Specify the desired text height and <enter>
5. Place the opposite corner.
6. Type the text.

**The apostrophe designates transparency. Transparent means you can use this command in the middle of another command. Thus we interrupted the Text command to run the spacetrans command and then you returned to the Text command.

EXERCISE 29

This exercise will give you more practice using Blocks, working with a scaled drawing and using the new 'spacetrans command.

INSTRUCTIONS:

A. Open your drawing **My Feet-Inches Setup.**

B. <u>STEP 1:</u> **Create the Blocks**
Draw the objects shown on page 29-4.
Select the "**Model**" tab and draw in **Model Space**. (Use the layers indicated.)

Change each one into a **BLOCK**. They should disappear as you create the Blocks.
(Refer to Lesson 28 for assistance)

<u>Do not start a new drawing, continue on to Step 2.</u>

C. <u>STEP 2:</u> **Drawing the Cabin**
Select the **"Qtr equals foot"** tab. (You should see your title block)
Next double click in the viewport to activate the Model space.
(Do not draw the cabin in Paper space)
Draw the **CABIN** shown on page 29-6.
Design your own Kitchen. Use appropriate layers.

Walls = 6" thick **Space behind doors** = 2"

Room Names = Text Height = 3/16" (Refer to 29-3 to choose a method)

Window sizes: In the Kitchen = 3 ft. In the Sleeping and Living rooms = 6 ft.

D. How to draw the dashed Electrical Lines.
A. Use Layer **Wiring**.
B. Draw the lines using **Arc** (Start, End, Direction)
C. Change the **Ltscale** to **.5** as follows:
<u>Globally:</u>
-Type at the command line: ***Ltscale <enter>***
-Enter new linetype scale factor <1.0000>:***type .5 <enter>***
OR
<u>Individually</u>: Select the arcs individually and use the Properties Palette

(Refer to page 27-12 for information on Linetype scales)

E. Dimension the drawing as shown (Refer to pg 27-11) Use dim style: Arch

F. **Save** as **EX-29** and **Plot** using Page Setup: **Qtr equals foot (All black)**

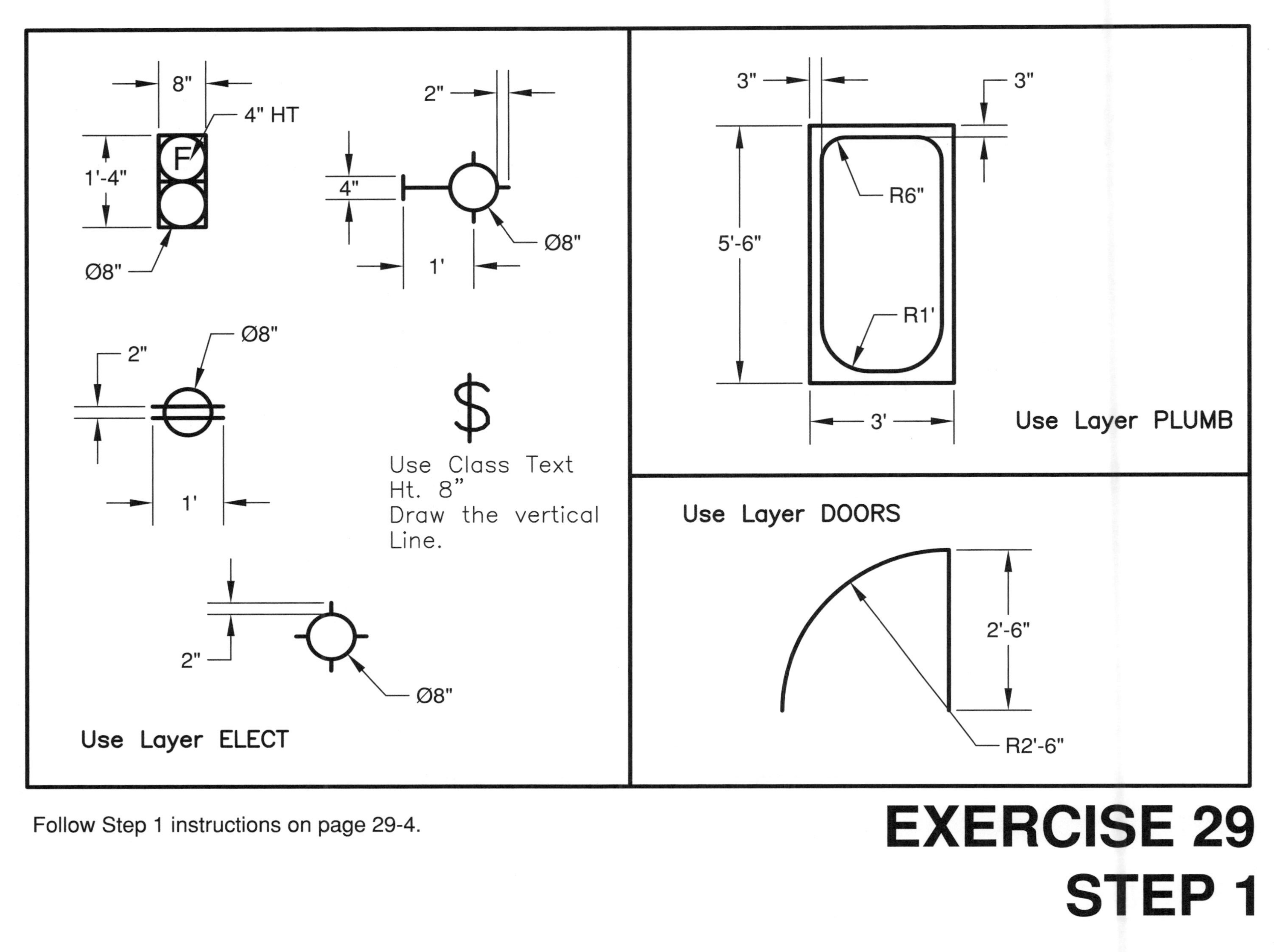

Follow Step 1 instructions on page 29-4.

EXERCISE 29
STEP 1

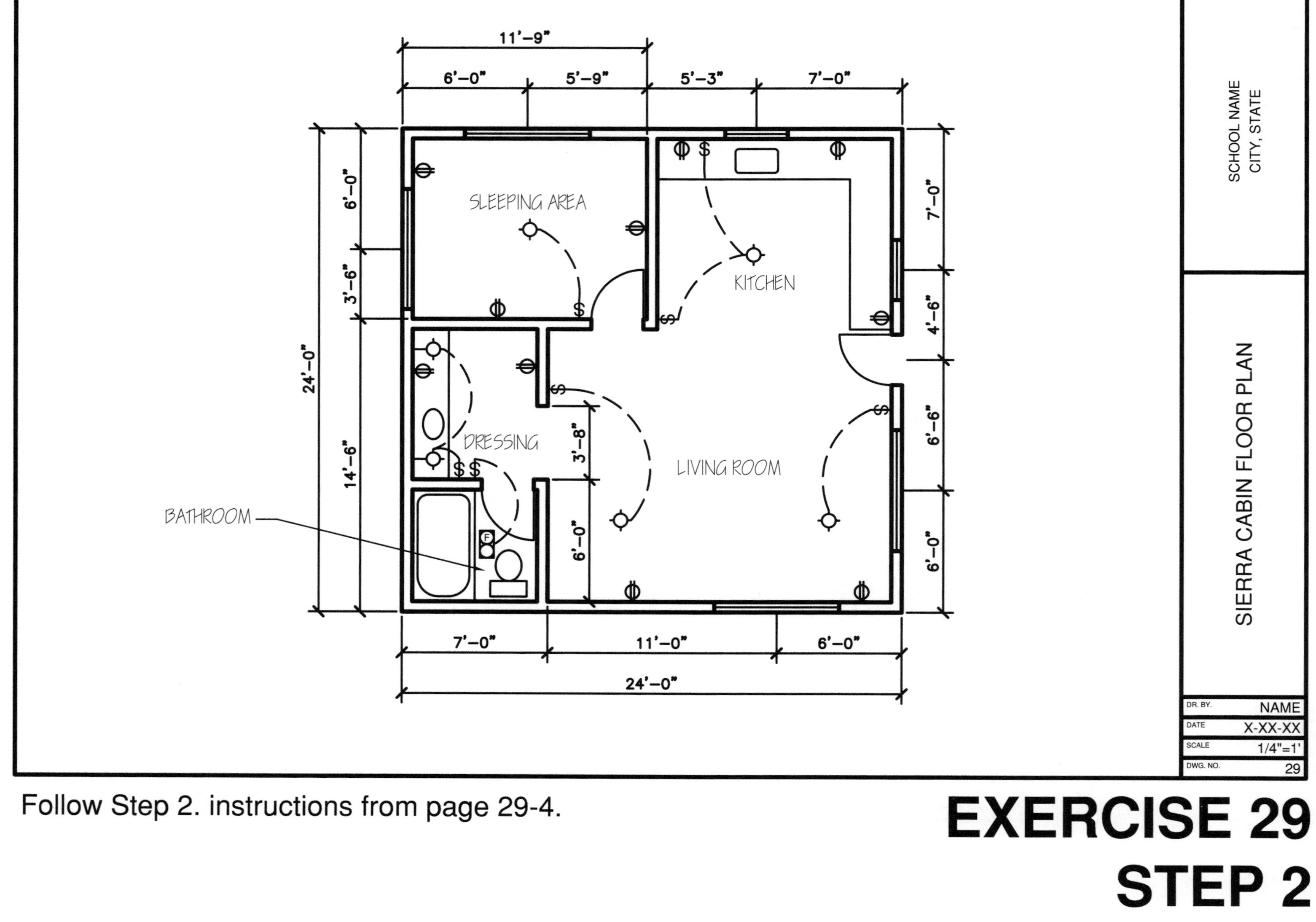

Follow Step 2. instructions from page 29-4.

EXERCISE 29
STEP 2

LEARNING OBJECTIVES

In this lesson we will review and practice:

1. "Polyline / Spline" commands. (Lesson 23)

LESSON 30

EXERCISE 30

INSTRUCTIONS:

A. Open drawing **My Decimal Setup**
Select the 11 X 17(1 to 1) tab. (You should see your title block)
Double click in the viewport area to get into model space.

B. **STEP 1.**
Draw the details shown on page 30-3. Do not dimension.
Draw the break line as follows:
   a. Draw the break line using **POLYLINE**.
   b. Change it to a curved line using:
   MODIFY / OBJECT / POLYLINE
      Select the Polyline
      Select **S**pline
   c. Draw the Hatch
   Use Hatch pattern ANSI 31, Layer = Hatch
   d. Thread dimensions below. Use Layer Threads.

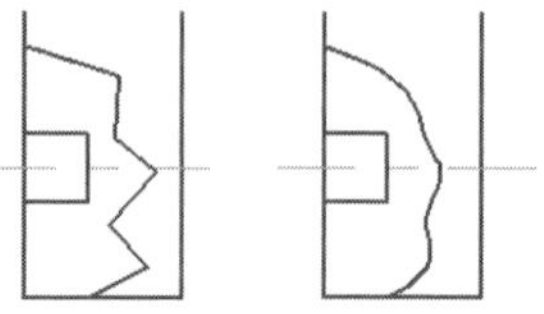

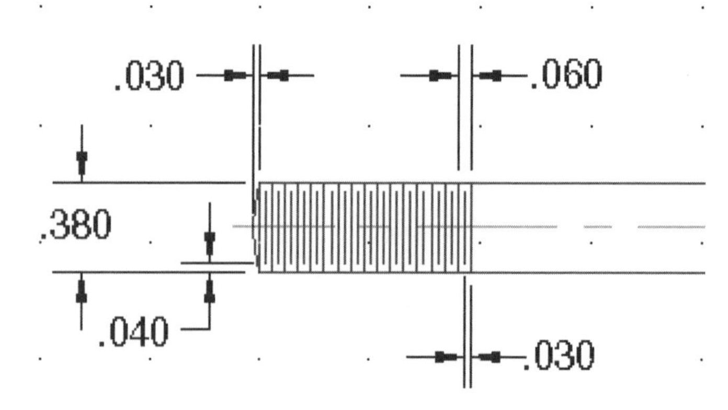

C. **STEP 2.**
Return to paperspace.
Draw the **PARTS LIST** (in paperspace) shown on page 30-4, STEP 2.
Make the changes to the title block.

D. **STEP 3.**
Move the parts into the proper locations to form the assembly shown.
The distance between the jaws is 1".
Draw the Balloons:
   a. Use Leader
   b. Use a .50 dia circle.
   c. Text ht. = .187

E. Save as **EX-30** and **PLOT** using Page setup **"11x17(1 to 1)All Black".**

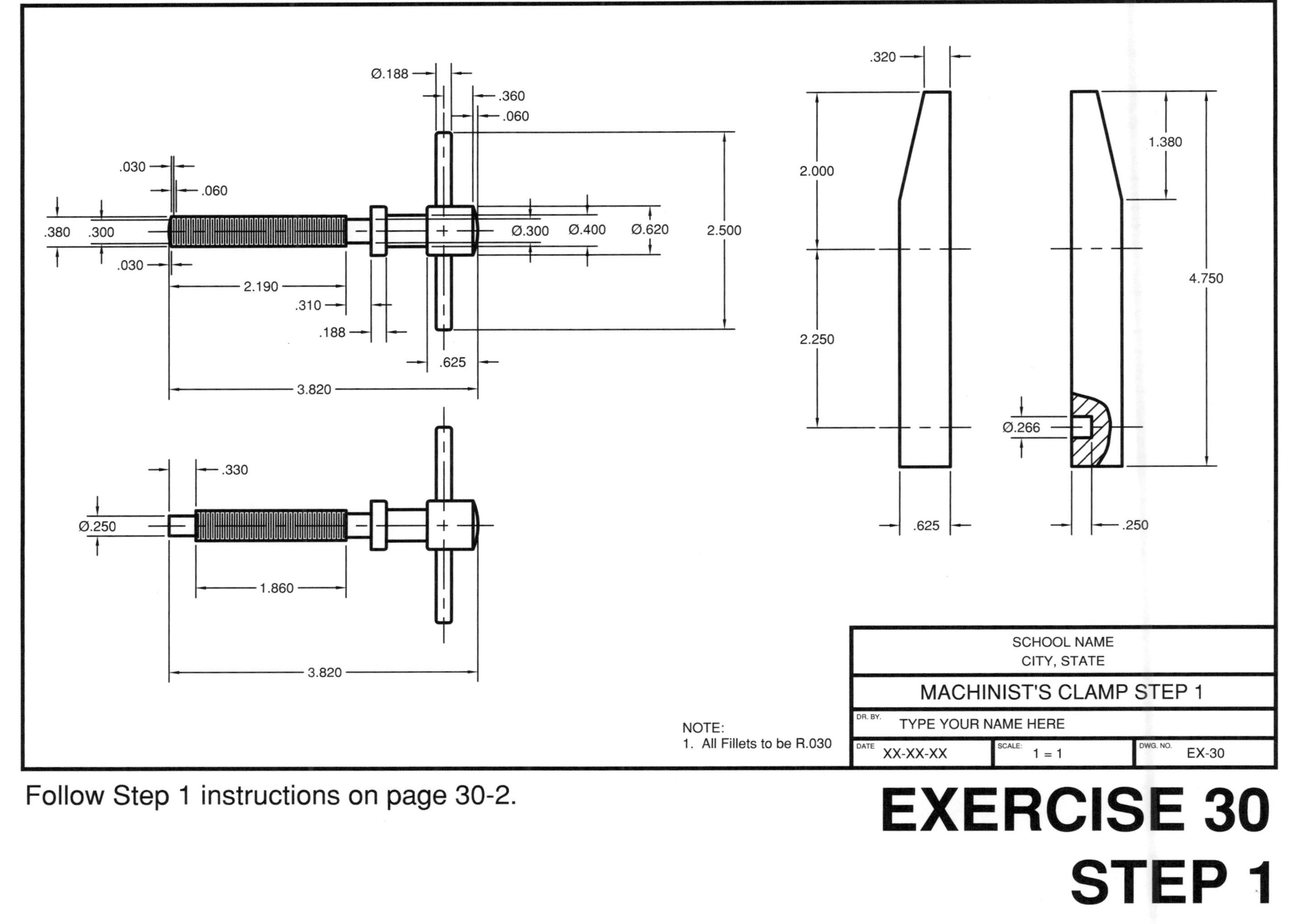

Follow Step 1 instructions on page 30-2.

EXERCISE 30
STEP 1

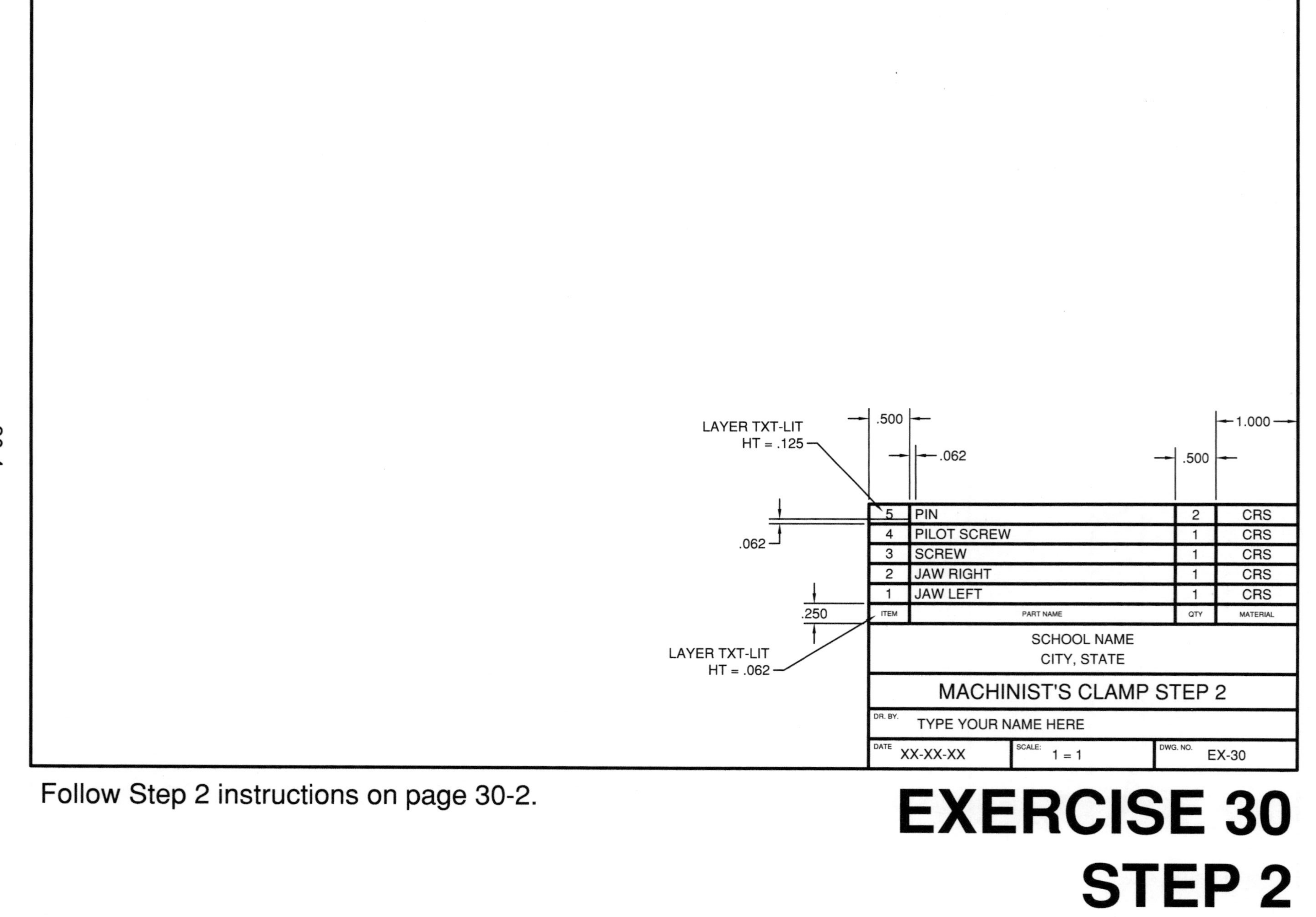

Follow Step 2 instructions on page 30-2.

EXERCISE 30
STEP 2

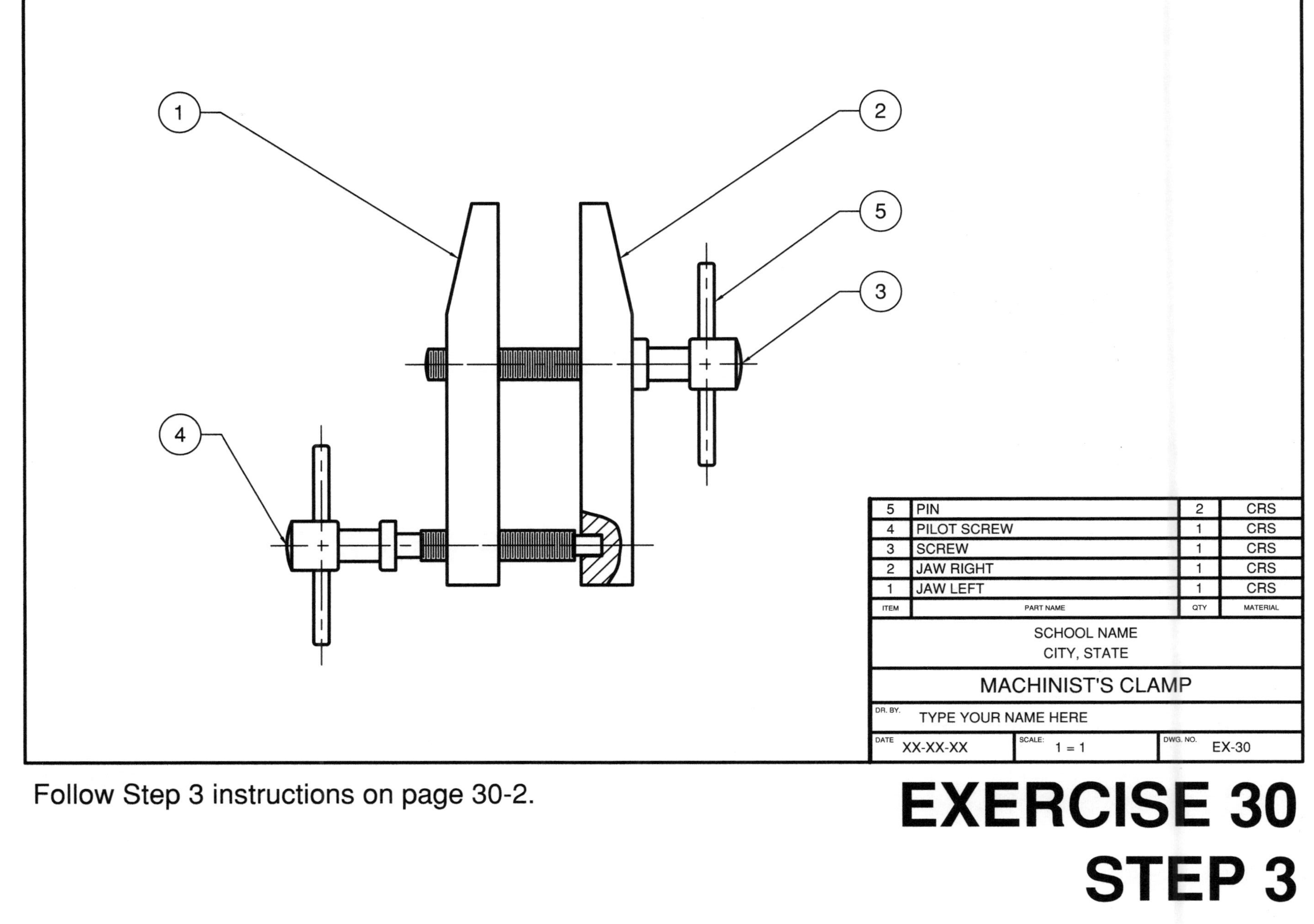

ITEM	PART NAME	QTY	MATERIAL
5	PIN	2	CRS
4	PILOT SCREW	1	CRS
3	SCREW	1	CRS
2	JAW RIGHT	1	CRS
1	JAW LEFT	1	CRS

SCHOOL NAME
CITY, STATE

MACHINIST'S CLAMP

DR. BY. TYPE YOUR NAME HERE

DATE XX-XX-XX | SCALE: 1 = 1 | DWG. NO. EX-30

Follow Step 3 instructions on page 30-2.

EXERCISE 30
STEP 3

NOTES:

APPENDIX A
Add a Printer / Plotter

The following are step by step instructions on how to configure AutoCAD for your printer or plotter. These instructions assume you are a single system user. If you are networked or need more detailed information, please refer to your AutoCAD users guide.

Note: You can configure AutoCAD for multiple printers. I suggest that you configure the plotter shown below to match the exercises in this workbook.

A. Select **File / Plotter Manager**
B. Select **"Add-a-Plotter"** Wizard

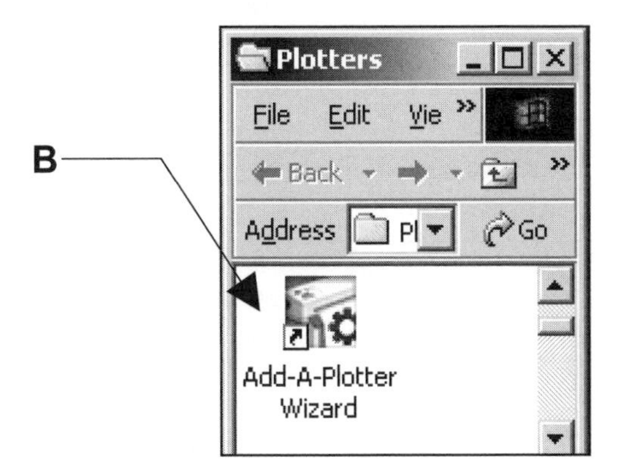

C. Select the **"Next"** button.

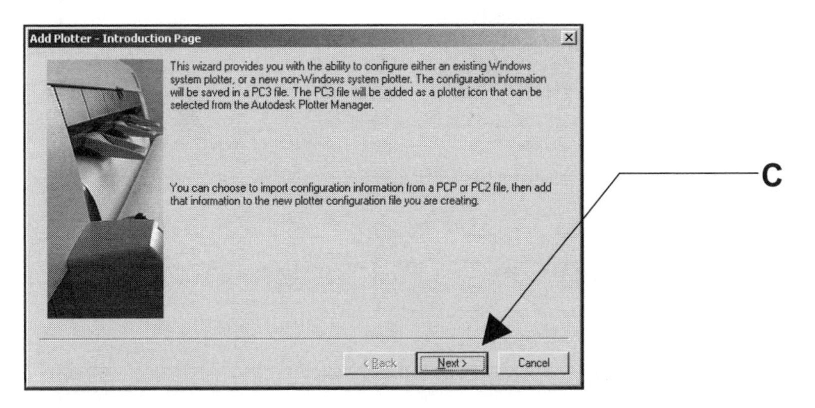

D. Select "**My Computer"** then **Next**.

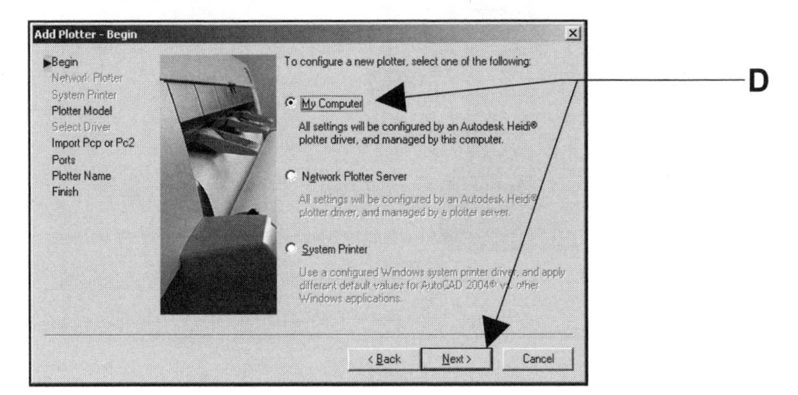

E. Select the **Manufacturer** and the specific **Model** desired then N**ext**.

(If you have a disk with the specific driver information, put the disk in the disk drive and select "Have disk" button then follow instructions.)

F. Select the **"Next"** box.

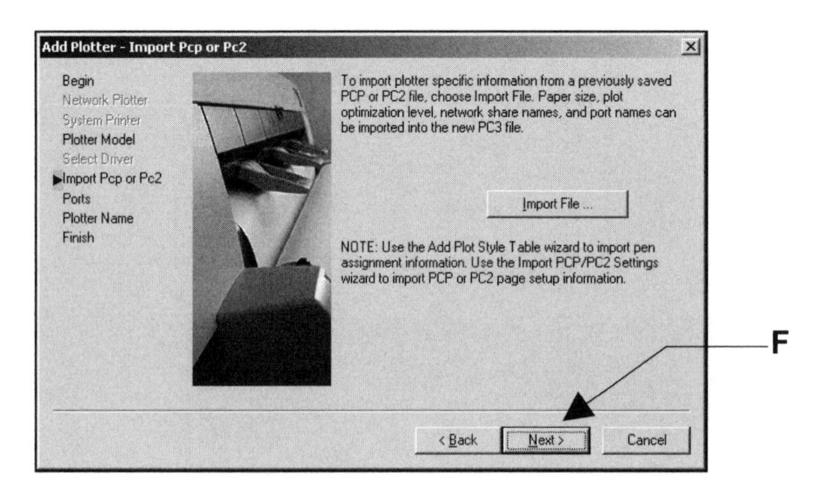

G. Select **"Plot to a port"** and **"LPT1"** unless your printer is attached to a Com port. Then select **"Next".**

H. The Printer name, that you previously selected, should appear. Then select **"Next"**

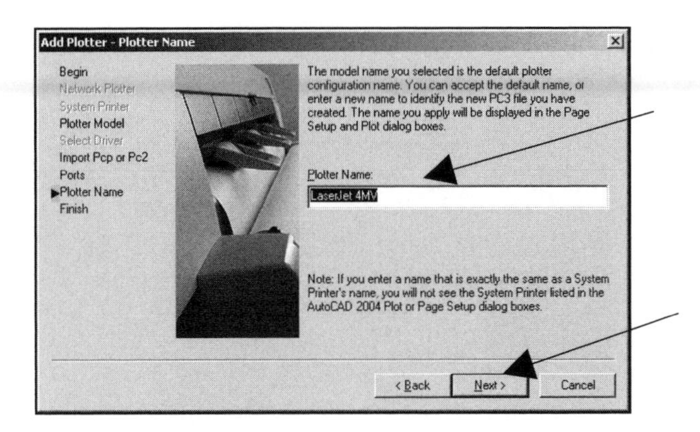

I. Select the **"Edit Plotter Configuration..."** box.

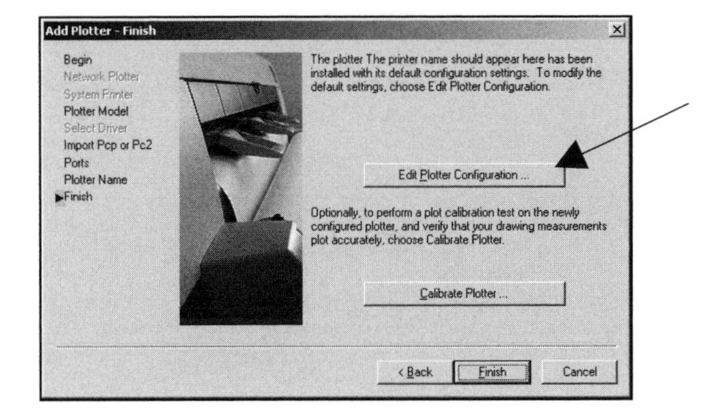

J. Select:

1. Device and Document Settings tab.
2. Media: Source and Size
3. Size: Ansi B (11 X 17 inches)
4. OK box.

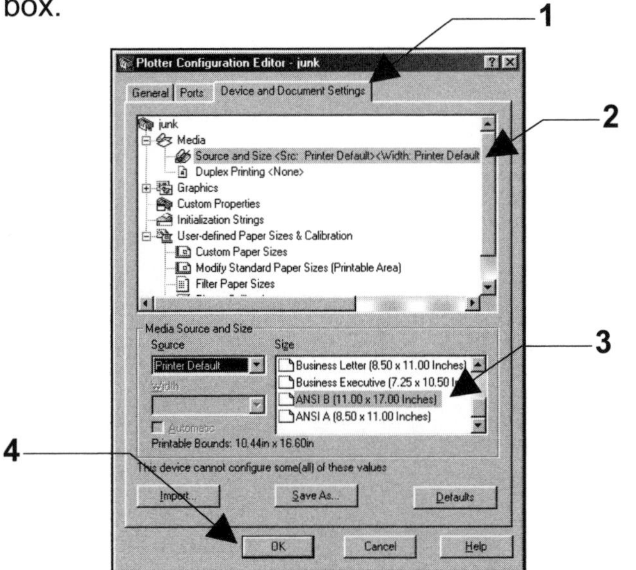

K. Select **"Finish".**

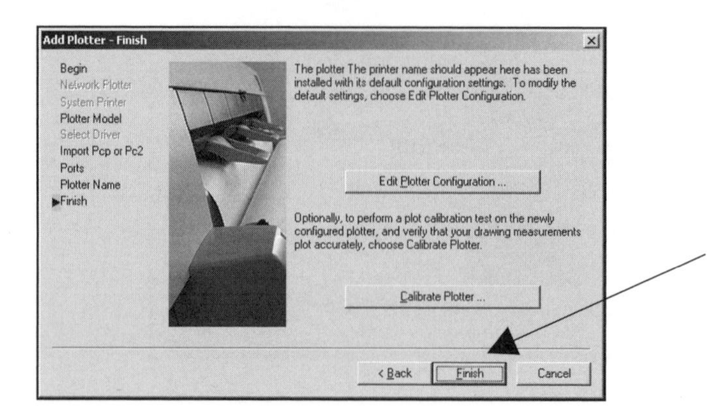

L. Now check the **File / Plotter Manager.** Is your printer / plotter there?

Plotters
File Edit View Favorites Tools
Back Search
Address Plotters Go
LaserJet 4MV.pc3
PublishToWeb JPG.pc3
PublishToWeb PNG.pc3
Locat 639 bytes My Computer

APPENDIX B
Dimension Style Definitions

When you select **Dimension / Style / Modify** the following dialog box will appear.

The following are descriptions for each setting within each section tab.

Lines and Arrows tab

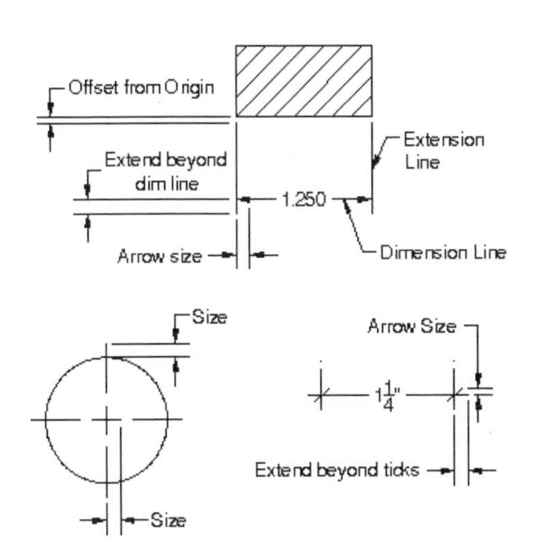

Dimension Lines

Color: Color of the dimension line.
Lineweight: Sets the width of the dimension lines.
Extend beyond ticks: Distance to extend <u>dimension</u> line beyond the <u>extension</u> line. Example above.
Baseline spacing: Spacing between baseline dimensions.
Suppress: Suppress means disappear. You may suppress individually or both.

Extension Lines

Color: Color of the extension lines
Lineweight: Sets the width of the extension lines.
Extend beyond dim. line: Distance to extend <u>extension</u> line beyond the <u>dimension</u> line. Example above.
Offset from origin: The distance between the object and the extension line. Example above.
Supress: You may suppress individually or both.

Arrowheads

1st : Sets the style for the symbol inserted for the first extension point
2nd: Sets the style for the symbol inserted for the second extension point and for <u>Diameter</u> and <u>Radius</u>.
Leader: Sets the style for the symbol inserted for Leaders.
Arrow Size: Sets the Size of Arrowheads. Example above.

Center Marks for Circles

Type: Select None, Mark or Line **Size:** Specify Size. Example above.

TEXT tab

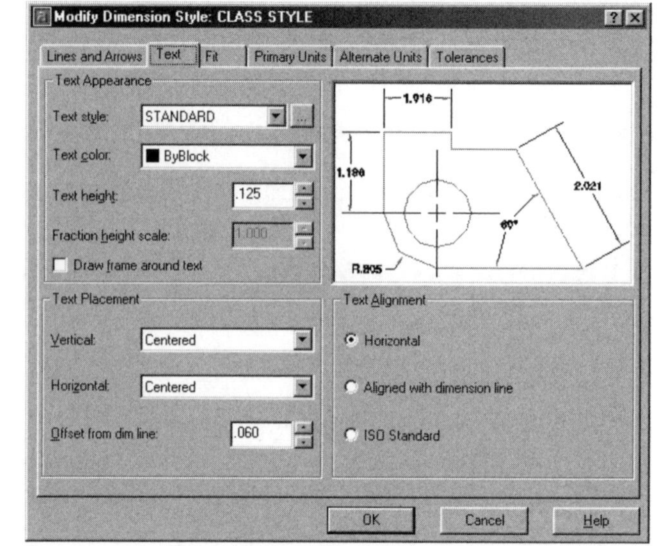

Text Appearance

Text Style: Select which text style to use for the dimension text. To create or change a style, select the […] button.

Text Color: Sets the color of the dimension text.

Text height: Sets the height.

Fraction Height Scale: Sets the size of fractions relative to the dimension text height. This is a factor not and actual height. Example: A setting of .50 would be half of the dimension text height.

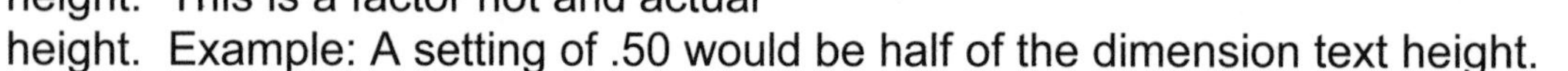

Draw Frame Around Text: If this box is checked, a box will be drawn around the text.

Text Placement

Vertical

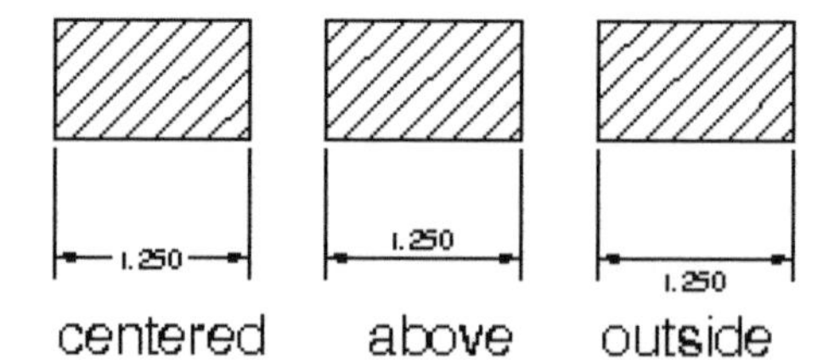

Centered: Centers dimension text between extension lines.
Above: Places dimension text above the dimension line.
Outside: Places dimension text on the side of the dimension line farthest away from the object.

Horizontal

At Ext Line 1: Moves text near first extension line.

At Ext Line 2: Moves text near second extension line.

1.250 1.250

At Ext Line 1 At Ext. Line 2

Over Ext Line 1: Places text over first extension line.

Over Ext Line 2: Places text over second extension Line

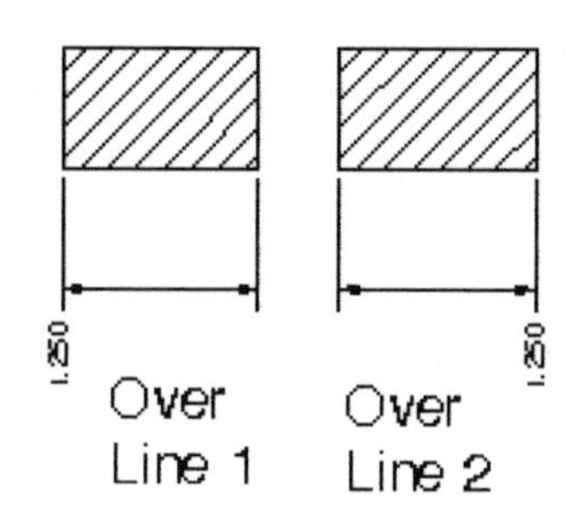

Offset from Dim Line: Sets the gap between the dimension text and the dimension line.

Text Alignment

Horzontal: Places dimension text horizontal.

Aligned: Aligns dimension text with the dimension line.

ISO Standard: Aligns text with the dimension line when text is inside the extension lines, but aligns it horizontally when text is outside extension lines.

FIT tab

Fit Options: Controls the placement of text and arrowheads based on the space available between the extension lines. If there is not enough room to place both text and arrows inside extension lines, you must choose what to move outside the extension lines.

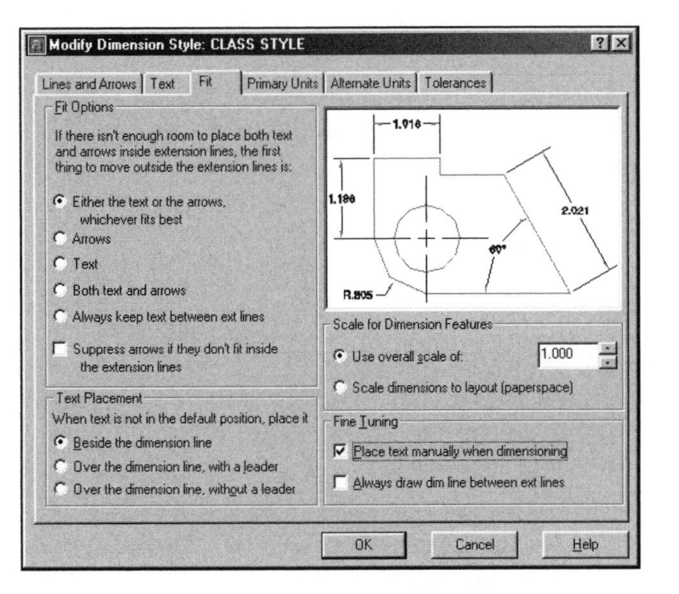

Text Placement: When dimension text is moved from its dim. style location, this setting controls the placement.

Scale for Dimension Features:

Use overall scale of: Set the factor for all dimension settings. Example: if this is set to 10 and the dimension text height is set to 1/8, the dimension text height would appear 1-1/4" ht. (10 x 1/8") The value does not change. ***Note****: this feature is left over from previous AutoCAD releases. It is not necessary now because of "Trans-spatial" dimensioning.*

Scale dimensions to layout (paper space): Calculates a scale factor based on the scaling of model space vs. paper space. ***Note****: this feature is left over from previous AutoCAD release. It is not necessary now because of "Trans-spatial" dimensioning.*

Fine Tuning

Place text manually when dimensioning: You control the placement of the text if this feature is ON. It is not automatic.

Always draw dim Line between Ext Lines: Draws the dimension line inside the extension lines even if the arrows are on the outside.

PRIMARY UNITS tab

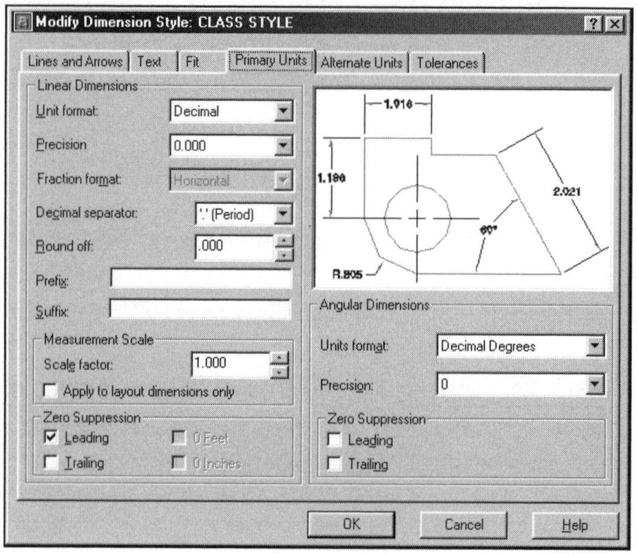

Linear Dimensions

Unit Format: Sets the units format for all dimensions except Angular.

Precision: Sets the number of decimal places in the dimension.

Fraction Format: Set the format for fraction to Horizontal, Diagonal or not stacked. Only available if Unit Format (above) is set to "Fractional".

Decimal Separator: Sets the style for the decimal separator to period, comma or space. Not available if Unit Format is set to Fractional.

Round Off: Sets rounding limits for dimensioning, such as .000 or 1/8. Rounds **up** to the nearest 3 place decimal or nearest 1/8".

Prefix: Add text in front of the dimension text. (**Flat for** 2.00)

Suffix: Add text after the dimension text. (1'-0" **Max**)

Measurement scale:

Scale Factor: AutoCAD multiples the dimension measurement by the value entered here. Example: If you draw a 1/2 inch line. Set this feature to 2. When you dimension the line, the dimension text will display 1. (2 X 1/2) If you set this feature to .50, when you dimension the line, the dimension text will display 1/4. (.50 X 1/2)
Apply to Layout Dimensions Only: Unnecessary now that we have Trans-spatial dimensioning.

Zero Suppression

The following two only work with decimals:
Leading: Controls the display of zeros before the decimal point. 0.50 = Off .50 = ON
Trailing: Controls the display of zeros at the end of the dimension. .500 = Off .5 = ON

The following tow only work with architectural:
Feet: Controls the display of zeros for feet. 0'-6" = OFF 6" = ON
Inches: Controls the display of zeros for inches. 6'-0" = OFF 6" = ON

Angular Dimensions

Units Format: Sets the units format for Angular. Does not affect Linear.
Precision: Sets the number of decimal places past the whole degree.
Zero suppression: Same as Linear.

Note: Alternate Units and Tolerance tabs are discussed in the "Advanced" workbook.

INDEX

T

U

V

W

X

Y

Z